Exploring the Cultural, Moral, and Technological Dimensions of Conflict

Khaled Al-Kassimi
https://orcid.org/0000-0003-1209-9493
American University in the Emirates, UAE

Tuba Işık
Istanbul Medipol University, Turkey

Muhammad Fahim
Monash University, Australia

Başak Gezmen
Istanbul Medipol University, Turkey

IGI Global
Scientific Publishing
Publishing Tomorrow's Research Today

Vice President of Editorial — Melissa Wagner
Director of Acquisitions — Mikaela Felty
Director of Book Development — Jocelynn Hessler
Production Manager — Mike Brehm
Cover Design — Jose Rosado

Published in the United States of America by
IGI Global Scientific Publishing
701 East Chocolate Avenue
Hershey, PA, 17033, USA
Tel: 717-533-8845 | Fax: 717-533-7115
Website: https://www.igi-global.com E-mail: cust@igi-global.com

Library of Congress Cataloging-in-Publication Data

(CIP Data Pending)
ISBN13: 979-8-3373-7589-2
Isbn13Softcover: 979-8-3373-7590-8
EISBN13: 979-8-3373-7591-5

British Cataloguing in Publication Data
A Cataloguing in Publication record for this book is available from the British Library.

Table of Contents

Detailed Table of Contents

This study examines the impact of digitalization on political communication processes within the context of social media platforms. Focusing particularly on the example of the 2024 U.S. presidential elections, it is argued that social media has evolved beyond merely facilitating interaction between voters and candidates, transforming instead into a new battleground where methods such as cyber warfare and media manipulation are employed. In this study, social media is approached as a platform where both domestic and global actors conduct information warfare, steer public opinion through disinformation, and directly interfere in electoral processes. In this context, it is emphasized that traditional forms of warfare are increasingly being replaced by strategic communication techniques conducted through digital channels. In conclusion, it is asserted that today, social media platforms are not only tools of communication but also digitalized forms of political competition and international intervention.

Technological innovation has continuously shaped the battlefield, dictating outcomes and redefining strategies. The evolution of warfare technologies includes key shifts such as the introduction of electronics and the rise of internet-dependent military systems in the 1990s. The reliance on digital infrastructure has transformed military operations, introducing vulnerabilities like cyberattacks targeting financial systems, intelligence, and weaponry. Media and social media platforms have expanded the landscape of warfare, enabling rapid dissemination of information and psychological operations powered by technology. This shift has facilitated the rise of unmanned systems, such as drones, which depend heavily on computer systems. Efficient cybersecurity frameworks are now essential to protect military and civilian digital assets, as data breaches can inflict profound economic and strategic damages. With the examples of past and recent past wars, particularly in the Indian subcontinent, this chapter explores these technological transformations and their implications for future conflicts.

This study dwells on the observation that war is inherently within the realm of human decision-making and that people can choose to be good or bad at any moment, for any duration, or in any action, which is generally overlooked. Generally, due to the excitement of the accompanying music, colors, allure of the subject, and other stimuli, many people fail to perceive this conflict and other types of war as part of narratives. The notions of "spare time and leisure" or "entertainment" attributed to media narratives might also cause people to fail to truly perceive physical or psychological struggle. Using a mixed-method approach, this study focuses on the symbolic analysis of life and death issues in the animated film Riverdance (2021) within the framework of "death and life" and their reflections in cultures. The findings put forward that beyond simply telling a story, the narrative focuses on narrating St. Patrick's Day in a way that facilitates marketing and popularization.

The study of war has long been confined to disciplines such as history, political science, and journalism. However, with the evolution of modern conflicts and the shifting global landscape, it is imperative that scholars also explore more unconventional or often overlooked mediums, including literature, art, and music—not just to study how civilians respond to war and the trauma it leads to, but also the intricacies of war itself. This study examines war and conflict narratives through a cultural lens, using Critical Discourse Analysis of Letters of Enchantment (2023), a fantasy duology by Rebecca Ross. Fantasy fiction, once perceived as an escapist genre, has recently developed into a medium capable of exploring serious socio-political themes, including warfare, propaganda, and trauma. This study explores the ways in which the duology portrays war through themes of media influence, artistic expression, censorship, and music suppression, drawing parallels with real-world historical and contemporary conflicts.

This chapter examines the representation of gender and identity in Indian war films, focusing on nationalism and militarism. Films Amaran (Rajkumar Periasamy, 2024) and Sita Ramam (Hanu Raghavapudi, 2022) place women at the centre of military narratives shaping historical understanding and soldiers' experiences. This chapter analyses the impact of gender relations on nationalism and patriotism, linking personal and national identity during conflict. Cross-cultural romantic relationships between Hindu soldiers and women from minority religions Christianity and Islam challenge and reinforce societal norms within tense militaristic contexts. Through Benedict Anderson's framework of nationalism, the research emphasises women's roles in military narratives, going beyond traditional romantic portrayals. This analysis is framed within popular media discourse on war and selected films and supplemented by secondary sources on media research to capture public perceptions. It also aims to deepen the understanding of how gender dynamics in Indian war cinema impact contemporary national identity.

This study investigates how cinema portrays the intrusion of war into domestic and private spaces, transforming them into battlegrounds for psychological and emotional struggles. Through a combination of psychoanalytic film theory and spatial analysis, the research examines how the traumatic legacies of war shape personal identities, intimate relationships, and physical spaces, even when removed from direct combat. Focusing on three films—The Zone of Interest (2023), The Pianist (2002), The Reader (2008),—the study explores how these narratives construct symbolic and visual representations of private spaces as sites of unresolved conflict. By analyzing mise-en-scène, character psychology, and narrative structure, the research demonstrates how war's echoes reverberate within memory, family dynamics, and spatial design, blurring the boundaries between the personal and the political. The Pianist and The Zone of Interest focus on the Holocaust's legacy, while The Reader examines post-war guilt and reconciliation.

This chapter examines how cinema constructs capitalist spatial orders at sea through Captain Phillips and A Hijacking. Using spatial theory, it introduces “cinematic territorialization” to show how film renders the ocean governable, securitized, and economically productive. These films offer contrasting views: Captain Phillips celebrates American power against racialized maritime threats, while A Hijacking exposes detached corporate logics of risk and negotiation. Linking this to a longer history of maritime myth-making, the chapter shows how cinema reproduces binaries of order/disorder and North/South. While acknowledging Somali perspectives, it focuses on how dominant film narratives naturalize capitalist power. Ultimately, it argues film actively shapes how we imagine ocean space, urging renewed critique as climate change and resource extraction open new oceanic frontiers.

This study analyzes how World War II–themed films and television productions from the 1990s to the present represent and deconstruct the archetypal hero–villain dynamic within the framework of traditional heroism. Centering on Saving Private Ryan, Band of Brothers, and The Thin Red Line, the discussion explores how these works both adhere to and challenge conventional heroic narratives. While Saving Private Ryan and Band of Brothers reinforce traditional heroic and villainous tropes through established heroism—thereby bolstering nationalist and patriotic discourses that legitimize American military endeavors and address past national traumas—The Thin Red Line offers a philosophical deconstruction of traditional heroism. Director Terrence Malick’s distinct style transcends Hollywood’s commercial mainstream, providing audiences with a contemplative lens on the nature of war.

The chapter critically examines how Hero and Villain archetypes materialize in war stories through historical, cultural and psychological lenses. Drawing from Jungian psychology, moral philosophy, media studies and social psychology, exploring how symbolic narratives transform complex geopolitical conflicts into stories of moral dichotomy between good and evil. It considers cultural and religious roots shape of monomyth along with concepts such as moral injury and virtuous violence, to investigate how propaganda, national identity, and collective memory uphold strict dualistic classifications, reinforcing ideological agendas. Through comparative examples, it disputes the oversimplified frameworks dominating wartime narratives. It seeks to demonstrate how simplistic stereotypes legitimize violence, shape historical memory and impact public understanding where moral framing shape consent. Media representations recycle these binaries, influencing opinions through familiar narrative tropes. The chapter calls for portrayals of human conflict with greater complexity and ethical awareness.

The goal of this study to investigate the atrocities against the Yazidi community, especially the women, during ISIS-led genocide. Using discourse analysis, this chapter examines Nadia Murad's memoir The Last Girl: My Story of Captivity, and My Fight against the Islamic State (2018) as a site of testimonial articulation that transcends personal trauma to represent the collective trauma. Furthermore, alongside the memoir, the chapter incorporates media appearances to analyse Murad's self-advocacy to navigate and challenge institutional mediation and inaction. The study uses testimonial theory and gender theory to understand testimony and the sexual slavery of Yazidi women as they were captured, traded and sexually violated. The study found that sexual enslavement decimated their individuality and connection to their culture while also denied their autonomy, narrative agency and expression. However, Murad's testimony served both personal and collective as it not only documented genocidal trauma, but demanded for recognition, accountability and justice.

The aim of the research is to identify the crisis pathway and examine the impact of crisis responses on the mass shooting in Belgrade, that occurred in May 2023. Official crisis responses and media reporting have been analyzed within the context of Situational Crisis Communication Theory, social impact, moral outrage, and crisis risk mitigation. Based on the document analysis, the flow and timeline of the crisis as

Preface

The essence of (post)-modernity is that nothing remains fixed. Everything is fluid. Conflict today is no longer overt or easily defined; it is diffused, mediated, and morally complex. As observed by Zygmunt Bauman in *Liquid Modernity*, "Precariousness, instability, vulnerability is the most widespread feature of contemporary life conditions" (2000:160). This edited volume, entitled *Exploring the Cultural, Moral, and Technological Dimensions of Conflict*, engages with the multifaceted nature of conflict in our contemporary world. No longer confined to traditional battlefields, conflict now unfolds on digital platforms, within cultural texts, and even in the intimate spaces of everyday life. By weaving together perspectives from technology, culture, morality, and media, this book reflects the complexity of modern conflict and the need to understand it beyond narrow disciplinary boundaries. In expanding the philosophical grounding of this volume, it is apt to recall intellectuals who have interrogated the cultural, moral, and technological dimensions of modern conflict from a philosophical worldview harmonizing the material with the metaphysical.

Herbert Marcuse (1964), in *One-Dimensional Man*, critiques the way technological rationality flattens human experience, reducing ethically complex questions to matters of mere efficiency. In the context of conflict, this rationality manifests in the normalization of surveillance, algorithmic warfare, and digital propaganda—tools that shape not only military strategies but also influence cultural narratives and moral perceptions. Several chapters in this volume engage directly with Marcuse's concerns, particularly in examining how media and technology are mobilized to legitimize violence or suppress dissent. In *The Question Concerning Technology*, Martin Heidegger (1977) warns that modern technology fundamentally "enframes" the world, reducing both human beings and nature to mere resources. This process of enframing is especially visible where remote-controlled systems, cyber operations, and the abstraction of violence lead to a marked dehumanization of conflict. Heidegger's analysis challenges us to reflect on the ways technological mediation not only alters the conduct of war but also its moral and existential dimensions. Several chapters in this book engage deeply with these concerns, examining how trauma, memory, and personal identity—from cinematic portrayals to digital testimonies—are distorted, shaped, and reshaped through technological lenses, thereby highlighting the profound impact technology has on our understanding of conflict.

Zygmunt Bauman's concept of "liquid modernity" (2000) radically reconfigures our understanding of social reality. Instead of viewing social structures as fixed or predictable, liquid modernity is characterized by fluid social structures, unstable identities, and elusive moral anchors. That is, Bauman contends that everything—identities, norms, and even moral reference points—has become fluid and unstable. In such a world, Bauman argues, war is no longer contained within traditional boundaries. Rather, it becomes diffused and often invisible, enacted through "war-machines" that operate across economic, informational, and cultural domains—that is, machine systems that push conflict through economic, informational, and cultural channels. The mechanisms operating these machines are not easily visible, yet they permeate daily life, entering domestic environments, cultural narratives, and digital platforms. This volume engages Bauman's ideas by exploring how conflict is mediated and represented through fantasy literature, survivor testimonies, and cinematic mythmaking. The representation of war in these forms highlights its ambiguity and the complex cultural work involved in making sense of violence. War, then, is not only a matter of physical confrontation but a phenomenon unfolding across multiple domains: personal identity, collective memory, and symbolic systems. The chapters in this book reflect this expanded and nuanced understanding, showing how conflict is experienced, remembered, and challenged in both individual and communal contexts. Thinkers like Marcuse, Heidegger, and Bauman underscore the necessity of philosophical engagement in a serious study of conflict and affirm that technology, culture, and morality are inseparable dimensions of the human condition in times of war.

This book represents a collaborative effort across countries, institutions, and academic fields. Each chapter offers an original exploration of how conflict manifests in new ways—through social media manipulation, cinematic narratives, testimonial literature, technological advancement, and the politics of memory. Together, they emphasize that the study of conflict is no longer restricted to the domain of Political Science or International Relations; rather, it is enriched by the lenses of communication studies, cultural analysis, philosophy, psychology, and literature. Our aim in curating this work has been to highlight the cultural, moral, and technological dimensions of conflict as inseparable elements of the human experience. Readers will encounter discussions that move from cyber warfare to mythmaking at sea, from Indian cinema to Holocaust memory, and from fantasy fiction to the lived trauma of genocide survivors. In doing so, the book hopes to broaden the reader's understanding of how war and violence are remembered, mediated, and reshaped in both collective and personal contexts. As editors of this volume, we are pleased to present a collection of scholarly works that engage with the multifaceted nature of conflict in our contemporary modern world. War and conflict have always shaped human civilization, but in today's Age of Technology, they are experienced not only on traditional battlefields but also within digital platforms, cultural texts, and personal spaces.

CHAPTER OVERVIEW

Chapter 1 examines the 2024 U.S. presidential elections as a case study in cyber warfare and media manipulation, showing how social media platforms have become digital battlegrounds for information warfare. **Chapter 2** focuses on technological innovation in warfare, tracing the evolution of digital and unmanned systems while highlighting vulnerabilities and cybersecurity needs in modern conflicts. **Chapter 3** shifts the analysis to cultural narratives, exploring how *Riverdance (2021)* uses symbolism to depict life, death, and cultural struggles through entertainment media. **Chapter 4** extends this cultural perspective by analyzing fantasy fiction as a site where music, art, and literature are mobilized to portray conflict and censorship, bridging the divide between escapism and political commentary. **Chapter 5** interrogates gender, nationalism, and identity in Indian cinema, particularly through films like *Sita Ramam* and *Amaran*, to show how women's roles are central to constructing national and military narratives. **Chapter 6** investigates the representation of domestic spaces in war films, showing how trauma transforms private life into contested terrain, with case studies ranging from *The Pianist* to *The Zone of Interest*. **Chapter 7** analyzes maritime conflicts through films like *Captain Phillips* and *A Hijacking*, revealing how cinema constructs binaries of power and reinforces capitalist spatial orders at sea. **Chapter 8** explores contemporary World War II films and series, highlighting how hero–villain dynamics are both reinforced and deconstructed across works such as *Saving Private Ryan* and *The Thin Red Line*. **Chapter 9** builds on archetypal studies of heroes and villains, applying Jungian and philosophical frameworks to show how oversimplified binaries legitimize violence and shape public perception of war. **Chapter 10** centers on Nadia Murad's testimony as a Yazidi survivor of ISIS atrocities, illustrating how personal trauma becomes collective memory while also demanding justice and accountability. **Chapter 11** provides a case study of the 2023 Serbia mass shootings, analyzing crisis communication strategies and the moral outrage that followed, with implications for public safety and state responses. **Chapter 12** closes the book with a philosophical reflection on Zygmunt Bauman's theory of postmodern wars, emphasizing dehumanization, technology, and the erosion of morality in contemporary conflicts.

In bringing these chapters together, we hope to offer more than a catalogue of conflicts and representations; we aim to illuminate the threads that connect them. Whether through the manipulation of social media, the ethical dilemmas of advanced technologies, or the cultural work of art and literature, each contribution illustrates the ways in which human societies confront, mediate, and attempt to make sense of violence. Conflict is never confined to battlefields; it permeates identities, memories, and even the moral fabric of communities. We also wish to acknowledge the diversity of perspectives represented in this book. Contributors hail from universities and institutions across Turkey, India, China, the United Kingdom, and the United Arab Emirates, among others. This international dialogue enriches the volume, reflecting that the study of war and conflict must remain global in scope. At the same time, many of the chapters are rooted in particular contexts—whether electoral politics in the U.S., mass-shootings, cinema in India, or the memory of genocide in Europe—underscoring the balance between universal patterns and localized experiences. Ultimately, this collection underscores the necessity of interdisciplinary scholarship when grappling with the cultural, moral, and technological dimensions of conflict. We invite readers to approach these chapters not simply as isolated studies, but as interconnected reflections on the challenges of our time. As editors, we hope this work inspires new conversations, encourages critical thought, and fosters an expanded understanding of the ways in which war continues to shape and unsettle the human condition. In closing, we turn to the enduring insight of one of the greatest—and arguably first—sociologists, Wali al-Din al-Hadrami (d. 1406), more famously known as Ibn Khaldun, whose reflections on the psychological dimensions of conflict remain strikingly relevant:

"Throughout history many nations have suffered a physical defeat, but that has never marked the end of a nation. But when a nation has become the victim of a psychological defeat, then that marks the end of a nation."

This insight reminds us that the deepest wounds of war are not always visible; they reside in spaces of memory, identity, and the moral fabric of communities. It is in these spaces that the true cost of conflict is measured. And it is here that this book begins its inquiry—into the cultural work of healing, the moral urgency of resistance, and the technological forces that shape how violence is remembered and understood.

Khaled Al-Kassimi

American University in the Emirates, United Arab Emirates

Tuba Isik

Istanbul Medipol University, Turkey

Muhammad Fahim
Monash University, Australia

Basak Gezmen
Istanbul Medipol University, Turkey

REFERENCES

Bauman, Z. (2000). *Liquid modernity*. Polity Press.

Heidegger, M. (1977). *The question concerning technology and other essays* (Lovitt, W., Trans.). Harper & Row.

Ibn Khaldun. (1967). *The Muqaddimah: An introduction to history* (Rosenthal, F., Trans.). Princeton University Press.

Marcuse, H. (1964). *One-dimensional man: Studies in the ideology of advanced industrial society*. Beacon Press.

Chapter 1

Cyber Warfare and Media Manipulation in the Age of Modern Conflicts: The Role of Social Media in the 2024 U.S. Presidential Election

Sevda Bozkurt
https://orcid.org/0000-0001-5882-0157
Gendermarie and Coast Guard Academy, Turkey

ABSTRACT

This study examines the impact of digitalization on political communication processes within the context of social media platforms. Focusing particularly on the example of the 2024 U.S. presidential elections, it is argued that social media has evolved beyond merely facilitating interaction between voters and candidates, transforming instead into a new battleground where methods such as cyber warfare and media manipulation are employed. In this study, social media is approached as a platform where both domestic and global actors conduct information warfare, steer public opinion through disinformation, and directly interfere in electoral processes. In this context, it is emphasized that traditional forms of warfare are increasingly being replaced by strategic communication techniques conducted through digital channels. In conclusion, it is asserted that today, social media platforms are not only tools of communication but also digitalized forms of political competition and international intervention.

DOI: 10.4018/979-8-3373-7589-2.ch001

INTRODUCTION

Throughout history, states have pursued their interests—security, economic gain, prestige, or ideology—through both peaceful and coercive means (Yalçınkaya, 2008, p. 17). When diplomacy fails, war emerges as the most forceful instrument of statecraft (Eslen, 2003, p. 13). War has been broadly defined as the organized use of force by states or alliances to impose their will on others (Varlık, 2013, p. 114; Clausewitz, 2011, p. 29). Legally, it involves state actors pursuing unresolved claims through national power elements, often in declared, prolonged, and intense forms (Akargün, 1993, p. 1–2; Keskin, 1998, p. 68).

From Hobbes's notion of war as an innate human condition (1997, p. 76–79) to early Mesopotamian conflicts and the moral legitimization of violence through Aquinas's just war theory, warfare has evolved alongside civilization (Sander, 2017, p. 33; Kurubaş, 2017, p. 18). Despite efforts to curtail it, war has remained a constant in human history—studies suggest 87 out of the last 100 years have witnessed conflict (Yılmaz, 2010; Herre and Roser, 2016). The roots of war are multifaceted. Realist theory sees war as a result of systemic imbalances and security dilemmas (Waltz, 2001, p. 205). Jomini (2013, pp. 17–18) similarly emphasizes the strategic and economic rationales behind warfare.

In tracing its evolution, analysts have classified war into four generations (Lind et al., 1989, pp. 22–26). First-generation wars featured organized infantry and linear tactics following the Peace of Westphalia (1648). The second generation introduced firepower and industrialization, evident in World War I's total and trench warfare (Parker, 1988; Van Creveld, 1991). The third generation, shaped by maneuver warfare and Blitzkrieg tactics, emphasized mobility and systemic collapse (US Marine Corps, 1997). Fourth-generation warfare marked a shift toward non-state actors, asymmetric strategies, and cognitive operations such as perception management and cyber warfare (Lind et al., 1989; Lind, 2004, p. 13–14). As Clausewitz aptly stated, "every age has its own kind of war." In our contemporary era, the nature of conflict has transcended classical definitions. Modern warfare is increasingly waged across political, technological, and informational domains.

Building upon this transformation, modern warfare has not only expanded its domain but also intensified its relevance in contemporary global politics. With the proliferation of digital technologies, the battlefield now includes social media platforms, data networks, and algorithmic environments. Modern conflicts are no longer fought solely with weapons and armies but with narratives, influence operations, and psychological tools aimed at shaping perceptions and manipulating behavior on a mass scale. In this evolving landscape, information dominance and cognitive superiority have become as critical as physical force, making modern warfare not just a matter of territorial gain but also one of ideological and informational control.

Within this framework, the present study focuses on the 2024 United States Presidential Election as a case through which to analyze the dynamics of modern warfare. This election is not merely a democratic process but also a critical arena where strategies of digital manipulation, cyber influence, and psychological operations unfold. The convergence of political campaigning with technological tools such as artificial intelligence, algorithmic targeting, disinformation campaigns, and social media warfare exemplifies how electoral processes have become fertile ground for contemporary conflict dynamics.

To investigate these dimensions, this study adopts a qualitative methodology grounded in case study and content analysis. The 2024 U.S. election will serve as the central case, allowing for an in-depth examination of the actors, narratives, digital tactics, and hybrid influence strategies deployed throughout the campaign process. Through this analytical lens, the study aims to shed light on the broader transformation of warfare—away from conventional military engagements and toward battles for perception, influence, and cognitive dominance in the digital age.

A War Without a Frontline: The New Dynamics of Modern Warfare

Modern warfare is not limited to classical armed conflicts between regular armies; rather, it is a multilayered form of struggle that spans diplomatic, economic, informational, psychological, and cyber domains. This type of conflict is characterized by a dynamic structure in which the geographic, physical, and cognitive boundaries of war become blurred, and confrontations are carried out both with and without weapons, involving both state and non-state actors (Jordan et al., 2008).

Modern wars are often conducted without formal declarations of war, using asymmetric and hybrid tactics. Unlike traditional wars with structured army confrontations, this type of warfare includes perception management, public opinion manipulation, cyberattacks, economic sanctions, and diplomatic encirclement (Renz, 2016; Gerasimov, 2016). Military strategies are no longer solely shaped by battlefield considerations but are increasingly aligned with broader political objectives.

The aim of modern warfare is not merely to destroy the enemy's military power or gain territory, as in classical warfare. Instead, the fundamental objective is to break the will of the enemy state or society, disrupt decision-making processes, and create strategic instability (Jordan et al., 2008, p. 9). This form of warfare seeks results primarily through psychological, cognitive, and institutional damage rather than physical destruction. As defined by David Jordan and his colleagues, success in modern warfare is measured not only by superiority on the battlefield but also by the capacity to neutralize the opponent's social structure, public opinion, decision-making elites, and information systems (Jordan et al., 2008). Therefore, controlling the flow of information, shaping perceptions, and achieving psychological dominance lie at the strategic core of modern warfare.

This reality is clearly reflected in Gerasimov's concept of "new generation warfare." According to Gerasimov, in modern conflicts, military force is used only as the final and fastest means; the primary objective is to destabilize the rival state internally beforehand through media, economic measures, and diplomacy (Gerasimov, 2016). In this context, modern war is a "process," the goal of which is to render the opponent's political decision-making mechanisms dysfunctional and to generate internal turmoil by undermining social cohesion. In conclusion, the ultimate aim of modern warfare is to go beyond physical destruction and achieve lasting superiority by disrupting internal order, breaking societal resilience, and producing strategic outcomes.

Although various factors contribute to the emergence of modern warfare, technological advancements play the most critical and decisive role. Innovations such as digitalization, artificial intelligence, cyber capabilities, and remotely operated unmanned systems have fundamentally transformed both the methods by which wars are conducted and the domains in which they take place (Horowitz, 2018; Jonsson, 2020). Today, warfare occurs not only in physical realms but also in cyberspace, on social media platforms, and within data infrastructures. With the increasing influence of non-state actors such as terrorist organizations, paramilitary groups, and hacker collectives, the classical symmetrical structure of war has collapsed; through hybrid war strategies, diplomacy, economy, media, and cognitive spheres have been simultaneously integrated into the battlefield (Lind et al., 1989; Renz, 2016). This new configuration blurs the boundaries between war and peace, civilian and military, reality and fiction. Consequently, the management of public perception and dominance over information have become as effective as direct armed struggle (Jordan et al., 2008). In this context, modern warfare can be defined as the form of conflict unique to the information age.

Furthermore, many wars today are waged without official declarations and within legal grey zones not clearly defined by international law. As Gerasimov noted, this allows uncertainty to become the defining characteristic of modern conflict (Gerasimov, 2016). Among all these components, technology stands out as the most critical determinant of the nature of conflict, functioning as the driving force behind the formation of modern warfare. Technology has profoundly reshaped both the methods and domains of warfare—particularly through the deployment of AI-supported weapon systems, unmanned aerial vehicles (UAVs), cyberattack capabilities, and deepfake technologies—which has led conflicts to shift from physical theaters to cognitive and digital dimensions. These tools demonstrate that modern warfare is no longer conducted solely on the battlefield, but increasingly through screens and algorithms.

Mass media and social media platforms, widely used by the public, are now regarded as the new frontlines of modern warfare. Due to the proliferation of technology and the expansion of global user bases into the billions, these platforms have become both arenas of warfare and direct instruments of conflict. Shaping public opinion through the production, dissemination, and manipulation of information is now of strategic importance, equal to military superiority. While traditional media has long been used to construct the legitimacy of warfare, in the conflict environments of the 21st century, social media platforms (e.g., Twitter/X, Facebook, Telegram) have transformed into dynamic and cost-effective tools instrumentalized for real-time propaganda, disinformation, psychological operations, and public opinion engineering. Jonsson (2020) describes this as a new war paradigm in which perception is more effective than armored divisions. These platforms are used to construct perceptions of victory or defeat, disrupt routine information flows, and establish psychological dominance, independent of physical outcomes on the ground.

Conflicts conducted via social media are particularly shaped by manipulative content, deepfake videos, bot accounts, troll networks, and targeted content dissemination. Russia's 2014 intervention in Crimea stands as a typical example demonstrating how social media tools can be used as strategically as conventional military assets (Renz, 2016). Similarly, in the Israel-Hamas, U.S.-Iran, and Ukraine-Russia conflicts, social media platforms have not only reported the wars but have also been employed as active components of the warfare itself. In modern conflict, media is no longer merely a medium that transmits information; it has become an actor that creates and shapes the reality of war.

Media and social media have become not only strategic objectives but also strategic tools in modern warfare. These instruments are employed both to gain international legitimacy and to influence the internal public opinion of rival actors, thereby transforming the frontlines of modern conflict from geographic to boundless, timeless, and multilayered domains. In this context, wars are no longer won or lost solely through weaponry, but also through words, images, and hashtags.

In conclusion, the concept of modern conflict reveals that warfare is not confined to tanks, rifles, and frontlines; rather, it denotes a form of struggle that is multilayered, irregular, and conducted at a cognitive level through digital technologies, cyber domains, artificial intelligence systems, and social media platforms. This new form of warfare is shaped not only by physical superiority but also by the control of information flows, management of perceptions, and the production of political legitimacy. In this regard, modern conflict transcends classical definitions of war, blurring the boundaries between war and peace, civilian and military, reality and fiction—thus constituting a new paradigm of conflict.

Today, modern warfare is no longer a singular phenomenon; it manifests in diverse forms such as hybrid warfare, cyber warfare, cognitive warfare, proxy wars, and fourth-generation warfare. Each type bears distinct characteristics in terms of execution methods, actors involved, and target domains. However, since this study specifically focuses on social media platforms—which have become decisive within the communication infrastructure of the digital age—it will primarily examine the concepts of hybrid warfare and cyber warfare. These two forms of conflict are directly connected to disinformation, perception management, digital propaganda, and cognitive intervention strategies conducted via social media, and are thus critically important for understanding the contemporary conflict environment.

The Changing Face of Modern Warfare: The Hybrid War Paradigm

Frequently encountered in academic and strategic literature through its derivatives such as "hybrid conflict" and "hybrid threat," hybrid warfare refers to a type of warfare that simultaneously utilizes special forces, irregular and regular troops, economic warfare, cyber operations, local insurgent support, information warfare, propaganda, and diplomacy (Glenn, 2009). In hybrid warfare, each of these elements operates in conjunction and mutual interaction, shaping the nature of war through flexible, multidimensional, and complex structures rather than fixed and predictable models.

The concept of hybrid warfare was systematically defined for the first time by Frank G. Hoffman in his 2007 work *Conflict in the 21st Century: The Rise of Hybrid Wars*. Hoffman argues that classical inter-state wars have been replaced by hybrid forms of conflict in which the lines between civilian and military domains blur, and where organized violence, terrorism, crime, and conventional warfare intertwine. In this context, hybrid warfare is described as a composite model wherein conventional military forces are employed in coordination with irregular tactics, non-state actors, asymmetric attacks, and criminal-terrorist activities (Hoffman, 2007, p. 29).

Various approaches to the nature of hybrid warfare further highlight its layered and intricate structure. According to Schroefl and Kaufman (2014, p. 7), hybrid warfare involves the simultaneous and coordinated use of regular and irregular military elements within the same operational environment. Josan and Voicu (2015, p. 50) expand on this view, defining hybrid warfare as a type of conflict in which not only regular military forces but also non-state actors are engaged. These actors operate alongside conventional military capabilities and employ unconventional methods such as terrorism, systemic violence, chaos-inducing strategies, cyberattacks, media manipulation, and economic interventions. From this perspective, hybrid warfare emerges as a multidimensional form of conflict, both in terms of the diversity of actors and the range of tools used.

Conversely, Lanoszka (2016) contends that hybrid warfare is not a technically new type of war but rather a strategic approach wherein states employ all instruments of national power—including military, diplomatic, economic, and informational tools—in an integrated manner to fulfill their foreign policy objectives. From this standpoint, hybrid warfare is not a defensive posture but rather an overtly offensive strategy aimed at destabilizing the target state or society. In sum, hybrid warfare represents the strategic convergence of conventional and unconventional instruments of power, rendering it a defining element of modern conflict and a key concept for understanding contemporary global security dynamics.

One of the primary objectives of hybrid warfare is to destabilize the political structure of a target country without formally declaring war, to overthrow the existing government, and to replace it with a political order aligned with the aggressor's interests (Güler, 2017). This illustrates that hybrid warfare is not merely a military strategy but also a tool for explicitly political objectives. To achieve these goals, actors engaging in hybrid warfare first conduct a detailed analysis of the target country's social fabric, ethnic and cultural fault lines, and political vulnerabilities. Based on this analysis, strategies are formulated to exploit societal sensitivities. In this process, intelligence-based strategic planning plays a critical role in the success of hybrid operations.

At the same time, this type of warfare extensively utilizes information technologies, communication infrastructures, and digital platforms, extending conflict beyond the physical realm into cognitive and social domains (Akın, 2019). However, advanced technological capabilities alone are insufficient to achieve success in hybrid warfare. Effective strategy must also take into account the historical background, strategic culture, geopolitical position, and economic structure of the target country to enable a holistic impact plan. Therefore, hybrid warfare seeks not only technological superiority but also superiority in information and the ability to manage social vulnerabilities. This transforms warfare from being limited to frontlines into a multidimensional struggle aimed at undermining the structural resilience of entire societies.

The concept of hybrid warfare aligns closely with the "new wars" theory, which Mary Kaldor (2013) defines based on the following key characteristics:

- The structure of actors has changed. New wars are no longer fought solely between regular state armies but involve a multiplicity of actors such as private security firms, mercenaries, warlords, paramilitary groups, and jihadist organizations.

- The nature of warfare has transformed under globalization. As wars become shaped by global flows, violence increasingly targets civilian areas, and the burden of conflict disproportionately falls on societies.
- State authority is eroding and violence is being privatized. These conflicts often emerge in environments with weak or dysfunctional state institutions, and the legitimate monopoly on violence becomes fragmented among privatized actors.
- Ideological foundations have been replaced by identity-based conflicts. New wars are frequently fought along ethnic, sectarian, or cultural lines, with identities being reconstructed through the very act of war.
- The aim is not territorial gain but psychological and political pressure. These wars prioritize the use of terror, fear, and instability to achieve political objectives. What is sought is not territorial control, but control over minds.
- The financing of conflict has shifted to global networks. Unlike traditional wars financed by state budgets, new wars are often sustained by international interest groups, illicit economic networks, and diasporic support, contributing to their prolonged nature.

In this context, the most distinguishing feature of hybrid warfare is the systemic use of violence by non-state actors, breaking the state's monopoly on legitimate force. Beyond conventional military means, hybrid warfare integrates political, technological, and societal tools to construct a multifaceted strategy against targeted states (Sadowski & Becker, 2010, p. 4). Hybrid warfare thus represents a complex and adaptive conflict model, waged not only on battlefields but also within information environments, economic structures, and the internal dynamics of society. Its ultimate goal is to weaken political authority, destabilize state institutions, and create chaos and power vacuums within the governing system.

Russia's 2014 intervention in Crimea is widely recognized in academic literature as one of the most concrete examples of hybrid warfare. During this intervention, while Russian Special Forces played a central role in the operation, other elements such as mercenaries, cyberattacks, economic pressure mechanisms, and media support were also utilized as integral components of the conflict. This multidimensional campaign demonstrated that hybrid warfare can be effectively implemented by states under asymmetric conflict conditions and revealed the defining characteristics of next-generation warfare. According to Josan and Voicu (2015, p. 51), these defining characteristics include:

1. Conducting military intervention without an official declaration of war,
2. Employing armed civilian elements in asymmetric conflict environments,

3. Blocking the military forces of the adversary indirectly through activist civilian groups, without engaging in direct confrontation,
4. Achieving military and psychological superiority through indirect and low-intensity methods,
5. Conducting simultaneous operations across land, sea, air, and cyber domains,
6. Shaping public perception by generating remote influence via mass media and communication tools.

This strategic approach enabled Russia to secure political and military advantages in its interventions in Crimea and Ukraine while avoiding direct responsibility and largely maintaining its international legitimacy. In this way, both the methods of implementation and the outcome-generating capacity of hybrid warfare have become visible through a striking example in the contemporary security environment.

Despite growing academic interest in recent years, hybrid warfare has not emerged as a novel type of conflict. Historically, the roots of this form of warfare can be traced back as far as the Peloponnesian War in the 5th century BCE (Mansoor, 2012, p. 2). In this context, Athens encouraged the Helots—an ethnically distinct group loyal to Athens—to revolt against Sparta in an effort to force the Spartans into peace negotiations. This example demonstrates the historical foundations of hybrid warfare by showing how asymmetric tools and non-state actors were used to achieve political objectives.

In more recent history, the 2006 war between Israel and Hezbollah and Russia's 2008 intervention in Georgia are considered contemporary examples of hybrid warfare strategies. Similarly, the Syrian Civil War is analyzed within this framework due to the employment of unconventional methods by both state and non-state actors (Bingöl, 2017).

Hybrid warfare is an inclusive form of conflict that can integrate all kinds of tools, methods, and actors to neutralize the adversary. Civilian and social elements that seemingly have no relation to warfare—such as criminal organizations, NGOs, terrorist groups, academic institutions, artists, journalists, and social media influencers—can each become components of this war. Due to its highly multidimensional nature, the events and processes involved in hybrid warfare are often misunderstood or misinterpreted (Gültekin, 2024).

A New Security Threat in the Digital Age: The Rise of Cyber Warfare

Throughout human history, war has manifested in various forms, adapting to the social, political, and technological conditions of each era. While early conflicts were waged with primitive tools, the development of weapons technology over time transformed warfare into a more destructive, organized, and multi-layered phenomenon. However, this transformation was not abrupt but rather the result of a centuries-long historical evolution. Within this historical continuum, the nature of warfare has undergone fundamental changes not only in terms of the tools used but also in the domains in which it is conducted and the structures it targets. In the contemporary era, cyber warfare has emerged as one of the most recent and striking links in this evolutionary chain, reshaping national security strategies. Unlike traditional forms of physical conflict, cyber warfare—driven by the dynamics of the digital age—targets information systems and societal order, thereby adding a new dimension to the phenomenon of war (Yenal & Akdemir, 2020, p. 415).

Cyber warfare, in its broadest sense, refers to digital offensive actions undertaken by one nation-state against the computer systems or network infrastructure of another, with the aim of causing damage, disruption, or functional failure (Clarke & Knake, 2014, p. 6). Within international doctrine, multiple definitions of cyber warfare can be found. According to Türkay (2013), cyber warfare is a method of conflict conducted via digital or technological means, encompassing activities such as the disabling of infrastructure, intelligence gathering, and propaganda dissemination. This definition highlights that cyber warfare is not merely a technical assault but also a multifaceted security threat with strategic and cognitive dimensions.

Taddeo (2012) defines cyber warfare as state-led attacks carried out to disrupt or control hostile elements, involving military-type operations in both physical and virtual environments at varying levels of intensity. Similarly, Shakarian et al. (2013) describe cyber warfare as a form of political action in cyberspace, conducted by state or non-state actors in response to perceived threats to national security and serving political objectives. This perspective underscores that cyber warfare is not only a technical operation but also an extension of national security strategies. In a similar vein, Beard (2013) defines cyber warfare as the aggregate of computer-based attacks conducted by one state against another state's public or civilian information systems using specialized software and technologies. Each of these definitions confirms that cyber warfare is regarded as a next-generation form of conflict aimed at both damaging technical systems and achieving strategic superiority.

Although there are various definitions of cyber warfare, a commonly accepted approach that has gained traction in the literature since the 2010s defines cyber warfare as cyberattacks conducted against a hostile state with the intent of inflicting damage comparable to that caused by physical warfare or rendering the digital infrastructure and critical computer systems of the target state inoperable (Açıkalın & Baykız, 2024, p. 204). This definition demonstrates that cyber warfare has evolved beyond being a mere threat in virtual space and has become a form of warfare with direct implications for national security and sovereignty.

While numerous state-sponsored cyberattacks have been carried out to date, one of the earliest and most notable examples is undoubtedly the Moonlight Maze operation. Moonlight Maze is considered one of the first large-scale state-sponsored cyber espionage operations recorded in the history of cyber warfare. Initiated in 1996 and continuing until 1998, this campaign targeted numerous strategic entities, including the U.S. Department of Defense, NSA, NASA, Department of Energy, and various universities and research institutions. The operation resulted in the theft of a substantial amount of sensitive data, including diagrams related to missile guidance systems, naval codes, and control systems, amounting to over five gigabytes of data. The fact that the attackers refrained from conducting attacks during working hours and Russian holidays reinforced suspicions that a state-sponsored actor linked to Russia was behind the operation. The U.S. government attributed the attack to a Russian-based cyber espionage group now known as "Turla." Moonlight Maze had repercussions beyond its technical impact, prompting a reassessment of U.S. cybersecurity strategies and leading to the development of new technologies and collaborative approaches to protect critical digital infrastructure (Açıkalın & Baykız, 2024, pp. 206–207). In this regard, the incident marked a turning point and set a precedent for future cyber threats.

The September 11, 2001 attacks on the World Trade Center, widely known as the "Twin Towers" not only reshaped global security paradigms but also profoundly influenced perceptions regarding the security of digital systems and cyberspace. Although the attacks were not cyberattacks in a direct sense, technical vulnerabilities such as the breach of Pentagon security systems, the disabling of air traffic radar, and the failure to receive hijacking signals from downed aircraft clearly exposed the fragility of advanced technological infrastructures. In this context, the 9/11 attacks are considered a technology-intensive act of terrorism. In their aftermath, concerns surrounding not only terrorism and national security but also cyber terrorism rapidly came to the fore. Public anxiety regarding the uncontrollability of information technologies and the loss of trust in digital systems heightened perceptions of threats in cyberspace and intensified discussions on the potential impacts of cyber terrorism. Particularly, the perceived invulnerability of American "impenetrable security networks" at the heart of post-Cold War hegemony was severely challenged by these attacks, drawing international attention and making the development of new cybersecurity policies inevitable (Holat, 2021, p. 112).

The United States is one of the countries most frequently targeted by cyber warfare. A noteworthy example occurred on April 24, 2013, when a hacker group of unknown identity took over the Twitter account of the Associated Press (AP) news agency and posted a fake message stating, "Breaking: Two explosions in the White House and President Obama is injured." Coming from such a credible source with over ten million followers, the message quickly caused public panic. Within just three minutes, the U.S. stock market experienced a sharp decline, resulting in an estimated $136 billion in economic losses (Özdemiroğlu, 2013). This incident clearly demonstrated that cyber warfare can cause direct harm not only to technical systems but also to a country's economic stability (Holat, 2021, p. 113).

Cyber warfare is not merely a technical mode of attack; rather, it constitutes a multilayered digital battlefield encompassing a wide range of methods and strategies. This form of warfare includes a broad spectrum of operations ranging from digital espionage and propaganda to system sabotage. According to Kara (2013, p. 40), cyber warfare involves methods such as espionage, manipulation, communication control, misinformation, and damaging systems through viruses. Core techniques include crashing systems via Trojan horses and malware, sabotage through cyber bombs, system lockouts, and infiltrating network structures to steal data or compromise information integrity. Additionally, disinformation campaigns conducted on social media platforms to influence public opinion highlight the cognitive and perceptual dimensions of cyber warfare. The common objective of these methods is to breach the target system's security in order to access, control, or incapacitate critical information. In today's cybersecurity environment, access to and control over information has become a new form of power projection, replacing the use of physical force (Holat, 2021, p. 111).

Emerging as a consequence of political tensions between states, cyber warfare has become a concrete security reality materialized through attacks targeting the strategic digital infrastructure and software resources of the information age. However, cyber warfare encompasses far more than the targeting of technical systems. The widespread use of the internet and the transformation of social media platforms into global communication networks connecting individuals, communities, and even states reveal that the effects of cyber warfare extend not only to infrastructure but also to human and societal domains. Social media, as a digital layer intertwined with individuals' everyday life practices, has evolved beyond a tool for communication and sharing to become a platform for perception management, psychological manipulation, and mass influence. In this context, the effects of cyber warfare surpass the physical realm and directly target individuals and communities on a psychosocial level. Therefore, analyzing the dimensions and effects of cyber warfare requires not only an examination of technical security processes but also an understanding of their societal and behavioral extensions (Çalışkan, 2023, p. 12). The penetration of cyber warfare into the realm of social consciousness has transformed social media platforms into not only communication tools but also strategic arenas for manipulation.

This transformation in the nature of warfare has also fundamentally altered traditional notions of strong and weak states. Today, having a large military force, including soldiers, tanks, aircraft, or heavy weapons systems, is no longer sufficient for a state to be considered powerful. Under current conditions, a state's strength is directly related to how well it adapts to technological developments, its resilience against cyber threats, and whether it can demonstrate deterrent capabilities and maintain a presence in digital environments. Damage that once required decades of conventional conflict can now be inflicted through a single computer at minimal cost and within a much shorter timeframe (Yenal & Akdemir, 2020, p. 416).

War in the Shadow of Digital Transformation: The Evolution of Manipulation, Propaganda, Disinformation, and Perception Management on Social Media

Historically, media has been among the most influential societal institutions in transforming how people access information and communicate. The emergence of media essentially began with the mass dissemination of written culture following the invention of the printing press in the 15th century. Johannes Gutenberg's development of printing technology broke the monopoly of elite groups such as the church and aristocracy over the production and distribution of knowledge. This process laid the groundwork for the construction of the modern public sphere, and in subsequent centuries, media emerged as a key tool in the democratization of society.

In the 18th and 19th centuries, the widespread circulation of newspapers and magazines, followed by the advent of radio and television in the early 20th century, transformed media into a visual and auditory medium capable of reaching broader audiences. Initially, these new communication technologies facilitated public access to information, contributed to awareness in areas such as education, culture, and public health, and promoted social integration by informing rural populations of central developments. Many broadcasting institutions operated under the principles of public service broadcasting, adhering to norms of impartial journalism and civic responsibility. As a result, media functioned as a vital instrument in nurturing democratic awareness and a culture of citizenship.

However, over time, this potential was instrumentalized and manipulated by various centers of power. Radio and television, in certain cases, became ideological apparatuses of the state. Especially in authoritarian regimes, media was used less for public benefit and more as a mechanism for social control and legitimization of political power. The Nazi regime achieved its most significant success in reaching the masses and steering public opinion through radio. During this period, radio became the central and most effective tool of official propaganda. Systematic broadcasts conducted by the Ministry of Propaganda aimed to exert cognitive control over the populace. To serve this purpose, low-cost, single-channel "Volksempfänger" (People's Receiver) radios were produced, capable only of receiving state broadcasts, and were mandatorily placed in a wide range of public venues—from restaurants and factories to schools and other communal spaces (Jowet & O'Donnell, 1999, p.

224). This ensured the constant exposure of the public to state radio programming, and media played a pivotal role in disseminating totalitarian ideology.

Similarly, in the ensuing years, television—especially during times of war and crisis—became central to state efforts in shaping public opinion and managing perceptions. One of the most striking examples of this occurred during the 1991 Gulf War. The U.S.-based CNN news network delivered live 24-hour coverage of the conflict to a global audience, demonstrating how television could evolve from a mere news source into a visual propaganda tool that shaped the reality of war. In these broadcasts, narratives emphasizing technological superiority and the notion of a "clean war" were prominently featured, while civilian casualties were either not shown or significantly downplayed (Kellner, 2009, p. 1). Thus, television played not only an informative role but also a strategic one in processes of political persuasion and legitimization.

By the 21st century, the process of digitalization had brought about a profound transformation in communication technologies. The most striking manifestation of this transformation has been the emergence of social media. With the development of Web 2.0 technologies, users ceased to be merely passive consumers of information and became active participants and content creators. Social media platforms structurally diverge significantly from previous forms of media. Content on these platforms circulates directly among users without undergoing editorial oversight, accuracy verification, or professional filtering processes. In fact, ordinary users with no institutional background or public reputation can, in some cases, reach audiences as large as those of long-established media outlets such as Fox News, CNN, or *The New York Times* (Allcott & Gentzkow, 2017, p. 211).

This development has, on one hand, strengthened democratic participation, facilitated access to information, and enabled the amplification of marginalized voices. Particularly during periods of crisis, social movements, or political campaigns, social media has served as a tool that facilitates individual mobilization and the formation of public opinion. Movements such as the Arab Spring, the Gezi Park protests, and #MeToo clearly demonstrate how social media can trigger collective action. Thus, social media has evolved beyond merely facilitating interpersonal communication to becoming a core component of social movements and political processes. Through networks, users can issue calls to action, gather participants, disseminate information, organize, and establish both dominant and counter-power spheres. In this respect, social media platforms stand out as privileged arenas where social and political mobilization takes place in the digital sphere (Castells, 2020, p. 195).

Despite these possibilities, the impact of social media on social and political life is not limited to its participatory, egalitarian, and emancipatory aspects. On the contrary, the structural characteristics of these platforms make them functional not only for public benefit but also as tools that are difficult to regulate, vulnerable to manipulation, and highly directive. In particular, the commercialization of user data, the algorithmic shaping of visibility, and the opaque practices of content moderation have transformed social media into a sphere of influence and power rather than merely a space for communication.

In this context, digital steering, misinformation, and polarizing content disseminated via social media generate new vulnerabilities across a wide spectrum—from individual preferences to collective decision-making processes. Therefore, social media embodies a paradoxical structure, presenting both opportunities and inherent risks. This contradiction becomes most evident when social media is instrumentalized not only to facilitate communication between individuals but also to influence how individuals think, make decisions, and develop behaviors.

A striking example of this manipulative potential is the Cambridge Analytica scandal that erupted in 2018. The scandal exposed the destructive impact of social media's manipulative capabilities on democratic processes. It revealed that the personal data of approximately 87 million users had been harvested from Facebook without consent and used to psychologically profile and target voters during the 2016 U.S. presidential election (Walsh, 2020, p. 850). This incident demonstrated that social media has evolved from a space of information sharing into a domain where algorithms and data analytics can shape individuals' political inclinations. As a result, social media has shifted from being a platform for free expression to a terrain where public opinion is reconstructed through invisible techniques of manipulation.

Before the rise of social media, traditional mass communication tools—particularly radio, television, and newspapers—served as highly effective instruments for shaping public opinion and conducting political campaigns. Due to their centralized structures and capacity to reach wide audiences, these media outlets were frequently utilized by political actors and centers of power to shape societal narratives. One of the most striking examples of the manipulation potential of traditional media is the 1929 campaign conducted by Edward Bernays—one of the founding figures in the history of public relations—for the Lucky Strike cigarette company.

During a period when smoking by women in public spaces was restricted by both social norms and certain legal regulations, Bernays designed a symbolic campaign linking smoking to women's liberation. As a relative of psychoanalyst Sigmund Freud, Bernays leveraged psychological persuasion techniques to frame smoking as a form of rebellion against male dominance. In this context, he collaborated with a group of feminist women, encouraging them to smoke in public during the Easter Parade. Bernays also strategically cooperated with the media, directing journalists to generate visuals and narratives aligned with the campaign's theme of "torches of freedom." The campaign effectively disseminated the image that 'free women smoke' throughout the media, significantly influencing public perceptions. As a result, the smoking rate among women, which stood at just 5% in 1923, rose to

12% by 1929 and eventually reached 35% in subsequent years (Gültekin, 2024, pp. 23–25). Through this media campaign framed around the rhetoric of women's freedom, the commercial objectives of the tobacco company were successfully obscured, offering a clear example of effective public manipulation by providing a false purpose to the public.

Today, especially through mass media and social media platforms, manipulation can be defined as the process of guiding and shaping the thoughts, perceptions, and behaviors of individuals, groups, or societies in alignment with the goals of particular interest groups. Manipulation may occur through the distortion or selective presentation of information and news in line with economic, political, religious, ethnic, or organizational interests (Karakaya, 2006, p. 176). In this context, actors who control the media frequently produce content that encourages the public to think in specific ways, often delivering these messages to mass audiences without their awareness. While traditional media can easily become a tool of manipulation in the hands of malicious actors, this manipulative potential has significantly increased in digital media given the technological capabilities of the present day. As Manovich (2001, p. 50) emphasizes, new media objects can be easily altered due to their susceptibility to algorithmic intervention; blurring in photos can be removed, colors and proportions can be modified, or objects can be deleted and replaced. As such, with the digitalization process, media has become programmable and well-suited for more sophisticated forms of manipulation.

Although visual manipulation has become more advanced and widespread thanks to digital technologies, such interventions are not exclusive to the digital age. In fact, examples of visual manipulation date back to periods long before the advent of digital tools. A particularly notable case from the Stalin era exemplifies this. Nikolai Yezhov, head of the Soviet secret police, arrested and executed many opponents during the period known as the Great Purge. However, in 1940, he himself was arrested on charges of disloyalty to the regime, tortured, and executed. After his execution, he was systematically removed from photographs taken with Stalin, earning him the nickname "the vanished commissar" in history (Stanger, 2018). This case illustrates that visual materials were manipulated for political purposes to reshape history and public perception long before digital technologies. In today's digital era, however, such manipulations have become far more rapid, widespread, and difficult to detect, sparking serious debates about the credibility of visual content.

Coxall (2013, p. 201) defines media manipulation as a process in which partisan media actors use various manipulative techniques to generate images or arguments that prioritize their own interests rather than those of their audiences. However, this process is not limited to manipulation alone; manipulation is often intertwined with other communication strategies such as disinformation, perception management, and propaganda, thereby creating a complex field of influence.

Jowett and O'Donnell (2006, p. 7) define propaganda as a deliberate and systematic process of influence that aims to shape the perceptions and manipulate the behaviors of a target audience to achieve outcomes desired by the propagandist. The historical roots of propaganda techniques trace back to Ancient Greece. Even in that early period, carefully crafted rhetoric was employed to influence public attitudes and behaviors, and deceptive tactics were utilized during times of war (Öymen, 2014, p. 111). Despite lacking modern communication tools such as newspapers, radio, and cinema, the Athenians effectively utilized the social and cultural institutions of their era to influence thought and opinion. Theaters, parliamentary sessions, courts, and religious festivals served as primary venues for propaganda; notably, playwrights used drama as a powerful tool to convey political, social, and moral messages (American Historical Association, n.d.).

Throughout history, the tools employed in propaganda activities have continually evolved, expanding the reach of such efforts in tandem with advances in communication. The impact of propaganda has grown stronger with the development of mass media, and today, it penetrates nearly every aspect of society (Çakı, 2018, p. 14). While the advancement of mass communication has broadened the scope and effectiveness of propaganda, technological progress has also enabled the use of new platforms for propaganda purposes. In this context, the rise of digitalization and internet-based communication tools has laid the groundwork for the increasing significance of social media as a propaganda medium. Unlike traditional propaganda tools—which were typically expensive and controlled by dominant power structures—social media platforms, due to their low or no cost, have made propaganda activities accessible to all, removing them from the exclusive domain of elite circles (Özdemir, 2015).

Moreover, social media transcends temporal and spatial limitations, enabling broader and more diverse audiences to be reached. Political actors, activists, and even states now use these platforms to mobilize domestic and international public opinion swiftly and effectively (Seo, 2014, p. 151). This shift has brought about a profound transformation in political communication. Politicians are now able to reach their constituents directly via social media platforms, sharing their thoughts, political agendas, and messages through microblogs without relying on traditional media outlets (Doğan, 2014, pp. 73–74). Today, propagandists often voice their messages on social media before doing so through mainstream news outlets. At the same time, the symmetrical communication capabilities provided by these platforms allow for immediate feedback from voters and opposition groups. This new communication environment marks a departure from an era in which politicians had to rely on journalists to reach the electorate—now, they can communicate directly through social media (Karagöz, 2018, p. 697).

The interactive and cost-effective communication opportunities offered by social media have further amplified the influence of propaganda conducted in digital environments. These platforms offer significant advantages in terms of political control mechanisms and create an open space accessible to all political actors. Citizens, politicians, political parties, opinion leaders, political advertisers, and manipulators alike converge on these platforms, which have become spaces where every kind of discourse circulates and where deep surveillance mechanisms are actively employed. While these environments may appear to support democratic debate, organized networks and interest groups systematically exploit them to steer political narratives. Indeed, numerous studies show that social media serves as a space for deliberation over public policy and politics (Woolley & Howard, 2019, pp. 8–9). However, this very function also transforms social media platforms into strategic tools for the execution of manipulative disinformation campaigns.

Disinformation is defined as false or misleading information that is deliberately produced and disseminated with the intent to deceive. Historically, the concept has been systematically used as a tool of propaganda, especially since the mid-20th century. The origin of the term is attributed to Soviet leader Joseph Stalin, who reportedly coined the term "dezinformatsiya" to make it appear as if it were of French origin. This deliberate naming strategy also aimed to create the impression that disinformation was a practice rooted in the West. Indeed, the 1952 edition of the Great Soviet Encyclopedia defines disinformation as "the dissemination of false information with the intent to deceive public opinion" (Encyclopedia, n.d.). Disinformation, which is produced intentionally to shape public opinion and incite social chaos, has a far more destructive and dangerous potential than misinformation spread without intent (Wardle & Derakhshan, 2017).

During the Cold War, disinformation became central to the psychological warfare between superpowers. In the bipolar world order, the Soviet Union systematically conducted disinformation campaigns under the rubric of "active measures." The Soviet Committee for State Security (KGB) invested significant efforts in producing fake documents and spreading conspiracies to discredit its adversaries. One of the most notorious examples is Operation INFEKTION in the 1980s, where the KGB successfully propagated the false claim that the AIDS virus had been developed by the United States—regarded as one of the most "successful" disinformation operations of the Cold War (BBC News, 2017). In response to the growing power of Soviet disinformation, the United States also changed its strategy in the early 1980s under the Reagan administration and began to develop mechanisms to expose and counter Soviet narratives (Schoen & Lamb, 2012, p. 26–27). Thus, the Cold War period witnessed the use of disinformation not only by the Soviet Union but also by the United States for both defensive and offensive purposes. This period marked a significant rise in awareness of disinformation in Western societies and led to the integration of the concept into international academic literature.

A striking example of the effective use of disinformation from the Cold War era is cited by Cole (1998, p. 172). In a horse race featuring one American and one Soviet horse, the American horse won while the Soviet horse came in second. However, the Soviet state news agency Pravda reported the event with the headline: "In an international race, the Soviet horse came in second," deliberately omitting the fact that there were only two competitors and using the term "international" to suggest a prestigious, large-scale event. While the statement about the Soviet horse finishing second is factually correct, the manner in which it was presented misled readers—demonstrating the manipulative nature of disinformation.

Historically, the media has been positioned as a key communication tool in democratic societies, tasked with informing the public, serving as a watchdog, and fostering social awareness. However, due to its wide-reaching influence, the media has always been subject to the pressures and influences of various interest groups, often shaping its content in line with the demands of economic and political power centers. This undermines the ideal of the media as a neutral instrument serving the public interest, turning it instead into a tool for reinforcing the power of dominant actors. Consequently, disinformation—based on the deliberate distortion of truth—has become a structural component of the media ecosystem. This is particularly visible in news content, which tends to be perceived as more credible by the public (Herman & Chomsky, 1999, p. 22).

In today's journalistic landscape, where the principles of impartial reporting are increasingly eroded, disinformation has become more complex. Critically assessing who produces news content, for what purposes, and through which processes has become a vital necessity in the field of communication (Herman & Chomsky, 1999, p. 22). In the digital media age, this vulnerability has deepened further: social media platforms, which reach billions globally and enable rapid dissemination, have become fertile ground for the spread of disinformation (Aydın, 2023, p. 2605). Moreover, the traditional practices of witnessing events, verifying information, and fact-checking in news production have been replaced by user-driven, real-time content creation on social media. This has led to the circulation of unverified information at unprecedented speeds. Fake content carrying the characteristics of disinformation not only shapes individual perceptions but also influences behavior and decision-making processes.

Today, disinformation on social media has become a significant global challenge affecting societies worldwide. During major social events, factually disconnected content and debates proliferate on these platforms, misleading large audiences. Individuals often unknowingly become victims or agents of disinformation. On X (formerly Twitter), for example, disinformation actors generate content related to trending topics, deploy hashtags, and disseminate it through multiple accounts, triggering the disinformation process. As participation increases, false information spreads rapidly and is sometimes amplified by high-profile accounts or news outlets, drastically increasing its reach (Ertem, 2019, pp. 61–62). Furthermore, popular culture figures, politicians, and social media influencers often share unverified content, thereby unintentionally contributing to the spread of fake news and enhancing the impact of disinformation.

During the COVID-19 pandemic, social media emerged as a critical medium for the dissemination of disinformation, with numerous unverified and non-scientific claims misleading large audiences. Misleading advice such as consuming high doses of vitamin C, drinking hot water, keeping the throat constantly moist, and using antibiotics—along with dangerously harmful recommendations like ingesting bleach or pure alcohol—revealed the severe dangers of disinformation on these platforms (Bernard, 2020). In this period of fear and uncertainty, individuals were often guided by false information circulating rapidly on social networks, leading them to adopt harmful practices. This demonstrated that disinformation does not merely create informational confusion but can also pose tangible threats to public health.

As observed in many similar cases, the spread of disinformation on social media during the pandemic reached such a critical level that even the companies managing these platforms found themselves compelled to take action against false information within their own systems. Consequently, social media companies began integrating anti-disinformation strategies into their corporate policies by establishing dedicated departments or collaborating with independent fact-checking agencies. For instance, the company META initiated a partnership with Reuters to verify the accuracy of content shared on Facebook and Instagram (Voice of America, 2020).

Considering that disinformation seeks to influence individuals' cognitive orientations through intentionally fabricated content designed to mislead and distort public opinion, a more comprehensive and strategic concept—perception management—has gained prominence. Unlike disinformation, perception management does not merely involve the spread of false information; it refers to a set of systematic and planned actions aimed at shaping the emotions, motivations, and cognitive processes of a target audience.

The term "perception management" was first introduced by the United States Department of Defense and defined as "actions to convey and/or deny selected information and indicators to foreign audiences to influence their emotions, motives, and objective reasoning, and ultimately the behavior of foreign governments, organizations, groups, and individuals" (Dictionary of Military and Associated Terms, 2020). While this definition emphasizes the behavioral outcomes of perception management, consciousness-oriented approaches define the concept as a strategic shaping of perceptions intended to generate consent and facilitate the achievement of specific objectives (Siegel, 2005, p. 118). Therefore, perception management is not limited to the manipulation of information; it represents a multidimensional process that targets individuals' mental and emotional states.

Perception management is defined as an interactive communication process aimed at shaping the attitudes and behaviors of the target audience in a specific direction. Structuring and directing messages in accordance with the audience's perceptual framework constitutes a key strategic approach within this process (Garfield, 2002, p. 33). The implementation of perception management typically involves four stages: first, attracting the attention of the target audience; second, sustaining that attention by delivering similar content; third, aligning the conveyed message with the audience's preexisting perceptions; and finally, repeating the message to ensure its retention and reinforcement (Callamari & Reveron, 2003, p. 15).

In this context, repetition plays a critical role in shaping public perception. As Le Bon (2014, p. 123) notes, a simple and comprehensible claim, when repeated frequently using the same wording, can eventually be accepted as truth without scrutiny. The persistent repetition of a message, regardless of its actual content, facilitates its entrenchment in public memory. As a result, while the source of the message may fade into the background, the content gains legitimacy through repetition and transforms into a perceived truth within the collective consciousness.

The concept of perception management is interpreted and defined in various ways depending on the context in which it is applied, the position of the actor, and the intended objectives. Within this framework, some approaches regard perception management as a manipulative tool aimed at deceiving and directing the target audience, while others see it as a legitimate process of influence within the context of strategic communication. From a marketing and persuasion-oriented perspective, perception management refers to a set of techniques aimed at guiding consumer behavior. At the individual level, these techniques are often employed—consciously or unconsciously—to persuade others, justify oneself, or maximize personal gain. States, on the other hand, utilize perception management as a tool to generate legitimacy, shape public opinion, and influence external actors at both the national and international levels. Similarly, corporations apply these techniques to gain competitive advantage or enhance their market influence, while political parties deploy them to shape voter behavior and undermine opposition. At the institutional level, such strategies are generally considered under the banners of "public relations" or "corporate communication." This indicates that perception management constitutes a multidimensional field encompassing disciplines such as communication, psychology, political science, and business administration.

Accelerated by technological advancements and globalization, the rapid circulation of information has further expanded the scope of perception management, embedding it at the core of contemporary social and political relations. These strategies, aimed at shaping the emotions and thoughts of the masses, are often implemented within different conceptual frameworks such as propaganda, public diplomacy, soft power, or marketing. However, at their core, they align with techniques that seek to persuade individuals and mobilize them toward specific objectives (Derman, 2021, pp. 64–65). In this regard, the media—particularly through visual and auditory content—emerges as a crucial instrument for influencing attitudes and behaviors on a large scale. The channels through which individuals access information directly affect the success of perception management efforts.

A historical examination reveals that perception management is not exclusive to the modern era; rather, it has served as an effective tool in military and political spheres since ancient times. In this context, the biblical story known as "Gideon's tale" offers a striking example of how perception management was applied in the field of warfare. According to the narrative, only 300 soldiers equipped with torches and trumpets were sent into the Midianite camp. The visual and auditory spectacle created an illusion of a much larger army, inciting panic among the enemy forces (Jowett & O'Donnell, 2011). This strategy can be considered an early example of establishing psychological superiority through perception rather than physical dominance.

A similar example is found in the Battle of Pelusium in 525 BC between Ancient Egypt and the Persians. The Persian army reportedly drew images of the goddess Bastet—sacred to the Egyptians—on their shields, and in some accounts, even carried live cats at the front lines. The Egyptian soldiers, reluctant to harm sacred animals, were demoralized and confused by this tactic. According to the ancient historian Polyaenus, the Persians used not only cats but also other animals considered sacred by the Egyptians—such as dogs, snakes, and sheep—effectively stopping the enemy's arrow fire (Elektratig, 2014). These examples illustrate that perception management has historically been crafted not only through verbal or written communication, but also through symbols, religious sentiments, and cultural values, becoming an inseparable element of warfare strategies.

Throughout history, perception management—especially in the military domain—has evolved into a comprehensive strategy encompassing propaganda and psychological warfare. It has served as a means not only to defeat enemy forces physically but also to control their minds and emotions. Over time, this strategy extended beyond the battlefield, influencing all layers of society. By the 21st century, the steering of perceptions had transformed into an intervention capable of affecting decision-making processes and even suppressing individuals' free will. In this transformation, the media has played a central role: messages disseminated through mass communication channels function as powerful mechanisms capable of reshaping individuals' beliefs, values, desires, and cultural identities (Utma, 2022, p. 61).

In the contemporary era, not only traditional mass communication tools but also new media environments introduced by digitalization are actively used within the scope of perception management. A wide range of topics—from political developments and economic crises to popular culture products and daily life practices—are extensively debated on social media platforms, holding the potential to influence public perceptions. Perception management has acquired a new dimension due to rapid technological advancements and the ability to disseminate information to large audiences within a short period. In the age of information, processes of persuasion are increasingly shaped by the opportunities provided by social media. Platforms such as Facebook, Twitter, and Instagram are not merely channels of communication, but powerful instruments that shape individuals' attitudes and behaviors (High & Solomon, 2011, p. 91). In these digital environments, access to information has become easier regardless of time and space, and content production and consumption have taken on an interactive character. However, the intensity and rapid circulation of information in these spaces often lead to deviations from objective reality; content is decontextualized and reconstructed before being presented to audiences. As a result, accessing accurate information is becoming increasingly difficult (Utma, 2022, p. 65).

Russia implemented a multilayered perception management strategy through media as part of its hybrid warfare approach during the 2014 Ukraine crisis. Especially with the aim of psychologically winning over the Russian-speaking population in Crimea, propaganda efforts were supported by narratives designed to legitimize Russia's military intervention both domestically and internationally. After Russian troops entered Crimea, Ukrainian television channels were shut down and replaced with pro-Kremlin Russian broadcasts, thereby exposing the public only to one-sided, manipulated information. During this period, social media was also actively employed; through fake blogs, troll accounts, and manipulated content, historical legitimacy claims such as "Crimea has always been Russian territory" were systematically disseminated. At the same time, the Ukrainian state was portrayed as a failed, impoverished, and radical entity; protesters were depicted as militant extremists, and the Ukrainian army as disorganized and demoralized. These strategies were used to win the loyalty of the Russian minority and to soften international reactions. Furthermore, the dissemination of fabricated or decontextualized content—such as footage of a child killed in Syria being falsely linked to the conflict in Ukraine—illustrates how the process reached the level of disinformation (Sönmez, 2022, pp. 505–507).

With digitalization, perception management has moved beyond being a limited strategy reliant solely on traditional media channels. Instead, it has gained a far more dynamic and widespread influence through social media platforms. Due to the interactive nature of these platforms, individuals have become not only targets but also producers of content, transforming perception management into a two-way and increasingly unpredictable process. In this regard, social media not only accelerates the flow of information but also plays a central role in shaping perceptions of reality. Consequently, the scope and impact of strategic communication activities have increased significantly compared to previous eras.

Reflections of Cyber Warfare on the Electoral Sphere: A Social Media-Centric Perspective on the 2024 U.S. Elections

Until the advent of the internet, mass communication tools operated within a highly institutionalized framework. These tools—such as newspapers, radio, television, magazines, and cinema—which are now considered traditional media, had a one-way flow of information and thus exerted direct influence on the social structure (Toplu, 2018, p. 355). The dissemination of public information, the construction of public agendas, and the visibility and legitimization of political actors were largely carried out through these media. In traditional media, practical elements such as on-camera appearances, rhetoric used in interviews, and body language were deemed essential in shaping the political identities of politicians. For this reason, messages to be broadcast and statements in interviews were not spontaneous, but were carefully designed through curated representations. As a result, politicians began to address not the public directly, but representations crafted in line with public expectations.

As its name implies, mass media—which conveys messages to the masses—enabled political actors to reach and influence audiences, yet rendered the public's role in this process largely passive. In this context, rather than focusing on informing the public, the media was primarily used to shape public opinion through strategically designed political agendas. Over time, media ceased to be a vehicle serving public interest and instead became a tool catering to the representational and legitimization needs of dominant actors. Hence, media communication drifted away from the ideal of participatory democracy and transformed into a hegemonic tool governed by a regulatory elite that created a filter between society and communication itself.

In the traditional media era, politicians' access to the public was largely subject to the approval of media owners, their editorial policies, and political orientations. Particularly, the presence of commercial media entities facilitated the manipulation of politics through media. News content was often shaped not by informational or political value but by emotional appeal. This undermined media independence and democratic oversight, ultimately depriving the public of the opportunity for impartial news consumption.

The technological developments in the late 1990s marked a pivotal moment, integrating all media and communication tools into a global, interconnected ecosystem accessible worldwide. These innovations not only initiated the transformation of media but also provided individuals with the ability to communicate in free and interactive environments, independently from institutional structures. The rise of internet technologies, particularly the Web 2.0 revolution, triggered the most significant rupture through the creation of social media platforms. These platforms enabled users not only to access information but also to produce, disseminate, and continuously modify content. This transformation eliminated the passive consumer position of the individual in the communication process, rendering them active participants in content creation. The shift in the individual's communicative role led to the evaluation of social media as a democratizing tool. Especially in terms of political participation and freedom of expression, this development promised a hopeful transformation. Consequently, the media landscape evolved into a freer communication platform where not only large media outlets but also individuals, communities, and small-scale entities could actively participate (Toplu, 2018, p. 355). The prominence of platforms such as Facebook, YouTube, and X (formerly Twitter) in various social and political movements—from the Arab Spring to the Occupy movement, Black Lives Matter, and feminist activism—serves as clear evidence of this trend.

Through social media, individuals have become agents in the communication process. As the interactive structure of these platforms and their opportunities became widely recognized, they quickly garnered mass attention and witnessed rapid increases in user numbers. The active participation of users on these platforms led to a decline in public interest in traditional media. In turn, traditional media institutions had to adapt to these developments to maintain their relevance. Nowadays, traditional communication channels like radio, television, and newspapers have begun to use social media platforms and prioritize direct interaction with their audience, valuing their feedback (Toplu, 2018, p. 359). Social media distinguishes itself from traditional media tools through its features such as speed, effectiveness, and real-time updates, as well as through its role in political participation, agenda-setting, and making societal demands visible. The interactive and dynamic nature of social media has granted institutions, just like individuals, a central role in strategic communication practices. Accordingly, many organizations today regard social media platforms as both a primary and indispensable element in shaping their communication strategies toward the public.

In its early days, social media was largely perceived as a space primarily used by younger populations for socializing and entertainment. For this reason, public institutions and political actors approached these platforms with caution and often failed to take them seriously. Initially, some politicians utilized social media during election periods to build personal image and support their campaigns, yet they overlooked its potential for direct communication with voters. In this early stage of political engagement with social media, the content largely consisted of traditional political discourse simply transferred to the new digital medium without significant adaptation.

In a short period, as social media platforms rapidly expanded their user base, they also became attractive spaces for institutional and political actors. What began as sporadic and election-focused use of social media by politicians evolved into more institutionalized and systematic practices. Politicians began to employ digital media consultants, make regular public statements via Twitter, and leverage the dynamic nature of the internet by streaming rallies and political advertisements through social media. As seen, political actors' attitudes toward social media have shifted in line with technological advancements and the increasing functionality of these platforms. These spaces, where individuals can directly express their societal demands and opinions, have gradually turned into significant tools for political actors to gauge public sentiment and shape their strategies accordingly. Consequently, social media platforms have transformed one-way communication between political actors, individuals, and society into an interactive communication structure, offering promising democratic potential.

Although political actors initially approached social media with skepticism and concern, over time, they developed strategies to exploit and even dominate the platform for their own interests. Thus, in addition to its democratic potential, social media has revealed its capacity to be utilized as an authoritarian tool for political control and mass manipulation. Particularly during election periods, social media becomes a highly active space where politicians deliver instantaneous messages, gauge public reactions, promote their campaigns, and target their opponents. At the same time, it has become a primary ground for manipulation, disinformation, propaganda, and perception management. When these communication strategies—long used in traditional media—began to be employed via social media, they became significantly faster, more widespread, and directly implementable. The lack of editorial oversight mechanisms, which were typical in traditional media, has facilitated the dissemination of unverified information, prioritized emotionally charged content, and enabled perception management through echo chambers with greater efficacy. As a result, it has become increasingly questioned whether political actors use these platforms to foster democratic participation or to establish digital dominance.

While social media is used effectively in many fields, it is in the political realm—especially during election periods—that it is employed most strategically and comprehensively. The most prominent examples of this trend can be observed in the U.S. presidential elections, where the active use of digital platforms and the global visibility of political competition have become particularly evident. Beginning with the 2008 election, the influence of social media has steadily increased through the 2016, 2020, and 2024 elections.

During the 2008 presidential campaign, Barack Obama became the first candidate to effectively use social media as a political tool. His use of these platforms to modernize campaign strategies, enhance social outreach, and energize his voter base marked a revolutionary shift in political communication (Genel, 2012, p. 30). By leveraging social networks in a centralized and effective manner, Obama pioneered the use of online technology in electoral politics (Borins, 2011, p. 85). According to Dostie-Goulet (2009, p. 418), Obama's strategic use of social networks in the 2008 U.S. presidential election led to a remarkable increase in youth political participation compared to previous elections. This demonstrated that social media platforms significantly boost young voters' motivation to engage in political processes. Similarly, Sotirovic (2004, p. 595) argues that the internet, when used to access political information and campaign news, has a positive effect on users, increasing their level of political awareness. Both scholars emphasize that digital platforms play a vital role not only in information transmission but also in encouraging participation and expanding democratic processes.

By 2016, the widespread adoption of artificial intelligence environments, the creation of new media ecosystems, and rapid technological developments had led to profound and swift changes in all domains, including political communication. Social media environments had become indispensable tools for political actors to reach target audiences, build images, share opinions and ideas, and announce events and activities (Doğan & Ertan, 2016, p. 139). Therefore, the 2016 U.S. presidential election marked a critical turning point in demonstrating the power of social media to influence political processes. In this period, social media evolved beyond being merely a communication tool and became an instrument for perception management and voter manipulation. For instance, during the 2016 U.S. presidential election, the Cambridge Analytica scandal involving Facebook founder Mark Zuckerberg revealed that algorithms and artificial intelligence were being used effectively to shape public opinion and played a pivotal role in the spread of manipulation and disinformation (İldem, 2021, p. 2). The widespread circulation of disinformation on social media during this election generated high engagement, leading to significant public opinion shifts and misleading outcomes.

In this context, not only social networks but also search engine algorithms played a critical role in the rapid dissemination of information and news to millions. These algorithms functioned as a form of digital gatekeeping by filtering users' searches according to their personal interests and presenting them as suggestions, thereby directing the flow of information (Uluk, 2018). Rather than official endorsement of a specific candidate, the heightened visibility and engagement in favor of Donald Trump during the 2016 election process stemmed from several interrelated dynamics. First, Trump's intense use of Twitter allowed him to bypass traditional media filters and set the public agenda directly through sensational and concise messages. Second, a significant portion of the disinformation circulating on Facebook and Twitter targeted Hillary Clinton, thus playing a crucial role in shaping voter perceptions in favor of Trump (Allcott & Gentzkow, 2017). Third, since social media algorithms tend to prioritize high-engagement content, Trump's provocative posts spread rapidly and gained viral traction. Lastly, within the framework of the Cambridge Analytica scandal, unauthorized collection of Facebook user data and the use of psychological profiling techniques to craft micro-targeted campaign messages helped mobilize voters in favor of Trump (Isaak & Hanna, 2018). Consequently, social media, in tandem with Trump's communication strategy, significantly influenced both the circulation of information and voter behavior.

The various scandals associated with social media platforms during the 2016 U.S. presidential election raised serious concerns among the public about the impartiality and reliability of digital media. In particular, the unauthorized collection of data from millions of users via Facebook and its use for political purposes by firms such as Cambridge Analytica increased doubts about the fairness of the election process. Additionally, findings that fake accounts, bot networks, and foreign actors conducted systematic disinformation campaigns to manipulate voters severely undermined public trust in social media platforms. These developments not only diminished users' individual trust in information but also cast doubt on the integrity of democratic processes.

The 2020 U.S. presidential election emerged as another turning point in which social media played a transformative role in both communication and information control within political processes. Due to the COVID-19 pandemic, in-person campaign activities were limited, prompting political candidates to use social media platforms as their primary communication channels. At this point, Joe Biden focused on paid advertising campaigns on platforms such as Facebook and Instagram, whereas Donald Trump adopted a direct and confrontational discourse through Twitter (Isaak & Hanna, 2018). Unlike previous elections, social media platforms took a more active regulatory approach; Twitter and Facebook added warning labels to Trump's misleading posts, intervening in the content (Kleis Nielsen & Graves, 2020). Consequently, Trump's campaign efforts were disrupted, and despite attempts to respond to the situation, his efforts proved ineffective. In contrast, the visibility of Biden's tweets increased on the platform, becoming a significant factor in his electoral success. These developments demonstrated that social media companies were no longer mere neutral communication infrastructures but had become regulatory powers and political actors directly intervening in the political process.

A significant development after the 2020 election further enhanced the impact of social media platforms on political processes. Twitter initially suspended Donald Trump's account temporarily, citing that his posts could threaten public order and encourage violence. Following the storming of the Capitol building on January 6,

2021, Twitter permanently suspended Trump's account, citing his role in inciting the incident. The platform justified the decision by assessing that Trump's rhetoric could embolden violent groups and endanger public safety (blog.x, 2021).

In 2022, a major change in Twitter's ownership structure occurred when Elon Musk acquired the platform and rebranded it as "X." This shift not only transformed the platform's corporate structure but also had substantial implications for public opinion formation, content moderation, and political discourse. Shortly after the acquisition, Musk conducted a poll to decide whether Donald Trump's personal account should be reinstated, sharing the results publicly. When 51.8% of participants supported reactivating Trump's account, Musk announced the decision using the Latin phrase "Vox Populi, Vox Dei," meaning "The voice of the people is the voice of God." After approximately two and a half years, Trump resumed posting from his "@realDonaldTrump" account and actively used the platform to conduct his 2024 election campaign. During the previous election, Trump had been unable to fully utilize Twitter's reach due to the platform's restrictions. Now, with Musk as the new owner of X (formerly known as Twitter), Trump gained a powerful ally in a domain where he had previously been limited (BBC News Turkish, 2022). This development provided Trump with advantages in a sphere where he had previously struggled to establish a presence.

In the previous election cycle, the platform that had blocked public opinion polls conducted by Trump supporters did not interfere with similar polls conducted again by Trump supporters in 2024. On the contrary, these tweets were disseminated and encouraged by Elon Musk, the new owner of the social media platform. In this context, Musk's statements following the failed assassination attempt on Donald Trump in July 2024 in Butler County, Pennsylvania, clearly revealed his political stance. Shortly after the incident, Musk publicly declared his official support for Trump's candidacy; later, in October 2024, during a town hall meeting in the suburb of Folsom, Philadelphia, he reaffirmed this support. In the same event, Musk urged voters to cast their ballots for Trump and emphasized that the upcoming elections would be a historical turning point "determining the fate of Western civilization." Moreover, he retweeted posts containing criticism of Harris, Trump's opponent in the elections, thus amplifying those critiques and contributing to their wider dissemination. These developments show that Elon Musk has become directly involved in Trump's election campaign, not merely as the owner of a social media platform, but also as an active political figure, particularly in strategically significant states such as Pennsylvania. This form of intervention has further deepened debates around the transformation of social media platforms from neutral digital spaces into determining actors in political processes (Reuters, 2024).

Under Musk's ownership, the X platform has become a more favorable environment particularly for right-wing media content creators. Having already experienced the impact of social media in previous elections, Donald Trump quickly adapted to the situation in the 2024 presidential race and made substantial use of social media participation. Trump strategically utilized each social media platform to reach different segments of the electorate, and his communication style on the X platform, in particular, set him apart from other political actors. For instance, he frequently shared links on his X account that directed followers to visual content on his Instagram page, but rather than posting the visuals directly, he used hyperlinks that guided viewers to the Instagram platform. Through such targeted links, Trump ensured that followers accessed more detailed and reinforcing information related to his posts (Doğan & Ertan, 2016, p. 148).

In his posts on X, Trump paid close attention to being simple, concise, and clear. His direct and colloquial language contrasted sharply with the distant and formal tone of traditional political discourse, generating a sense of closeness between him and voters. Furthermore, unlike other candidates who emphasized "we," Trump frequently used the pronoun "I," thereby reflecting an ego-centered leadership model. In addition to his self-aggrandizing statements, he also adopted a mocking and belittling tone toward his opponents. This suggests a deliberate rhetorical strategy aimed at elevating his personal charisma while discrediting rivals. What is striking, however, is the extraordinary public reaction to such behavior, which, although bordering on incivility and generating a polemical environment, garnered substantial engagement. The high number of likes and retweets for such mocking, humorous, and derogatory posts indicates that Trump's confrontational tone and rhetoric allowed him to interact far more effectively with audiences (Doğan & Ertan, 2016, p. 148).

Another important element of Trump's social media strategy is his ability to integrate elements of popular culture and adapt to the dynamics of digital platforms. Political actors now align themselves with platform-specific trends and humorous language to capture the attention of their audiences. As a result, politics has increasingly intertwined with the forms of popular culture. Politicians are now not only seen as representatives but also as figures within digital culture, seeking fame, visibility, and agenda-setting power. For instance, during the 2024 elections, Trump launched his TikTok account and shared his first video from a UFC event, presenting himself as a fun, energetic, and "cool" leader to younger voters. The video, which quickly garnered millions of views, was interpreted by young users as a portrayal of Trump's off-stage persona. Such content, produced from Trump's official account, not only increased his recognition among young voters as a politician, but also positioned him as a figure of the digital age. This contributed to generating sympathy toward Trump among "independent" or "undecided" young voters in particular.

One of the most prominent aspects of the 2024 presidential elections was the use of artificial intelligence. During the campaign period, the candidates tested the potential of artificial intelligence in political communication in unprecedented ways. Donald Trump, for example, shared an AI-generated image on his personal X account depicting Kamala Harris as a communist dictator. He also manipulated public perception of celebrity endorsements by circulating an AI-produced image suggesting that Taylor Swift supported his candidacy. On the other hand, Harris's campaign faced allegations that a photograph of a crowd used in one of her rallies was AI-generated. Although these allegations were later refuted, the size of campaign crowds became a symbolic battleground for both sides (The SAIS Review, 2024). While these examples show that AI is being creatively utilized for political propaganda and perception management, current studies indicate that such content still has a limited effect on changing voter behavior.

The interactive, fast, and cost-effective nature of social media makes it an indispensable tool during election campaigns for candidates seeking to reach target audiences, express themselves, and rally support. Among the various social media platforms, the prominence of X in the 2024 elections was no coincidence. A 2024 study by the Pew Research Center revealed that X significantly surpassed other platforms in terms of political content engagement. While 59% of X users cited political content as one of the main reasons for using the platform, this figure was 36% for TikTok, and only 26% for Facebook and Instagram. X's longstanding role as a venue for news, commentary, and political debate, along with its users' confidence in expressing political opinions, reinforced its appeal. Most X users stated that one of the reasons they share political content is their sense of freedom in expressing their views on the platform. This highlights the importance of digital political spaces where users are not merely passive observers but active participants. Furthermore, 74% of X users reported that at least some of the content they encounter is political, compared to 52% on Facebook, 45% on TikTok, and 36% on Instagram. These differences reflect not only user preferences but also the influence of algorithmic structures and content delivery mechanisms in shaping the visibility of political content (Pew Research Center, 2024). Consequently, X stands ahead of other platforms in terms of both political engagement and modes of expression, establishing itself as a primary arena for political communication for both individuals and political actors. As the importance of digital media in election campaigns grows, so too does the budget allocated to these channels. In the 2024 election cycle, digital advertising accounted for an increasing share of total political advertising expenditures (Emarketer, 2024), demonstrating the growing significance of these platforms in political communication.

Recent research highlights how X has amplified the visibility and engagement of pro-Trump content while also reshaping itself around Elon Musk's worldview and political stance (University of Oregon School of Journalism and Communication, 2024). Echo chambers, made possible by personalization algorithms, have contributed to this dynamic. These algorithms, which restrict content diversity and amplify homogenous viewpoints, pose a significant threat to democracy. Moreover, echo chambers have accelerated the spread of misinformation and fake news, while normalizing undesirable behaviors.

During the 2024 U.S. presidential elections, social media and digital platforms remained at the center of debates concerning foreign influence. Countries such as Russia, China, and Iran attempted to intervene in the U.S. electoral process through these platforms. Russia, in particular, carried out a comprehensive campaign to influence American voters via social media and digital tools. These efforts included generating and disseminating fake videos, hiring right-wing social media influencers to promote Kremlin-friendly narratives, and creating fake websites that mimicked credible American news outlets. According to Microsoft, a Russian propaganda group known as "Storm-1516" targeted Vice President Kamala Harris with AI-generated fake videos during the campaign. Among these was a fabricated video claiming Harris injured a young man in a 2011 traffic accident. Other examples included fake content alleging the destruction of ballots, bomb threats at polling stations on election day, and sexual misconduct accusations against Minnesota Governor Tim Walz. Although these videos were later confirmed to be false, they were widely disseminated on X and viewed millions of times (Miller, 2024). Reports also indicated that Russian actors paid American right-wing influencers to spread Kremlin propaganda, while also creating highly realistic fake news websites to boost the credibility and reach of such disinformation (Huo, 2024). By leveraging new technologies, including AI-generated voice recordings and multimedia, Russian operatives were able to spread more persuasive forms of disinformation. These activities aimed to create social tension, foster distrust, exacerbate societal divisions, and erode public confidence in democratic institutions.

China was one of the countries that intensified its social media-based influence operations during the 2024 U.S. presidential elections and sought to affect political discourse within the United States. A research company identified that, as part of a Chinese-origin influence campaign dubbed "Spamouflage," fake online accounts were created to pose as Americans and manipulate voter groups. These accounts published content on controversial topics already dividing American society—such as abortion, homelessness, support for Ukraine, or U.S. foreign policy toward Israel—on popular social media platforms like TikTok, Facebook, YouTube, and X, with the aim of fostering polarization. Although these accounts were eventually exposed as fake and known to have shared AI-generated content, they nonetheless reached over a million views, demonstrating that they had a tangible impact (Bond, 2024). Rather than overtly supporting any one candidate, China appeared more inclined to portray American democracy as dysfunctional. To that end, it reportedly conducted smear campaigns and negative propaganda against candidates whose agendas contradicted Chinese interests (Miller, 2024).

Another country that stood out for its negative influence in the 2024 U.S. elections through the use of social media and digital platforms was Iran. Iran carried out both direct cyberattacks and indirect propaganda via social media. The most significant scandal involved the hacking of email accounts belonging to Donald Trump's campaign team and the subsequent transfer of stolen documents and confidential materials to the opposing political camp (Gerstein & Woodruff Swan, 2024). It was revealed that Iran had established at least eight English-language fake news websites, which it used extensively for propaganda purposes throughout the election period. These websites tailored their content to specific target audiences and frequently published AI-generated disinformation articles (Chopra, 2024). Posts that mocked Trump, ridiculed candidates, or touched on sensitive issues such as religion or LGBTQ+ rights attracted significant attention on social media. Iran's multilayered operations targeting the U.S. elections were part of an information warfare strategy aimed at escalating existing tensions in American society and undermining the democratic nature of the electoral process.

With the impact of digital transformation, social media platforms have evolved beyond being mere tools of interpersonal communication. They have become strategic arenas for political campaigns, public opinion formation, and even global power rivalries. As evidenced in the case of the 2024 U.S. presidential elections, social media has surpassed its role as a communication tool to become a primary space where politics is practiced. These platforms serve as arenas in which political actors engage directly with voters, shape campaign strategies, disseminate discourse, and target opponents. At the same time, they have emerged as multilayered battlegrounds vulnerable to foreign interference.

Within this context, social media has become a stage not only for competition among domestic candidates but also for global influence struggles. The efforts of countries such as Russia, China, and Iran to influence American voters through AI-supported disinformation, fake accounts, and manipulative news networks have expanded the concept of electoral security from the physical safeguarding of ballot boxes to include digital defense. Consequently, social media possesses a dual character: while it has the potential to enhance democratic processes, it also offers avenues for manipulation and exploitation.

During the 2024 election process, the increasing prominence of the X platform (formerly known as Twitter) further solidified its central position in the production and consumption of political content. Elon Musk's managerial influence over the platform has revealed that social media ownership represents not only economic but also political power. Musk's overt support for Donald Trump and the alignment of platform policies in favor of this political stance demonstrate that social media companies are no longer neutral digital infrastructures, but rather politicized actors. This development raises serious concerns regarding the principle of equality in democratic competition and increases skepticism about the neutrality of digital environments.

In light of these developments, social media platforms have become indispensable arenas in modern electoral processes. These media channels, which significantly influence voter decision-making, also serve as stages for emotional manipulation, algorithmic distortions, and echo chambers that create ideologically closed circuits. In this context, social media should be understood not merely as a tool of communication, but also as a digital front that deepens social polarization, hosts information warfare, and shapes political consciousness.

In conclusion, social media platforms in contemporary democracies are not only a digitalized extension of political competition, but also the stage for a new generation of conflict and power struggles. The 2024 U.S. presidential election has clearly illustrated that these platforms are principal arenas not only for domestic political actors, but also for global powers engaging in cyber intervention, disinformation, and propaganda warfare. In this regard, traditional instruments of war are increasingly being replaced by perception management and information wars conducted through digital means. The goals of generating societal polarization, supporting or discrediting political figures, and even undermining trust in democratic institutions are now being pursued not through physical confrontation, but via digital strategies. Therefore, social media should be regarded not only as an interactive tool that facilitates political participation, but also as a new battleground capable of generating influence, intervention, and digital dominance.

CONCLUSION

The transformative effect of digitalization on political processes has reshaped the structural dynamics of contemporary democracies. One of the most striking examples of this transformation is the central role that social media platforms now play in political participation, public opinion formation, and the conduct of election campaigns. Social media is no longer merely a space for individual communication but has become a hybrid battlefield where both national and international actors engage in strategic struggles. In this context, the theoretical approaches presented throughout the study clearly demonstrate that social media has evolved into a kind of "postmodern battlefield" in which the competition for political power is waged on digital grounds.

The theoretical framework addressing media effects theories, political communication strategies, and perception management techniques reveals that social media platforms have transformed from passive tools into domains of active intervention. In particular, analyses based on Habermas's theory of the public sphere indicate that instead of creating an ideal democratic forum, social media has taken on a function that deepens polarization within algorithmically shaped echo chambers. The personalized delivery of political content to individuals not only influences voter preferences but also manipulates perceptions of reality and renders democratic decision-making processes vulnerable to manipulation.

The 2024 U.S. presidential election, examined in detail in the fourth chapter, serves as a field-based reflection of this theoretical framework. This election process revealed that social media platforms now go far beyond the functions of traditional media outlets and have become the primary front in political competition. Candidates' focus on digital platforms in their campaign strategies proves that interactions on social media lead directly to political consequences. The discourse wars conducted via platforms such as X (formerly Twitter), TikTok, Instagram, and Facebook have not only shaped voter behavior but have also become one of the main forces in determining public opinion. During this period, not only domestic political actors but also global powers have been active in digital arenas. Findings indicating that countries such as Russia, Iran, and China interfered in the U.S. elections show that social media has become not only a domestic political tool but also an integral part of foreign policy and security strategies. This situation points to the evolution of concepts like "cyberwar" and "information warfare" beyond their classical meanings toward struggles for digital hegemony. Particularly, interventions carried out through AI-generated fake accounts, deepfake videos, and disinformation content now threaten election security not in physical terms but in digital domains.

Another key issue emphasized throughout the study is the politicization of social media ownership. Elon Musk's interventions on X have shown how platform policies can be shaped in favor of or against certain political actors. This development signals a new architecture of power in which platform owners, like the users themselves, have become political actors. Thus, social media platforms are transforming into centers of power that, unlike traditional media outlets, are not subject to regulatory oversight but have a far wider impact.

All these findings demonstrate that while social media offers significant opportunities for democratic political participation, it also entails serious risks. The manipulability of voter behavior in digital environments, the spread of false content, the systematic manipulation of public opinion, and the erosion of transparency in election processes require that social media platforms be evaluated not merely as communication tools but as strategic arenas of contestation.

In conclusion, this study shows that social media platforms are not only the venues where the pulse of politics is taken but also the stages of digitized forms of electoral security, political manipulation, and global competition. Politics is now conducted not just in public squares or on television screens but also within digital ecosystems shaped by algorithms; and traditional methods of warfare are

REFERENCES

Açıkalın, Ş. N., & Baykız, T. (2024). Değişim ve dönüşümün eşiğinde siber savaş. *Muhafazakâr Düşünce Dergisi*, *20*(67), 200–220.

Akargün, S. E. (1993). *Harp Tarihi Ders Kitabı-1.* Kara Harp Okulu Matbaası.

Akın, A. (2019). Stratejik Analiz. https://www.stratejikanaliz.com/hibrit-savas/

Allcott, H., & Gentzkow, M. (2017). Social Media and Fake News in the 2016 Election. *The Journal of Economic Perspectives*, *31*(2), 211–236. DOI: 10.1257/jep.31.2.211

American Historical Association. (n.d.). The story of propaganda. https://www.historians.org/resource/the-story-of-propaganda/

BBC News. (2017). Putin's 'chef', the man behind the troll factory. https://www.bbc.com/news/world-europe-39419560

BBC News Turkish. (2022). Donald Trump'ın Twitter hesabı yeniden açıldı, https://www.bbc.com/turkce/articles/c0jg0vqk5p7o

Beard, M. (2013). Cyberwar and just war theory. Applied ethics, risk, justice and liberty, 1-12.

Bernard, N. (2020) 'Corona' günlerinde her 10 kişiden altısı yalan haber paylaşıyor. https://www.diken.com.tr/podcast-corona-gunlerinde-her-10-kisiden-altisi-yalan-haber-paylasiyor/

Bingöl, O. (2017). Hybrid war and its strategic implications to Turkey. *Gazi Akademik Bakış*, *11*(21), 107–132. DOI: 10.19060/gav.379591

Bond, S. (2024, September 3). China is pushing divisive messages using fake U.S. voters, NPR. https://www.npr.org/2024/09/03/nx-s1-5096151/china-tiktok-x-fake-voters-influence-campaign

Borins, S. (2011). Online Adaydan Online Başkana. In Devran, Y. (Ed.), *Seçim Kampanyalarında Geleneksel Medya.* İnternet ve Sosyal Medyanın Kullanımı, Başlık Yayın Grubu.

Çakı, C. (2018). *Propaganda. M. Karaca ve C. Çakı içinde.* İletişim ve Propaganda, Eğitim Yayınevi.

Çalışkan, A. (2023). Siber savaş: Bilgi krizi mi yoksa güvenliği mi? *SAVSAD Savunma ve Savaş Araştırmaları Dergisi*, *33*(1), 1–32. DOI: 10.54078/savsad.1188851

Callamari, P., & Reveron, D. (2003). Chine's Use of Perception Management. *International Journal of Intelligence and CounterIntelligence*, *16*(1), 1–15. DOI: 10.1080/713830380

Castells, M. (2020). İnternet Galaksisi: İnternet, İş Dünyası ve Toplum Üzerine Düşünceler, (T. Asrak Hasdemir, E. Öztürk, H. Çakmak, A. Yıldız, S. K. Özdemir, B. T. Yıldırım, M. D. Aydın, N. Yıldız, A. Aydın, A. A. Yıldız, & A. E. Yıldız, Trans.). Phoenix Yayınevi.

Chopra, A. (2024). Iran 'news' sites, hackers target Trump ahead of US election. AL-MONITOR. https://www.al-monitor.com/originals/2024/09/iran-news-sites-hackers-target-trump-ahead-us-election

Clarke, R. A., & Knake, R. K. (2010). Cyber war: The next threat to national security and what to do about it.

Clarke, R. A., & Knake, R. K. (2014). Cyber war. Old Saybrook: Tantor Media, Incorporated.

Clausewitz, C. V. (2011). *Savaş Üzerine* (Koçak, S., Trans.). Doruk Yayıncılık.

Cole, R. (1998). *The Encyclopedia of Propaganda*. M.E. Sharpe.

Coxall, M. (2013). *Human Manipulation: A Handbook*. Cornelio Books.

Derman, G. S. (2021). Perceptıon Management in the Media. *International Journal of Social and Economic Sciences*, *11*(1), 64–78.

Dictionary of Military and Associated Terms. (2020). Perception Management. https://www.dtic.mil/doctrine/jel/doddict/index.html

Doğan, A., & Ertan, T. (2016). Siyasal İletişim Aracı Olarak ABD Başkanlık Seçimlerinde Twitter'ın Kullanımı: Hillary Clinton ve Donald Trump Örneği. *Atatürk İletişim Dergisi*, *11*, 139–159.

Doğan, İ. (2014). Propaganda Aracı Olarak İnternet: Kayseri İli Merkez Seçmeni Üzerine Bir Alan Araştırması, [Unpublished graduate dissertation]. Selcuk University Social Sciences Institute, Konya.

Dostie-Goulet, E. (2009). Social Networks And The Development Of Political İnterest. *Journal of Youth Studies*, *12*(4), 405–421. DOI: 10.1080/13676260902866512

Elektratig (2014). Did Cambyses Have Cats Nailed to His Soldiers' Shields? Elektratig. https://elektratig.blogspot.com/2014/01/did-cambyses-have-cats-nailed-to-his.html

Emarketer. (2024). 2024 Political Ad Spending Will Jump Nearly 30% vs. 2020. Emarketer. https://www.emarketer.com/press-releases/2024-political-ad-spending-will-jump-nearly-30-vs-2020/

Encyclopedia. (n.d.). Disinformation. https://encyclopedia.pub/entry/29164

Ertem, Y. E. (2019). Sosyal Medyada Dezenformasyon. [Unpublished graduate dissertation]. Marmara University Social Sciences Institute, İstanbul.

Eslen, N. (2003). *Tarih Boyu Savaş ve Strateji*. Q Matris Yayınları.

Garfield, A. (2002). The Offence of Strategic Influence: Making the Case for Perception Management Operations. *Journal of Information Warfare*, *1*(3), 30–39.

Genel, M. G. (2012). Siyasal İletişim Kampanyalarında Sosyal Medyanın Kullanımı (12 Haziran 2011 Seçimleri "Twitter" Örneği, The Turkish Online Journal of Design. *Art and Communication, TOJDAC*, *2*(4), 23–31. DOI: 10.7456/10204100/005

Gerasimov, V. (2016). The value of science is in the foresight: New challenges demand rethinking the forms and methods of carrying out combat operations. *Military Review*, *96*(1), 23–29. https://www.armyupress.army.mil/portals/7/military-review/archives/english/militaryreview_20160228_art008.pdf

Gerstein, J., & Woodruff Swan, B. (2024. Three Iranians charged in hacking of Trump presidential campaign. Politico. https://www.politico.com/news/2024/09/27/iranians-charged-trump-campaign-hack-00181454

Glenn, R. W. (2009). Thoughts on "Hybrid" Conflict. Small Wars Journal, https://smallwarsjournal.com/wp-content/uploads/2022/02/188-glenn.pdf

Güler, H. (2017). Rusya Araştırmaları Enstitüsü. http://www.rusen.org/hibrit-savas-nedir/

Gültekin, M. (2022). Hibrit Savaş, Sosyal Medya ve Propaganda. İslami Analiz. https://islamianaliz.com/makale/12496108/mucahit-gultekin/hibrit-savas-sosyal-medya-ve-propaganda

Gültekin, M. (2024). *Algı yönetimi ve Manipülasyon: Kanmanın ve Kandırmanın Psikolojisi*. Pınar Yayınları.

Herman, E. S., & Chomsky, N. (1999). *Medya Halka Nasıl Evet Dedirtir*. Minerva Yayınları.

Herre, B., Guirao, L. R., & Roser, M. (2016). War and Peace. https://ourworldindata.org/war-and-peace

High, A. C. & Solomon, D. H. (2011). Locating Computer-Mediated Social Support Within Online Communication Environments.

Hobbes, T. (1997). Leviathan. Touchstone.

Hoffman, F. G. (2007). Conflict in the 21 st Century: The Rise of Hybrid Wars. Potomac Institute for Policy Studies. https://www.potomacinstitute.org/images/stories/publications/potomac_hybridwar_0108.pdf

Holat, O. (2021). Yeni Medya ve Siber Savaş Kavramları Bağlamında Stuxnet Saldırısı Örneğinin İncelenmesi. *Abant Kültürel Araştırmalar Dergisi*, *6*(11), 105–121.

Horowitz, M. C. (2018). Artificial Intelligence, International Competition, and the Balance of Power. *Texas National Security Review*, *1*(3), 36–57.

Huo, J. (2024). Foreign İnfluence Efforts Reached A Fever Pitch During the 2024 Election. NPR. https://www.npr.org/2024/11/09/nx-s1-5181965/2024-election-foreign-influence-russia-china-iran

İldem, T. (2021). Dezenformasyona Karşı Toplumsal Dayanıklılık. Ekonomi ve Dış Politika Araştırmalar Merkezi. https://edam.org.tr/wpcontent/uploads/2021/03/210305-Dezenformasyon-Ildem.pdf

Isaak, J., & Hanna, M. J. (2018). User Data Privacy: Facebook, Cambridge Analytica, and Privacy Protection. *Computer*, *51*(8), 56–59. DOI: 10.1109/MC.2018.3191268

Jomini, A. H. (2013). *Savaş Sanatı* (Büyükonat, A. T., Trans.). Doruk Yayımcılık.

Jonsson, O. (2020). Putting Modern War in Perspective, in Oscar Jonsson (Ed.), Modern Warfare: New Technologies and Enduring Concepts. Stockholm Free World Forum. https://www.academia.edu/44001543/Modern_Warfare_New_Technologies_and_Enduring_Concepts

Jordan, D., Kiras, J. D., Lonsdale, D. J., Speller, I., Tuck, C., & Walton, C. D. (2008). *Understanding Modern Warfare*. Cambridge University Press.

Josan, A., & Voicu, C. (2015). Hybrid Wars in the Age of Asymmetric Conflicts. *Review of the Air Force Academy*, *1*(28), 49–52.

Jowet, G. S., & O'donnel, V. (1999). *Propaganda and Persuasion*. Sage Publications.

Jowett, G., & O'Donnell, V. (2006). *Propaganda and Persuasion*. Sage Publications.

Kaldor, M. (2013). In defence of New Wars. *Stability*, *2*(1), 1–16. DOI: 10.5334/sta.at

Kara, M. (2013) Siber Saldırılar, Siber Savaşlar ve Etkileri, [Unpublished graduate dissertation]. İstanbul Bilgi University.

Karagöz, K. (2018). Post-Truth Çağında Yayıncılığın Geleceği. *Trt-Akademi, 3*(6), 678–709.

Karakaya, S. (2006). Türk Medyasında Haberin Manipülasyonu ve Dezenformasyon. In Türkoğlu, N. (Ed.), *Medya Okuryazarlığı* (pp. 176–178). Marmara Üniversitesi İletişim Fakültesi.

Kellner, D. (2009). The Persian Gulf TV War Revisited. https://pages.gseis.ucla.edu/faculty/kellner/essays/gulftvwarrevisited.pdf

Keskin, F. (1998). *Uluslararası Hukukta Kuvvet Kullanma: Savaş, Karışma ve Birleşmiş Milletler*. Mülkiyeliler Birliği Vakfı Yayınları.

Kleis Nielsen, R., & Graves, L. (2020). News You Don't Believe: Audience Perspectives on Fake News. *Journalism Studies, 21*(6), 826–841. DOI: 10.1080/1461670X.2020.1745667

Kurubaş, E. (2017). Dünya Politikasının Dönüşümü ve Uluslararası İlişkiler Çağı. In Kardaş, Ş., & Balcı, A. (Eds.), *Uluslararası İlişkilere Giriş* (7th ed.). Küre Yayınları.

Lanoszka, A. (2016). Russian Hybrid Warfare and Extended Deterrence in Eastern Europe. *International Affairs, 92*(1), 175–195. DOI: 10.1111/1468-2346.12509

Le Bon, G. (2014). *Kitleler Psikolojisi* (Can, H., Trans.). Tutku Yayınevi.

Lind, W. S. (2004). Understanding Forth Generation War. U.S.Army Combined Arms Center. https://www.hsdl.org/?view&did=482203

Lind, W. S., Nightengale, K., Schmitt, J. F., Sutton, J. W., & Wilson, G. I. (1989). The Changing Face of War: Into the Fourth Generation. *The Marine Corps Gazette, 73*(10), 22–26.

Lind, W. S., Nightengale, K., Schmitt, J. F., Sutton, J. W., & Wilson, G. I. (2001). The Changing Face Of War: Into The Fourth Generation. *The Marine Corps Gazette, 85*(11), 65–68.

Manovich, L, (2001). The Language of New Media. Massachusetts Institute of Technology.

Mansoor, P. R. (2012). Introduction: Hybrid Warfare in History. in Peter R. Mansoor ve Williamson Murray (Ed.), Hybrid Warfare: Fighting Complex Opponents from the Ancient World to the Present. Cambridge University Press.

Miller, M. (2024). Influence Campaigns From Iran, China, Russia Ramping Up Ahead of Elections, Microsoft Finds. Politico. https://www.politico.com/news/2024/10/23/influence-iran-china-russia-elections-microsoft-00185075

Öymen, O. (2014). *Bir Propaganda Silahı Olarak Basın.* Remzi Yayınevi.

Özdemir, İ. (2015). Propaganda Aracı Olarak Sosyal Medya. T.C Republic of Türkiye Ministry of Interior. https://www.icisleri.gov.tr/arem/propaganda-araci-olarak-sosyal-medya

Özdemiroğlu, P. (2013). Sahte Tweet'ler Bizi Yıldıramaz, Milliyet Gazetesi, https://www.milliyet.com.tr/ekonomi/sahte-tweet-ler-bizi-yildiramaz-1698125

Parker, G. (1988). *The Military Revolution: Military Innovation and the Rise of the West, 1500–1800.* Cambridge University Press.

Pew Research Center. (2024). How Americans Navigate Politics on TikTok, X, Facebook and Instagram. https://www.pewresearch.org/internet/2024/06/12/how-americans-navigate-politics-on-tiktok-x-facebook-and-instagram/

Renz, B. (2016). Russia and "Hybrid Warfare". *Contemporary Politics, 22*(3), 283–300. DOI: 10.1080/13569775.2016.1201316

Reuters. (2024). Philadelphia Prosecutor Sues Elon Musk Group to Stop $1 Million Lottery for Voters. https://www.reuters.com/legal/philadelphia-da-sues-elon-musk-pac-stop-1-million-lottery-voters-cnbc-reports-2024-10-28/

Sadowski, D., & Becker, J. (2010). Beyond the "Hybrid" Threat: Asserting the Essential Unity of Warfare. *Journal Article*, *7*(1), 13.

Sander, O. (2017). *Siyasi Tarih: İlk Çağlardan 1918'e* (33rd ed.). İmge Kitapevi.

Schoen, F., & Lamb, C. J. (2012). *Deception, Disinformation, and Strategic Communications: How One Interagency Group Made a Major Difference.* National Defense University Press.

Schroefl, J., & Kaufman, S. J. (2014). Hybrid Actors, Tactical Variety: Rethinking Asymmetric and Hybrid War. *Studies in Conflict and Terrorism*, *37*(10), 862–880. DOI: 10.1080/1057610X.2014.941435

Seo, H. (2014). Visual Propaganda in the Age of Social Media: An Empirical Analysis of Twitter İmages During the 2012 Israeli–Hamas Conflict. *Visual Communication Quarterly*, *21*(3), 50–161. DOI: 10.1080/15551393.2014.955501

Shakarian, P., Shakarian, J., & Ruef, A. (2013). *Introduction to Cyber-Warfare: A Multidisciplinary Approach.* Morgan Kaufmann Publishers – Elsevier.

Siegel, P. C. (2005). Perception Management: IO's Stepchild? *Low Intensity Conflict & Law Enforcement.*, *13*(2), 117–134. DOI: 10.1080/09662840500347314

Sönmez, G. (2022). Rusya'nın 2014 Ukrayna Krizi'ndeki Hibrit Savaş Stratejisi. *Gazi Akademik Bakış*, *15*(30), 489–511. DOI: 10.19060/gav.1136936

Sotirovic, M., & McLeod, J. M. (2004). Knowledge as Understanding: The Information Processing Approach to Political Learning. In Kaid, L. L. (Ed.), *Handbook of Political Communication Research* (pp. 357–394). Lawrence Erlbaum Associates.

Stanger, M. (2018, Mayıs 2). 6 People Who Were Literally Erased From History. Business Insider. http://uk.businessinsider.com/people-who-were-erased-from-history-2013-12/#nikolai-yezhov-joseph-stalins-head-of-secret-police-1

Taddeo, M. (2012). Information Warfare: A Philosophical Perspective. *Philosophy & Technology*, *25*(1), 105–120. DOI: 10.1007/s13347-011-0040-9

The SAIS Review. (2024). Social Media, Disinformation, and AI: Transforming the Landscape of the 2024 U.S. Presidential Political Campaigns. Johns Hopkins University. https://saisreview.sais.jhu.edu/social-media-disinformation-and-ai-transforming-the-landscape-of-the-2024-u-s-presidential-political-campaigns/

Toplu, M. (2018). Kamu Politikalarının Oluşumu ve Uygulanmasında Medya/Sosyal Medya'nın Rolü. *Journal of Communication Theory & Research.*, *46*, 346–376.

Türkay, Ş. (2013). Siber Savaş Hukuku ve Uygulanma Sorunsalı. *Journal of Istanbul University Law Faculty*, *71*(1), 1177–1227.

Twitter Inc. (2021). Permanent suspension of @realDonaldTrump. Twitter Blog. https://blog.twitter.com/en_us/topics/company/2020/suspension

Uluk, M. (2018). *Hakikat Sonrası Çağda Yeni Medya ve Yalan Haber*. Dorlion.

University of Oregon School of Journalism and Communication. (2024). Media Influence on Politics: 7 Election Trends in 2024. https://journalism.uoregon.edu/news/media-influence-on-politics

US. Marine Corps. (1997). Warfighting. https://www.marines.mil/Portals/1/Publications/MCDP%201%20Warfighting.pdf

Utma, S. (2022). *Medyada Gerçeklik Algısının Dönüşümü: Medya ve Dezenformasyon*. İKSAD Publishing House.

Van Creveld, M. (1991). *The Transformation of War*. The Free Press.

Varlık, A. B. (2013). Savaşı Tanımlamak: Terminolojik Bir Yaklaşım. *Avrasya Terim Dergisi*, *1*(2), 114–129.

Walsh, J. P. (2020). Social Media and Moral Panics: Assessing The Effects Of Technological Change On Societal Reaction. *International Journal of Cultural Studies, 23*(6), 840–859. Advance online publication. DOI: 10.1177/1367877920912257 PMID: 40479194

Waltz, K. N. (2001). *Man, the State and War: A Theoretical Analysis* (3rd ed.). Columbia University Press.

Wardle, C. & Derakhshan, H. (2017). Information Disorder: Toward an Interdisciplinary Framework for Researchand Policymaking, Council of Europe Report DGI 09.

Woolley, S., & Howard, P. (2019). Computational Propaganda Worldwide. in Samuel Woolley ve Philip Howard (Ed.), Computational Propaganda Political Parties, Politicians and Political Manipulation on Social Media (p. 3-20). Oxford University Press.

Yalçınkaya, H. (2008). *Savaş Uluslararası İlişkilerde Güç Kullanımı*. İmge Kitabevi.

Yenal, S., & Akdemir, N. (2020). Uluslararası İlişkilerde Yeni Bir Kuvvet Çarpanı: Siber Savaşlar Üzerine Bir Vaka Analizi. ÇAKÜ. *Journal of Institute of Social Sciences, 11*(1), 414–450.

Yılmaz, M. E. (2010). *Savaş ve Uluslararası Sistem*. Nobel Yayın Dağıtım.

Chapter 2
Rapid March of Technology:
Transforming the Way Future Wars Will Be Fought

Debadyuti Dasgupta
Independent Researcher, India

Tannistha Dasgupta
https://orcid.org/0000-0002-9973-261X
VIT-AP University, India

Renuprasad Patki
https://orcid.org/0000-0002-8698-2847
VIT-AP University, India

ABSTRACT

Technological innovation has continuously shaped the battlefield, dictating outcomes and redefining strategies. The evolution of warfare technologies includes key shifts such as the introduction of electronics and the rise of internet-dependent military systems in the 1990s. The reliance on digital infrastructure has transformed military operations, introducing vulnerabilities like cyberattacks targeting financial systems, intelligence, and weaponry. Media and social media platforms have expanded the landscape of warfare, enabling rapid dissemination of information and psychological operations powered by technology. This shift has facilitated the rise of unmanned systems, such as drones, which depend heavily on computer systems. Efficient cybersecurity frameworks are now essential to protect military and civilian digital assets, as data breaches can inflict profound economic and strategic damages. With

DOI: 10.4018/979-8-3373-7589-2.ch002

the examples of past and recent past wars, particularly in the Indian subcontinent, this chapter explores these technological transformations and their implications for future conflicts.

INTRODUCTION

Technological innovations have long shaped the dynamics of warfare, often determining the fate of battles and empires. From the ingenious deployment of bows and arrows by Genghis Khan in his conquest of Asia, to the transformative use of gunpowder cannons in the naval siege of Tenochtitlan, and the precision-guided bombs that revolutionized modern counterterrorism efforts in Afghanistan, advancements in technology have consistently redefined the nature of conflict (Buzan & Hansen, 2009). The evolution of warfare reflects a shift from direct physical confrontations to the implementation of ranged weapons, to the strategic deployment of unmanned systems, and the emergence of cyber warfare. This progression highlights technology's role in shaping military strategies and redefining global security (North Atlantic Treaty Organization [NATO], 2020; Olsen, 2010). The use of such technology has brought about changes and prompted the adaptation of various doctrines in response to the introduction of different weapon systems in a field where technology constantly evolves. Military science and weapon engineering had a historical stagnancy till the Renaissance period. However, with the advent of gunpowder, a drastic change was forced in the techniques of warfare. The introduction of gunpowder demonstrated that the effective range of weapon systems could be significantly extended through advancements in small arms technology, fundamentally altering the nature of warfare. With the development of higher-quality explosives, the dynamics of the battlefield shifted as controlled-explosion weapons were mounted on mobile platforms such as vehicles and ships, once again transforming the strategic landscape and tactical approaches of warfare. The introduction of aeroplanes also transformed the battlefield, as these aircrafts were capable of delivering explosives directly onto enemy trenches and positions and leaving opposing forces without effective means of defence (Roy, 2022; Olsen, 2010). Nations soon developed interceptive strategies and established their air forces to counter these aerial threats, thereby shifting the domain of warfare into the aerial sphere. With the arrival of aircraft, defence systems were adapted to counter aerial threats, which led to advancements in military equipment, particularly in surveillance and early warning detection technologies. This marked the entry of electronics and semiconductor technologies into the domain of military advancements. Besides, the emphasis on stealth represented another significant shift in warfare, a transformation that began with the introduction of submersible war vessels and military aircraft. Submarines equipped with stealth technology significantly contributed to the advancement of underwater weapon systems, thus shifting the domain of warfare from surface operations to undersea combat (Gabriele, 2024). This transformation was supported by specialized developments in underwater communication technologies, which enabled effective control and coordination below the surface.

The transformation of technology transfer in warfare was introduced in the mid-1970s, marking a pivotal moment with the widespread integration of electronic systems into military operations. This era saw a significant improvement in the use of electronics for aircraft, nuclear arsenals, and weapon guidance systems, heralding a gradual shift towards automation. By the 1990s, advanced militaries increasingly relied on digital technologies, particularly computers and the internet, for critical activities such as military intelligence, economic transaction data management, and safeguarding sensitive military hardware information (Potzsch, 2015). The end of the Cold War further intensified these developments, as both sides (the Western Bloc led by the United States and the Eastern Bloc, led by the Soviet Union) engaged in cyber operations like hacking attacks, which foreshadowed the growing significance of cyberspace in contemporary conflicts. This evolution persisted into the 21st century, with cyberspace emerging as a critical infrastructure subject to persistent threats. High-profile cyber-attacks targeted financial institutions, global economic databases, and national economic systems, often linked to proxy wars involving nation-states and non-state actors (Hakimi et al., 2015). Such activities have driven the development of advanced cybersecurity frameworks designed to protect critical national and international infrastructures.

In an era defined by rapid technological advancement and digital integration, the intersection of cybersecurity and national defence has become increasingly critical. The growing dependence on digital identification systems, inadequate regulation of access to citizens' data have created serious security-related vulnerabilities and threats. Sensitive information is increasingly exploited for large-scale economic disruption by professional hackers associated with military intelligence agencies, terrorist organizations, and untraceable private entities. In addition to uncontrolled access to public data, the dependency of modern weaponry on computer-based technologies, such as the growing adoption of drones and unmanned aerial systems, has significantly heightened the risk of cyber intrusions, making security threats more pervasive and immediate (Castrillo et al., 2022; Durfey & Sajal, 2022). These systems, while transformative, remain highly susceptible to cyber threats, with malware and hacking posing significant risks to operational readiness and military effectiveness. Therefore, "the issues related to safety, security and privacy must be taken into consideration" (Castrillo et al., 2022, 1).

Transformation of Technology

The potential for technology to transform conventional warfare suggests a notable shift in the role of employing manpower in combat. The extensive shift in weapon technology has been attributed to the rapid development of drones, which are also known as Unmanned Aerial Vehicles (UAVs). Drones are aircraft operated without a human pilot onboard. They are controlled remotely by a human operator or autonomously through Artificial Intelligence (AI) based software and pre-programmed flight paths. UAVs use a combination of GPS, sensors, and communication systems to navigate and perform tasks (Grigore & Cristescu, 2024). Since the 1990s, the advancement of drone technology was only limited to the US, which has focused on safeguarding human lives in combat. The first drones were used for minor tasks of either helping the infantry or completely used for surveillance or for replacing the role of an Air Force pilot for surveillance (Elish, 2017). At present, drones are used in warfare for precision strikes, intelligence gathering, and real-time surveillance, which enables militaries to engage targets with minimal risk to soldiers. They are also used for logistical support, such as delivering supplies in combat zones and assessing damage after a strike. Some drones are capable of hovering in place for extended periods and can either engage at the most strategic moment or return to base via an alternate route if the mission is aborted or no enemy presence is identified (Grigore & Cristescu, 2024; Ambodo, Saputro, & Komariah, 2021).

In recent times, the tactical significance of advanced drone systems has been demonstrated in their dual roles of reconnaissance and direct strikes (Johnson, 2024). With the integration of advanced robotic systems and control technologies, the traditional concept of human targets in warfare may shift. While operators may no longer require conventional physical military training, they still undergo specialized and continuous training to develop the cognitive skills, quick decision-making, and technical proficiency required to remotely control advanced battlefield systems. Consequently, both offensive and defensive forces would likely need to first address the challenges posed by these unmanned systems, such as drones and robots, before engaging with human or other targets (Nag, 2023). The ongoing Russia–Ukraine war has offered an alarming preview of this evolving nature of warfare, in which emerging technologies are redefining the culture and dynamics of the battlefield. The superior development in computer programming has contributed to this transformation. For instance, during the Russia-Ukraine conflict, Russian forces began using First-Person-View (FPV) drones more actively within their units for surveillance and also to support ground infantry, artillery, and combat drones (Ibrahim, 2024). As Ibrahim (2024) explains, "FPV drones were used initially for battlefield intelligence, surveillance and reconnaissance (ISR)" have now emerged as a threatening weapon system" (p. 1). The shift of the attitude towards the use of drones has changed due to such rapid innovation in technology, with the possibility of finding such weapon systems to be expendable. In the case of FPV drones, there is no human risk as these drones have a minimum range of 2 kilometres. The FPV drone pilot is in a much safer environment compared to the boots on the ground during combat. Such drones, mostly used either for self-destructive assaults or for surveillance and provide human life a sense of comfort and safety while having the capability to cause severe damage to the enemy.

The development of unmanned technologies, including robots with reasoning capabilities, AI, and humanoid or animal-like structures equipped with onboard weapons, may significantly alter battlefield dynamics by performing tasks traditionally carried out by humans. The massive deployment of drones has also resulted in the escalation of the usage of AI to make such devices operate themselves according to predictable and unpredictable situations (Kuhn & Williams, 2025; Ibrahim, 2024). The early development of AI was closely linked to military surveillance systems. At present, similar technologies are embedded in drone operations. AI now supports decision-making processes, including target identification and engagement. AI, when integrated into the control systems of unmanned aerial vehicles (UAVs), enables coordinated group behaviour and autonomous decision-making in the deployment and operation of onboard weapon systems against designated targets. The control systems can be fed with the desired input and can be left to operate autonomously without human interference or control. According to Ibrahim (2024), "In the near future, the incorporation of AI in FPV drones will allow them to operate with greater autonomy, rendering jamming and other electronic countermeasures obsolete" (p. 1). Unmanned technologies have also played a significant role in the recent 2025 India-Pakistan conflict, demonstrating their growing importance on the modern battlefield.

At present, with the Indo-Pak conflict of May 2025, the doctrine of the Indian Armed Forces in the deployment of its forces has seen exceptional changes in countering Pakistan in the span of 10 years. Considering an aerial scenario, till 2019, the Indian response to Pakistan has been mainly through the use of fighter jets. UAVs were largely limited to surveillance activities in conflicts starting from the beginning of the 21st century (Kolhatkar, 2024; Gupta & Dubey, 2024; Kaura, 2020). But in the 2025 conflict, both nations used drones extensively, with Pakistan targeting mostly Indian civilian areas while India targeted the Pakistan Air Force bases, air defence systems and missile bases. Night drone attack exchanges continued on a massive scale, which were controlled by pilots in monitor rooms, and these drones crossed the border for these attacks. The Pakistan Air Force bases and air defence systems, which were not attacked after the 1971 war, were subjected to heavy bombing both by the fighter jets and by drones (Al Jazeera, 2025; British Broadcasting Corporation [BBC] News, 2025; Peri, 2025; Roy-Chaudhury, 2025). This shows an adaptation of these new technologies to work with human presence on the ground as well. This evolving technological landscape could reshape strategic and tactical approaches in warfare, though the full implications of such changes remain to be fully understood (Kuhn & Williams, 2025).

The targeted drone attacks, followed by missile firing, were followed by the dispatch of the Indian Air Force fighter-bomber aircraft. The Indo-Pak conflict of 2025 witnessed significantly fewer casualties compared to previous engagements between India and Pakistan. Thus, indicating a marked shift in the nature and scale of hostilities. According to some of the leading Indian newspapers and media houses, Indian casualties are low after the first day of the exchange. However, the Indian casualties were more on the civilian population compared to the military targets as Pakistan's strikes affected the Indian civil population of Jammu area, while Pakistani casualties were more on the military side, with Indian casualties being low after the first day of fire exchange (The Hindu, 2025; The Times of India, 2025; New Delhi Television Network [NDTV], 2025). This was perhaps due to the extremely precise guided weapons fired from Indian drones or from mobile missile launching platforms that fired at high accuracy, hitting "terrorist" targets and launchpads, including on Pakistani military and air force bases, causing the maximum damage without damaging the nearby civilian infrastructure like residential areas and schools. It may be observed that the Indian Air Force conducted air strikes following the successful neutralization of Pakistani air defence systems by Indian drones. Simultaneously, Indian air defence units intercepted and destroyed incoming Pakistani drones, reflecting a coordinated and strategic use of both offensive and defensive aerial capabilities (NDTV, 2025). This transformation of weapons suggests a shift in focus: whereas attacks once targeted human operators or enemies directly, the priority now is to first engage or disable drones and unmanned systems before moving on to human targets. The development of drone technology has also contributed to the rise of the concept of swarm drone warfare. The deployment of such swarm drones was seen very much in this conflict (BBC News, 2025). Swarm drones, deployed in groups by Pakistan were effectively neutralized through active anti-aircraft gunfire (The Times of India [TOI], 2025).

The historical use of wire-controlled tracked bots, such as the Goliath, a popular name for the German demolition vehicle used during World War II to destroy bridges or tanks, set an early precedent in remote-controlled warfare. This concept re-emerged in the 1990s with the development of Unmanned Ground Vehicles (UGVs) and mobile platforms equipped with small arms, which could be remotely operated by humans from a safe distance (Szakacsi, 2016; Olsen, 2010). The next innovation in this self-destructive disposable technology was transferred onto drones. Since these drones can be piloted or controlled by computer systems, with no risk of human losses, the concept of suicide drones (also known as Kamikaze drones) came into existence (Grigore & Cristescu, 2024; Ambodo, Saputro, & Komariah, 2021). These drones vary significantly in size and function, ranging from long-range unmanned aerial vehicles capable of carrying substantial explosive payloads to compact, agile first-person-view (FPV) drones designed for precision strikes or close-quarters operations. Normally, the design of these drones is based on the explosive carrying chamber, and such small-scale drones of both FPV scale and of bigger scales have been seen to be deployed in the Russia-Ukraine war (Antebi & Adar, 2024). Even if the payloads are small, "these payloads are capable of killing a person in a direct strike and can cause extensive damage in closed areas, such as vehicles, cargo trucks, tanks and armoured personnel carriers" (Antebi & Adar 2024, p. 2). Targets of such drones may include tanks, artillery gun positions and even a single combat soldier. The reliance on such weapon systems not only reduces risk to human lives but also establishes these drones as bigger targets for a defender. This transition toward the future of warfare may lead to an increased deployment of such weapon systems, both to safeguard human lives and to serve as a deterrent before the utilization of conventional and cost-intensive military assets.

The presence of drones and unmanned ground vehicles equipped with mounted infantry, anti-armour, and anti-aircraft weapons indicates that ground warfare has undergone a significant transformation with the latest technological advancements, and it continues to evolve in response to emerging battlefield dynamics. In the present or future wars, the composition of battlefields may range from the use of infantry to the deployment of drones, or a combination of both. Besides, the standard tactics and operational procedures of ground-based human infantry have already begun to change and are expected to continue evolving (Roy, 2022; Olsen, 2010). The fundamental structure of defences, including the positioning of fire bases, traps, and emplacements, undergoes corresponding adjustments in response to the deployment of such weapon systems. In the future, perhaps the military would rely more on unmanned technologies, and as a result, comprehensive military tactics, along with military doctrines, would be compelled to undergo significant revision.

Electronic Warfare

Electronic warfare represents an often-overlooked aspect of modern warfare that could have significant implications in modern warfare. One phenomenon that requires attention is the Electromagnetic Pulse (EMP), which occurs naturally near the surface of the sun due to solar flares. Artificially generated EMPs can damage or destroy electronic components by disrupting the circuits of any functioning mechanical or electronic system (Pereira & Kunkolienkar, 2013). If employed strategically, these devices could potentially incapacitate a city's electronic infrastructure and severely affect a nation's economy (Pereira & Kunkolienkar, 2013; Johnson, 2024). In the past, invading armies aimed at capturing ports, economic centres, and markets to disrupt an enemy's access to vital resources. However, with the advancement of technologies capable of generating large-scale EMPs or transient electromagnetic disturbances, these economic hubs could be made inoperable without the need for direct military invasion, thus, effectively halting the flow of essential goods and supplies into a port or a nation.

Such actions, carried out without the formal declaration of war, could lead to significant economic setbacks for the targeted civilian population with the possible aim of inducing mass starvation (Pereira & Kunkolienkar, 2013). These kinds of EMP weapons, which are battle-tested, have been seen in the recent Russia-Ukraine war, where such FPV drones are shot down by the use of EMP weapons, cutting off the power of these drones at a close range (Nguyen, 2023). The next phase in the evolution of such weapons involves developing non-nuclear missiles and bombs that can be launched with precision from mobile weapon systems. These launch platforms may include fighter-bombers, naval vessels, or missile-equipped trucks designed to transport and deploy armaments closer to the frontline (Nasui, 2023). The rapid advancement of mobile and man-portable weapon systems has driven a shift toward non-contact warfare. This evolution has extended the battlespace across multiple domains, often targeting areas where the adversary lacks the capability to respond effectively. As the operational landscape continues to evolve, the integration of EMP technologies is prompting military strategists to rethink both offensive and defensive doctrines.

Electronic warfare emerges as a decisive factor in shaping the effectiveness of drone operations on the modern battlefield. Drones are known for their ability to evade detection and reduce the need for ground troops, but Calcara et al. (2022) state that drones are vulnerable to sophisticated measures such as jamming, spoofing, and electronic countermeasures. Besides, the rise of EMP-capable weaponry has led to the rapid development of countermeasures, like jamming the radar, continuous wave immersion, and electromagnetic shielding (Choi et al. 2020; Chen, Mao, & Wu, 2016). The vulnerabilities indicate that drones cannot operate independently or guarantee offensive dominance; instead, they require integration with other assets, including robust command, control, and communications infrastructure. With the examples from recent conflicts (2019-2021) in Libya, Syria, and Nagorno-Karabakh, Calcara et al. (2022) argue that the presence of effective electronic warfare systems and air defence systems can significantly blunt the impact of drone operations, as drones require highly skilled personnel and advanced technology to succeed. They conclude that the mastery of electronic warfare tactics and technologies is critical for any military seeking to employ drones or defend against them in contemporary and future conflicts. Therefore, the increasing sophistication of these technologies suggests that future conflicts may be characterized by rapid, high-impact disruptions where the distinction between military and civilian targets becomes increasingly blurred. This highlights the need for international dialogue and clear guidelines on the use of electromagnetic weapons to help prevent escalation and avoid unintended consequences in a world which increasingly shaped by technological advancement.

As Nagle (2023) states that "With weapons systems undergoing constant refinement, we ought to be able to, at the same time, to refine them in the moral dimension" (p. 23). The ethical concerns around electronic warfare are deeply tied to the risks it poses to civilians and the uncertainty about accountability. In the recent conflict between India and Pakistan, civilians living in border areas of both countries are severely affected, as attacks on civilian infrastructure cause widespread panic, anxiety, and even loss of life. A major concern is that civilian systems like power grids or hospitals could be accidentally affected. As technologies like artificial intelligence and cyber tools evolve quickly, it becomes harder to apply traditional rules of war. It is also difficult to tell apart military and civilian targets in the digital world, which increases the chance of unintended harm. Since these attacks (e.g. cyber-attacks) can be launched easily and do not always cause immediate physical damage, they might seem less serious, but they can still cause major disruption and even lead to dangerous escalation. There is need for urgent updates to international laws and ethical guidelines required for evolving technology.

Psychological Warfare

One of the most important theoreticians on the subject of psychological warfare, Lineberger states that "Psychological warfare is the planned use of propaganda and other psychological actions having the primary purpose of influencing the opinions, emotions, attitudes, and behavior of hostile foreign groups in such a way as to support the achievement of national objectives" (1954, p. 40). This foundational definition by Linebarger underscores the strategic and systematic nature of psychological warfare. By referring to the "planned use", Linebarger makes it clear that such operations are not improvised or incidental; rather, they are methodically designed and executed in coordination with broader military or political campaigns. Psychological warfare operates with specific objectives and draws upon intelligence, sociocultural understanding, and psychological insight to achieve its ends. The inclusion of both "propaganda and other psychological actions" signals that psychological warfare encompasses more than verbal or written communication. It includes a wide array of psychological tactics, such as the dissemination of rumours, staged events, disinformation campaigns, and manipulative visual media, all aimed at influencing perceptions.

After World War II, breakthroughs in medicine paved the way for psychology to grow as a field, gaining new insights and broader recognition in understanding human behaviour and mental health. Military intelligence has historically prioritized the systematic study and manipulation of adversary cognition, recognizing the human mind as a critical domain of conflict. The effort to influence or disrupt an opponent's mental processes has driven the development of non-kinetic strategies. These are commonly classified as forms of cognitive or psychological warfare. These methods have involved spreading carefully chosen materials and organising brainwashing efforts with the aim of either weakening existing beliefs or introducing new ways of thinking (Ibrahim, Rhode, & Daseking, 2023; Katoch, 2023). These actions attempt to change how individuals or groups behave by weakening their unity or encouraging people to join and cooperate within the targeted communities. The evolution of these techniques reflects a technological and conceptual shift in the conduct of warfare, wherein the manipulation of belief systems and ethical orientations is used as a tool for achieving strategic objectives without direct physical engagement (Regalado, 2024).

Military intelligence has long seen the mind as an important domain, studying how enemies think and finding ways to influence their decisions. The development of non-kinetic warfare has focused more on carefully changing how enemies make decisions by drawing on progress in psychology and brain science (Deppe & Schaal, 2024). This highlights that shaping beliefs and values can reach strategic goals more effectively than traditional military power. Historical initiatives such as the CIA's MKUltra program (1953–1973) exemplify early attempts to weaponize psychological principles through coercive interrogation techniques and pharmacological manipulation (Claverie & Du Cluzel, 2022). Present-day approaches involve advanced forms of cognitive warfare by using carefully crafted misinformation, tailored online content, and brain-related technologies to influence how groups think and act (Deppe & Schaal, 2024). These programmes work in two ways: breaking down old beliefs in opposing groups and introducing new values that support their goals. The implementation of these strategies is grounded in three interlinked pillars. First, predictive behavioural modelling- it analyzes cultural, historical, and psychological data to map and anticipate decision-making patterns of the target group (Claverie & Du Cluzel, 2022). Second, information ecosystem manipulation is employed to shape public perception by controlling media narratives and saturating social networks to influence opinions and behaviours. Finally, engaging in neurocognitive interference, although still largely theoretical in military contexts, explores emerging technologies designed to disrupt or modify neural processing pathways. The aim is to potentially alter cognitive functions and influence human responses at a biological level. The shift from physical battles to psychological engagement blurs the distinction between combatants and non-combatants, since whole populations are vulnerable to influence campaigns. This raises complex ethical questions, especially about people's ability to choose freely and where persuasion ends and manipulation begins in the flow of information (Deppe & Schaal, 2024).

A notable illustration of North Korea's covert operations can be observed in its systematic abductions of Japanese and South Korean nationals. These abductions, often carried out along coastal regions by North Korean naval operatives, involved the forcible transfer of individuals into the North, where they were subjected to ideological indoctrination. In some cases, these individuals were later redeployed into their countries of origin to carry out espionage missions on behalf of the North Korean regime (Hughes, 2003). Historical records show this tactic was part of a wider strategy of asymmetric warfare during the Cold War, backed by the Soviet Union. These efforts helped gather intelligence and sometimes involved terrorist acts in South Korea and Japan (Kim, 2003).

Hence, such power utilization for war is something that is one of the resorts for covert warfare against a target nation. A similar approach has been employed by the USSR during its time, and the same is employed by the Chinese in the reduction of its Muslim population in the Xinjiang province. As the Chinese communist ideologies clash with the traditional values of Islam, Muslim youth in Xinjiang are pressurized to abandon their faith and adopt loyalty to the state. Similarly, as argued by Greitens, Lee, and Yazici (2020), the Chinese government monitors and controls the general population through strict surveillance, using technology, credit scores, and citizenship ratings to track any actions against the state.

A clear example of modern strategic thinking is how nations deliberately avoid full-scale wars. Instead of choosing direct military action, which often leads to high political, economic, and human costs, indirect methods are often adopted. One of the most critical examples that can be seen regarding this case is during the use of media in the 1971 Indo-Pak war on the Eastern front. With low morale in the Pakistani military, Indian media published news about paradropping a Brigade of Paratroopers near Dhaka. This news reached the Pakistani Eastern Army commander, causing the military to surrender earlier than expected, however, only a Battalion was dropped, not a Brigade. Jacob (1997) suggests that the enemy commander's demoralized mental state and the circulation of paradrop photos from an Indian army archive (supposedly used as fake news), may have contributed to Pakistan's perception of the situation, potentially influencing the decision to surrender.

Military intelligence agencies have historically played a critical role in influencing national leaders, infiltrating foreign military structures, intelligentsia, universities, and businesses to serve the interests of another nation (Pecht & Tishler, 2015). This approach has its roots in military intelligence, where gathering information about the enemy goes hand in hand with building relationships with key people inside their country. Information can be used to manipulate public perception of the nation's capabilities, opinions about ongoing national problems, and can be used to generate a movement either against a government or against that nation's armed forces. Initially, these operations relied only on human operatives, but from the 20th century, technology has come into play. The role of the internet has been phenomenal here as it allows operatives or human assets within an enemy nation to be recruited to capture images of sensitive documents from secure military facilities. These images, often containing critical information, can then be sent through WhatsApp or email in just seconds. The internet has made it a matter of seconds to transfer vital data to leak information. With the rise of modern cyberspace and electronic warfare, the necessity for conventional warfare may diminish. These tools allow external powers to manipulate political leaders and destabilize internal governance without direct military engagement. By exploiting these advanced capabilities, entire nations or their leaders may be influenced to act against their national interests, potentially undermining their sovereignty and creating internal strife (Pecht & Tishler, 2015; Verma, 2022).

A nation's critical systems include essential infrastructures like power, communication, finance, transport, water, healthcare, and defence networks. The future of warfare is likely to focus increasingly on the control, disruption, and destruction of a nation's critical systems. This shift suggests a move toward non-contact warfare, where human involvement is minimal, and the emphasis is placed on soft civilian targets and institutions. This strategy may aim to transform an adversarial nation into an ally and eliminate the potential for future threats without the need for traditional combat. If necessary, conventional warfare could still be pursued, but it would likely be preceded or accompanied by these non-traditional methods of influence and disruption.

The development of non-conventional weapons systems may further reshape the landscape of warfare. For instance, long-range ballistic nuclear missiles, such as those potentially launched from submarines, or guided sniper rifle bullets intended for targeted assassinations, represent a shift toward precision in military operations (Chen et al., 2023). Also, small surveillance drones equipped with explosive payloads could become essential as they allow covert operations and intelligence gathering. Such technologies, which enable precise strikes on high-value targets, may significantly alter the course of conflicts. These developments, combined with unconventional surveillance techniques, could provide critical intelligence to military units and may reshape conventional warfare dynamics by lowering the reliance on large-scale deployments of personnel and equipment (Chen et al., 2023; Kilcullen, 2019). Warfare has clearly changed over time, with most of its history shaped by physical battles using the technology available up to the 20th century (Verma, 2022; Kilcullen, 2019). As we have entered the 21st century, it has become increasingly apparent that military power is not solely dependent on the destruction of infantry battalions or armoured columns to diminish a nation's defensive or offensive capabilities. Contemporary strategies may involve cyberattacks targeting communication channels, which can disrupt the functionality of both manpower and equipment. In some cases, rather than engaging in mass aerial bombardments, nations can deploy malware through cyberspace to cripple an economy or destabilize key infrastructure (Pecht & Tishler, 2015). This is reflected in Verma (2022), where he stated that "The future battlefield environment appears to be diffused, diverse, and disruptive" (p. 57).

Social Media and Media Influence

The integration of social media into contemporary society has profoundly influenced the dynamics of modern warfare, like in the destabilization of governments and regimes. Since its widespread adoption beginning around 2010, social media has been extensively utilized to influence national elections and to systematically shape public perceptions (Chaudhari & Pawar, 2021; Farkas & Schwartz, 2018). In addition to its socio-political impact, social media has been used as a tool for propaganda, similar to the methods employed in psychological operations (Mabima, 2018). According to Rashi and Schleifer (2023), "Modern propaganda is no longer meant to modify ideas, but rather to provoke action. It is no longer designed to change adherence to a doctrine, but rather to make the individual cling irrationally to a process of action. It is no longer a method for transforming one's opinion but rather is developed to engender an active and mythical belief (p. 200). They emphasize that modern propaganda targets behaviour rather than belief with the aim to drive people to act without rational thought. Instead of persuading individuals through logic, it promotes emotional commitment to a cause or narrative.

Demuyakor (2021) states that "the new, emerging media can also propagate grassroots propaganda, which is simply a form of open resistance against anybody oppressively exercising his or her power" (p. 6). For instance, the recent deployment of propaganda warfare in cyberspace was observed during the political crisis in Bangladesh in 2024. Bangladesh, a nation with a history of political assassinations and military coups, witnessed the downfall of its previous prime minister, Sheikh Hasina. Her government reportedly fell due to the strategic use of social media propaganda and persuasive communication (Hossain, Hasan & Wasantha, 2025; Jahan, n.d.). Opposition groups, including anti-Awami League factions and Islamist fundamentalists, utilized Facebook reels to highlight narratives of police brutality. These visuals were crafted to incite public sentiment against Sheikh Hasina's administration, further fuelling anti-Indian sentiments due to Hasina's historical alliances with India and her government's pro-Indian policies, which were deeply unpopular among Islamist forces in Bangladesh (Tripathi, 2024). The circulation of viral reels depicting police brutality during the crackdown on student protests, featuring footage of shootings and the hijacking of protestors by Islamist militants, was strategically synchronized with sentimental music and widespread memes. The campaign successfully shifted public opinion against Sheikh Hasina and the Awami League government, which led to widespread political unrest. The current interim government of Bangladesh, under the guidance of Muhammad Yunus, may serve as a strategic asset, particularly in its reported efforts to establish a listening post to monitor the geopolitical influences of both China and India (Amin, 2025).

Besides the influence of social media, the mainstream media has placed the China-Pakistan Economic Corridor (CPEC) as a central focus in South Asian geopolitics, specifically in the struggle over narratives between India and Pakistan. In their study, Murtaza and Saud (2025) state that countries construct opposing narratives that serve their respective national agendas. The Pakistani media presents CPEC as "a transformative initiative that promotes development and regional connectivity"; on the other hand, the Indian counterparts portray it as "a geopolitical threat that raises sovereignty concerns and increases China's economic and strategic influence in the Indian Ocean" (p. 249). Their study analyzed television news, newspaper articles, talk shows, and social media trends and found that state-driven narratives, misinformation, and selective media representations shape public opinion and intensify nationalistic sentiment. Based on information from mainstream media, social media plays a significant role in amplifying these narratives and deepening diplomatic tensions. Murtaza and Saud (2025) suggest that there is a need for improved media literacy and journalistic standards to promote a more informed understanding of CPEC's implications. On a similar line, Thankachan, Thomas, and Menon (2025) conducted a thematic analysis of case studies on the Mumbai (2008), Uri (2016), and Pulwama (2019) terrorist attacks. They studied news reported in Indian and Pakistani print media to explore biases and the reliance on stereotypical reporting in national newspapers. They examined how *The Hindu* and *Dawn,* Indian and Pakistani newspapers respectively, which are positioned as agents of both peace and conflict, shaped propaganda narratives of war and reconciliation. They argue that the print media in India and Pakistan reinforce nationalist narratives to influence public perception and sustain bilateral tensions. They highlighted "the media's tendency to exacerbate conflicts rather than work towards proposing solutions" (p. 170). The study suggests a meaningful change in how the media tells stories and urges a shift toward more thoughtful and positive reporting that can help build peace and understanding between the two nations, which are historically in conflict.

Fahim and Islam (2024) studied the extent of dedicated airtime to the Second Karabakh War (2022) by Eastern and Western media and the use of lexico-semantic manipulation, euphemisms, and dysphemisms in their discourse. The study identified a pro-Armenian bias in the Western media coverage, whereas Eastern channels adopted varied positions. Fahim and Islam (2024) state that "when reporting on war crimes, media bias can contribute to the dehumanization of one side while defending or justifying the actions of the other" (p. 91). Their study concluded that media bias is often shaped by journalists' ideological beliefs, national interests, and the broader goals of a country's foreign policy.

Similarly, the role of media has been exponential in the recent conflict in the Indian subcontinent. The background of the conflict and the portrayals of the events by the media are elaborated in this section. In May 2025, tensions between India and Pakistan erupted into a brief but intense military confrontation following a deadly terrorist attack in the town of Pahalgam, which killed 26 people. The town is a popular tourist destination which is located in Indian-administered Kashmir. India immediately attributed the attack to Pakistan-based insurgent groups, which revived India's long-standing accusations that Pakistan harbours and supports militants operating in Kashmir (Ministry of External Affairs [MEA], 2025). Over the next few days, both nations launched military strikes and exchanged artillery fire across the Line of Control (LoC). They also carried out retaliatory diplomatic and trade measures; however, both sides ultimately agreed to a ceasefire on May 10.

During the May 2025 conflict, Indian and Pakistani mainstream media were heavily criticized for spreading misinformation and panic. Kathju (2025) reported that television channels aired doctored visuals, recycled footage from unrelated conflicts (like Gaza and Lebanon), and even broadcast video game clips, falsely presenting them as live war updates. This misinformation reached alarming levels, with some channels falsely claiming attacks on Pakistani cities, captures of enemy soldiers and intensified public fear. Independent digital news/media platforms like *Alt News* and *The Quint*, in their attempt to debunk claims found that AI-generated videos and audio deepfakes targeting public perception were rampant in both India and Pakistan and "each country was directing propaganda outwards, but rather inwards- towards its citizens" (Kathju, 2025, p. 1). News studios transformed into theatrical 'war rooms' with sirens and Computer-Generated Imagery (CGI) graphics, which prompted India's Ministry of Home Affairs to issue an advisory against such dramatizations. The use of sensationalism, unverified content, and patriotic fervour over responsible reporting led to widespread disillusionment and panic. This damages media credibility, erodes public trust, and causes long-term psychological harm, which highlights the negative impact of sensationalist reporting during national crises.

Information warfare between the two countries was seen in the disparity in the claims made regarding air strikes, as both claimed to have caused severe damage by reciprocal drone and missile strikes on the military bases of the other. The rapid dissemination of such claims through social media platforms and 24-hour news channels significantly amplified the spread of conflict-related misinformation among people. This created parallel narratives of military success within both nations' digital ecosystems. However, according to The New York Times (2025), the analysis of the satellite images revealed India demonstrated more measurable success in targeting Pakistani military facilities like the Bholari air base with confirmed damage to hangars and runways. Whereas evidence for Pakistan's retaliatory strikes on the claimed attack on India's Udhampur base remained questionable because there was no conclusive visual proof, and the satellite image showed no damage to the airbase.

Economic Warfare

Economic warfare is a very old form of warfare that has existed for a long time. However, it has evolved significantly, and it remains relevant today. A relevant historical example can be seen in the trade war waged against India during its colonization. This was done by imposing taxes and tariffs that undermined India's small-scale industries and by promoting British imports to dominate Indian markets. The aim here was to ultimately pave the way for control over the Indian territory by the British colonizers (Awasthy, 2025). Similarly, during World War I, the Allied Powers imposed sanctions on the nations affiliated with the Axis Powers (Chohan, 2024). A similar pattern emerged during World War II, when the United States imposed an oil embargo to curb Japanese expansionism in Asia. In the following decades, economic sanctions became a key tool of U.S. foreign policy, especially against nations aligned with the Soviet bloc, such as North Korea. It also targeted Iran after the overthrow of the Shah, largely because of Iran's anti-American stance (Scarpino & Trainer, 2024). This approach was also evident during the Taliban's resurgence in Afghanistan, where the U.S. imposed sanctions to limit the Taliban's financial resources and income. Sanctions were likewise implemented against Russia as a means of economic pressure. Recently, the United States raised tariffs on Chinese goods as high as 145 per cent to intensify economic pressure on China and signal a new phase of economic warfare. Upon returning to office in 2025, President Trump imposed a universal 10 per cent tariff on all Chinese imports. He introduced a policy of "reciprocal taxation," which matched tariffs and trade barriers that other countries placed on the US, including a 34 per cent reciprocal tariff aimed specifically at Chinese goods (Clarke, 2025). These moves reflect economic battles aimed at countering perceived unfair trade practices.

In several instances, efforts to weaken an adversary through tariffs or trade wars have not always achieved their intended outcomes. When powerful nations or the international community impose revenue-blocking measures, often led by the U.S., targeted states often search for alternative strategies. At the same time, other countries may try to take advantage of the gaps left in global trade. Economic competition has grown as advances in merchant shipping, road networks, and air transport allow faster responses to trade restrictions. To apply or resist geopolitical pressure, states may launch infrastructure projects that connect airports and river ports to reinforce trade routes and weaken rivals. These advances in aviation and transport have made economic pressure a more appealing option for states that want to avoid direct military conflict. For many, it acts as an early move in a wider strategic contest (Chohan, 2024). The contest for control over key ports, such as those in Sri Lanka and Bangladesh, illustrates how naval merchant shipping and port access can influence regional power dynamics. For example, China and India's involvement in Bangladeshi ports has reportedly reduced US influence and prompted the U.S. to strengthen its presence in South Asia and counter China's growing naval and economic power (Bose & Ray Chaudhury, 2023).

These economic measures can promote corruption, destabilize economies, adversely affect global market confidence, and sometimes fail to decisively alter the political behaviour of targeted countries. The indirect harms of economic warfare thus often extend beyond the intended impact on enemy governments and pose complex ethical and strategic challenges for policymakers.

Redefining the Battlefield of Tomorrow

Looking ahead, future warfare may involve the implantation of malware or the use of phishing attacks to extract sensitive information, which could be exploited for specific strategic objectives. Contemporary society has witnessed a transformative evolution from the widespread adoption of smartphones to the emergence of sophisticated wearable technologies. Pearl, Intriligator, and Liu (2025) state that "One of the most significant trends in wearable technology is the seamless integration with a host of emerging technologies, including AI, the Internet of Things (IoT), augmented and virtual reality (AR/VR), and blockchain" (p. 12), thus representing a significant shift in how humans interact with digital systems. In future, wearable devices are anticipated to evolve into biologically integrated systems, including prosthetic replacements and ingestible electronics equipped with onboard computational capabilities. However, this technology is not without limitations. For example, the *Strava* fitness application unintentionally disclosed sensitive U.S. military sites through its global heatmap functionality. This event highlights the need to implement strict limitations on data sharing and to conduct thorough impact assessments to prevent similar breaches (Pearl, Intriligator, & Liu, 2025). Additionally, such advancements pave the way for pervasive surveillance mechanisms, particularly when deployed to monitor individuals of strategic importance like high-ranking military personnel, government officials, weapons engineers, or influential businessmen who are involved in framing national policies. Current methodologies for extracting sensitive data from these individuals often involve phishing attacks, viruses or malicious software applications that transmit data from their digital devices.

The next phase of attacks could be hacking attacks that destroy wearables or sensitive information to hinder the usability of the data. For example, any high-value military target is usually connected to intranet connection that is restricted solely to government or military use. However, this intranet can still be vulnerable to hacking attempts by allowing unauthorized access. Once breached, it can be subjected to surveillance, tapped for information, or even manipulated to disable or destroy specific systems connected to it. For example, during the Indo-Pakistan conflict of 2025, the Facebook Page of Karachi Port Trust was hacked and was shown as 'Temporarily Closed' with a document published stating that the Indian Navy had destroyed facilities in the Karachi port. This was later attributed to unauthorized cyber activity targeting the official Facebook page of the Karachi Port Trust, using the images of strikes in Gaza from the year 2020, which resulted in the mass dissemination of misleading information (Outlook, 2025).

Furthermore, the involvement of business entities may also play a powerful role, where their investments in foreign nations could be employed to weaken a nation's defence, intelligence, or policy decisions. This could also extend to financially supporting internal separatist movements to create further disruption. Such activities have been observed in cases where Chinese companies have acquired shares or entire firms involved in manufacturing defence equipment and software for the U.S. military, to potentially gain access to sensitive technologies and transmit this information abroad (Joshi, 2022; O'Connor, 2019). Moreover, financial support for espionage activities is not confined solely to businesses; non-governmental organizations (NGOs) and charitable entities, often with less scrutiny, can also be utilized for surveillance and intelligence-gathering purposes. These organizations may enable outsiders to gain access to foreign governments or military systems to influence decisions to serve external interests instead of the priorities of the host nation.

The domain of space emerged as another frontier for military power when Germany developed and tested the V-2 rocket between 1936 and 1944, which is a major technological leap that also opened the door to space exploration. After World War II, the United States and the Soviet Union entered a fierce competition to use space for military purposes. They focused on satellites for communication and intelligence gathering. The Soviet Union launched Sputnik in 1957, the first artificial satellite, followed by the United States establishing NASA in 1958. Both countries declared to display dramatic shows of strength, including nuclear tests, but these plans were abandoned due to risks, global concerns, and the danger of militarizing space (Geppert, Brandau, & Siebeneichner, 2018). Instead, the step that was taken was to start space programs by individual nations in the name of science and progress. However, the main idea behind this was launching satellites into space for both communication and surveillance. This technology proved highly valuable for military intelligence operations because satellite imagery provided critical insights into enemy positions and movements. As satellite technology advanced, the precision and reliability of data collection significantly improved. Similarly, communication satellites became essential, since many emerging technologies depended heavily on consistent and secure satellite-based communication channels (Muszynski-Sulima, 2023). This dependency indicates that the potential destruction or incapacitation of satellites could result in the severe disruption of a nation's entire communication infrastructure. Given the critical role and relative vulnerability, satellites have increasingly become attractive targets for adversarial actions like cyberattacks and hacking that aim to compromise or disable their functionality. This domain of warfare is in both the realm of cybersecurity and space warfare. Given these considerations, protecting satellite infrastructure remains crucial for national security in the changing landscape of modern conflict.

Historical analysis reveals that the domain of warfare has undergone significant transformation, particularly since the early 20th century. Initially confined primarily to traditional ground, naval, and aerial theatres, modern warfare has expanded into multiple domains, including cyber and space. This shift has been driven by advancements in technology that provide strategic and tactical advantages. As warfare continues to evolve toward greater automation, there is a great rise in the use of drones and UGVs. It is anticipated that future conflicts will predominantly involve automated weapon systems (Antebi & Adar, 2024). Also, with global rules and deterrence strategies reducing the chance of nuclear war, the focus of warfare in the physical world is now shifting toward advanced automated systems, like drones and robots. Hence, combat may slowly be restricted only to the deployment and destruction of such unmanned mobile weapon systems.

As much as we have seen in the development of drones, which almost do similar tasks carried out by a conventional air force, we can expect the same to be done during ground operations. A steady shift can be observed in the complete automation of armoured vehicles, with onboard weapon systems operated entirely by AI, thereby reducing human involvement in armour operations. According to Ibrahim (2024), "As AI becomes more affordable and accessible, these capabilities are likely to become standard features in FPV drones" (p. 1). Autonomous drones mark a major change in warfare, narrowing the distinction between drones and missiles. Unlike traditional missiles, which are costly and few in number, AI-powered drones can be made cheaply in large quantities. These drones can form swarms of self-guided weapons that could overwhelm even the strongest defences. Although the Russia-Ukraine war has exhibited a steady mixture of battlefields ranging from World War I era trench warfare to futuristic drone warfare, the slow arrival of armed robots can reduce human dependence in warfare. Therefore, while automated systems may play a larger role in future warfare, human involvement will remain crucial in conventional battles. Technology transfer in future conflicts is expected to focus on non-contact methods, while territorial aggression may still involve conventional warfare, supported by modern unmanned technologies (Verma, 2022; Kilcullen, 2019). In non-contact warfare over cyberspace, attacks would spread through hacking. These could target websites, destroy documents and data, steal information, infiltrate civil and defence networks, and implant viruses to disrupt systems. Military software and space programs could also be rendered useless, and both humans and AI might work individually or together on these methods of warfare.

As the range of available weapons grows, several countries are also working on developing biological weapons. With genetic engineering advancing quickly, the ability to alter and create new viruses could change warfare dramatically, potentially making traditional weapons less relevant. A nation could spread new strains of viruses and control the distribution of vaccines, thus using disease as a tool to target specific populations. Viruses have long been preferred in the development of biological weaponry, as viral infections are difficult to cure and new strains can be easily created either through mutations or genetic engineering (Balali-Mood, Moshiri, & Etemad, 2013).

Hence, without the destruction of property, future wars with biological weapons being deployed to attack a nation's population without the use of any kind of conventional weaponry. Even without the complete wiping out of a population, biological weapons can be seen to take a complete hit on the economy of a nation through the steps a nation or a population has to take. As the future of warfare progresses into new domains like hacking, cyber expansion, biological weapons, and shifting media narratives, the world's financial centres could also become targets of destruction. As technology progresses, there is a likelihood of increasing attacks on human intelligence and the mind itself. Since in the current times, individuals remain connected to their devices and wearables through the internet, there may be a gradual progression toward greater reliance on constant connectivity. Consequently, the potential for exploitation in this domain of warfare may differ from traditional forms. Entire ecosystems connected to the internet and the devices could be used to spread misinformation, change the perception of people, and relationships between nations or communities, thus leading to the possibility of war being initiated by twisting narratives and changing people's perception of reality. High-scale malware assaults could also be aimed at a targeted population to propel them towards financial loss, data breaches, and data compromises to progress further attacks. Social media could play a significant role in shaping public perception to either undermine or support a government, incite aggression, and target individuals within minutes. It can be anticipated that in this domain of cyber warfare, the duration of conflicts could be exceptionally brief. As technology progresses and new domains of human access and usage emerge, it may be argued that the frontlines of wars and conflicts will extend into these domains as well.

REFERENCES

Al Jazeera. (2025). Have India and Pakistan started a drone war? https://www.aljazeera.com/news/2025/5/8/have-india-and-pakistan-started-a-drone-war

Ambodo, T., Saputro, G. E., & Komariah, U. N. (2021). Multi-Agent Drone as Loitering Munition System to Support Military Operations Carrying Out Policy Mandate. *International Journal of Arts and Social Science*, *4*(5), 2581–7922.

Amin, A. (2025). Sino-Indian rivalry in Indo-Pacific and Bangladesh's geo-strategic hedging in response to its security threats. *Discover Global Society*, *3*(1), 1–20. DOI: 10.1007/s44282-025-00144-1

Antebi, L., & Adar, O. (2024). *FPV Drones: From the Ukrainian Battlefield to the Middle East*. Institute for National Security Studies.

Awasthy, V. (2025). Financial imperialism and British colonial power in India: An interwar history of economic strategies and fallouts. In Mandala, V. R. (Ed.), *Exploring power and authority in Indian history across the ages* (pp. 217–243). Springer., DOI: 10.1007/978-981-96-1679-4_12

Balali-Mood, M., Moshiri, M., & Etemad, L. (2013). Medical aspects of bio-terrorism. *Toxicon*, *69*, 131–142. DOI: 10.1016/j.toxicon.2013.01.005 PMID: 23339855

BBC News. (2025). India and Pakistan drone conflict: What we know so far. https://www.bbc.com/news/articles/cwy6w6507wqo

Bose, S., & Ray Chaudhury, A. B. (2023). Crossing currents: India-China contention and the Teesta conundrum in Bangladesh. *Observer Research Foundation*. https://www.orfonline.org/expert-speak/crossing-currents-india-china-contention-and-the-teesta-conundrum-in-bangladesh

Buzan, B., & Hansen, L. (2009). *The Evolution of International Security Studies*. Cambridge University Press. DOI: 10.1017/CBO9780511817762

Calcara, A., Gilli, A., Gilli, M., Marchetti, R., & Zaccagnini, I. (2022). Why drones have not revolutionized war: The enduring hider-finder competition in air warfare. *International Security*, *46*(4), 130–171. DOI: 10.1162/isec_a_00431

Castrillo, V. U., Manco, A., Pascarella, D., & Gigante, G. (2022). A Review of Counter-UAS Technologies for Cooperative Defensive Teams of Drones. *Drones (Basel)*, *6*(3), 65. DOI: 10.3390/drones6030065

Chaudhari, D. D., & Pawar, A. V. (2021). Propaganda analysis in social media: A bibliometric review. *Information Discovery and Delivery*, *49*(1), 57–70. DOI: 10.1108/IDD-06-2020-0065

Chen, P., Mao, C., & Wu, G. (2016). *Shielding effectiveness assessment of enclosure by means of EMP excitation*. In *2016 Asia-Pacific International Symposium on Electromagnetic Compatibility (APEMC)* (Vol. 1, pp. 273–275). IEEE. DOI: 10.1109/APEMC.2016.7523032

Chohan, U. W. (2024, July 15). Modern economic warfare and its implications. *SSRN*. https://ssrn.com/abstract=4894995

Choi, S., Kwon, O.-J., Oh, H., & Shin, D. (2020). Method for Effectiveness Assessment of Electronic Warfare Systems in Cyberspace. *Symmetry*, *12*(12), 2107. DOI: 10.3390/sym12122107

Clarke, J. (2025, May 24). What tariffs has Trump announced and why? *BBC News*. https://www.bbc.com/news/articles/cn93e12rypgo

Claverie, B., & Du Cluzel, F. (2022). Cognitive warfare: The advent of the concept of 'cognitics' in the field of warfare. In B. Claverie, B. Prébot, N. Buchler, & F. Du Cluzel (Eds.), *Cognitive warfare: The future of cognitive dominance* (pp. 1–7). NATO Collaboration Support Office. https://hal.science/hal-03635889

Demuyakor, J. (2021). The propaganda model in the digital age: A review of literature on the effects of social media on news production. *Shanlax International Journal of Arts. Science and Humanities*, *8*(4), 232–239. DOI: 10.34293/sijash.v8i4.3598

Deppe, C., & Schaal, G. S. (2024). Cognitive warfare: A conceptual analysis of the NATO ACT cognitive warfare exploratory concept. *Frontiers in Big Data*, *7*, 1452129. Advance online publication. DOI: 10.3389/fdata.2024.1452129 PMID: 39552633

Durfey, N., & Sajal, S. (2022). A comprehensive survey: Cybersecurity challenges and futures of autonomous drones. In *Proceedings of the 2022 Intermountain Engineering, Technology and Computing Conference* (pp. 1–6). IEEE. DOI: 10.1109/IETC54973.2022.9796881

Elish, M. C. (2017). Remote Split: A History of US Drone Operations and the Distributed Labor of War. *Science, Technology & Human Values*, *42*(6), 1100–1131. https://www.jstor.org/stable/26580327. DOI: 10.1177/0162243917731523

Fahim, M., & Islam, M. N. (2024). Mapping the Air Time of Eastern & Western Media on Conflict and War: A Comparative Study of BBC, DW, TRT and Al Jazeera on the Coverage of Second Nagorno-Karabakh War & the Aftermath. *Communicatio Socialis*, *37*(1), 79–98.

Farkas, J., & Schwartz, S. A. (2018). Please like, comment and share our campaign! How social media managers for Danish political parties perceive user-generated content. *Nordicom Review*, *39*(2), 19–33. DOI: 10.2478/nor-2018-0008

Gabriele, J. (2024). *Underwater communication technologies for submarines: Experimentation and innovations* [Honours thesis, University of New Haven]. University of New Haven Digital Commons. https://digitalcommons.newhaven.edu/honorstheses/80/

Geppert, A. C., Brandau, D., & Siebeneichner, T. (2018). Militarizing Outer Space Astroculture, Dystopia and the Cold War. *Spaceflight*.

Greitens, S. C., Lee, M., & Yazici, E. (2020). Counterterrorism and preventive repression: China's changing strategy in Xinjiang. *International Security*, *44*(3), 9–47. DOI: 10.1162/isec_a_00368

Grigore, L., & Cristescu, C. (2024). The Use of Drones in Tactical Military Operations in the Integrated and Cybernetic Battlefield. *Land Forces Academy Review*, *29*(2), 269–273. DOI: 10.2478/raft-2024-0029

Gupta, A., & Dubey, M. A. (2024). *Reforms in Indian military since 2014* [Research report]. ResearchGate. https://www.researchgate.net/publication/380034381_Reforms_in_Indian_Military_Since_2014

Hakimi, M., Reid, N. L., & Witten, S. (2015). An introduction: Adapting to a rapidly changing world. *American Society of International Law Proceedings*, *109*, 1–7. DOI: 10.5305/procannmeetasil.109.2015.0001

Hossain, M. A., Hasan, M. M., & Wasantha, N. (2025). The power of social media in driving political change: A case study on Bangladesh's regime change in 2024. *International Journal of the Sociology of Law*, *2*(1), 257–275. DOI: 10.62951/ijsl.v2i1.405

Hughes, C. W. (2003). North Korea: Coming in from the cold? *Journal of East Asian Studies (Seoul)*.

Ibrahim, A. (2024). Employment of FPV Drones in Russia-Ukraine War: Lessons and Future Outlook. *Modern Diplomacy*. https://moderndiplomacy.eu/2024/12/14/employment-of-fpv-drones-in-russia-ukraine-war-lessons-and-future-outlook/

Ibrahim, F., Rhode, S., & Daseking, M. (2023). Cognitive & psychological warfare. *The Defence Horizon Journal*. https://tdhj.org/blog/post/cognitive-psychological-warfare/

Jacob, J. F. R. (1997). *Surrender at Dacca: Birth of a Nation*. Manohar Publishers.

Johnson, E. (2024). How Far Does the Russian Military Presence Stretch? *Florida Political Chronicle, 31*(2).

Joshi, M. (2022). *China's military-civil fusion strategy: The US response and implications for India*. Observer Research Foundation. https://www.orfonline.org/research/china-s-military-civil-fusion-strategy-the-us-response-and-implications-for-india

Kathju, J. (2025). Indian media fuels panic with disinformation. *Al Jazeera Media Institute*. https://institute.aljazeera.net/en/ajr/article/indian-media-fuels-panic-disinformation

Katoch, P. C. (2023). Brain warfare. *SP's Naval Forces*. https://www.spsnavalforces.com/experts-speak/?id=606&h=Brain-Warfare

Kaura, V. (2020). India's Pakistan policy: From 2016 'surgical strike' to 2019 Balakot 'airstrike'. *The Round Table*, *109*(3), 317–327. DOI: 10.1080/00358533.2020.1760499

Kilcullen, D. (2019). The evolution of unconventional warfare. *Scandinavian Journal of Military Studies*, *2*(1), 61–71. DOI: 10.31374/sjms.35

Kim, Y. (2003). Kidnapping Politics in East Asia. *Journal of East Asian Studies (Seoul).*

Kolhatkar, A. J. (2024). Escalation and air power. *Journal of Defence Studies*, *18*(2), 101–120. https://www.idsa.in/wp-content/uploads/2024/08/18-2_AJ-Kolhatkar-07.pdf

Kuhn, U., & Williams, H. (2025). Governing the impact of emerging technologies: Actors, technologies, and regulation. *The Journal of Strategic Studies*, *48*(3), 559–570. DOI: 10.1080/01402390.2025.2472213

Linebarger, P. M. A. (1954). *Psychological warfare* (2nd ed.). Duell, Sloan and Pearce.

Mabima, J. (2018). *Social networking sites as a tool of psychological operations: A case study. SSRN.* https://doi.org/DOI: 10.2139/ssrn.3261039

Ministry of External Affairs. (2025). *Statement by Foreign Secretary: OPERATION SINDOOR* [Press release]. Government of India. https://www.mea.gov.in/Speeches-Statements.htm?dtl/39473/Statement+by+Foreign+Secretary+OPERATION+SINDOOR

Murtaza, G., & Saud, A. (2025). India's Fifth-Generation Warfare against Pakistan and the Role of Indian Media in Shaping Public Perceptions of the China-Pakistan Economic Corridor (CPEC). *Journal of Social Sciences Review*, *5*(1), 249–256. DOI: 10.62843/jssr.v5i1.487

Muszyński-Sulima, W. (2023). Cold War in space: Reconnaissance satellites and US-Soviet security competition. *European Journal of American Studies, 18*(2). Advance online publication. DOI: 10.4000/ejas.20427

Nag, H. (2023). *Drone Warfare Capacity Building in Securing India's National Interests*. Vivekananda International Foundation.

Nagle, W. J. (2023). *Morality and modern warfare*. Highlyy Publishing LLP.

Nasui, R. C. (2023). An Upgrade to the Gepard System. Possible Usage of the CA-94. *AFASTUD, 2023*, 28.

NDTV. (2025). India-Pakistan tension live update: Operation Sindoor, Pahalgam terror attack, Jammu and Kashmir, Rajasthan, blackout in Punjab. https://www.ndtv.com/india-news/india-pakistan-tension-live-update-operation-sindoor-pahalgam-terror-attack-jammu-and-kashmir-rajasthan-blackout-in-punjab-8366542

New York Times. (2025). India and Pakistan Talked Big, but Satellite Imagery Shows Limited Damage. https://www.nytimes.com/interactive/2025/05/14/world/asia/india-pakistan-attack-damage-satellite-images.html

Nguyen, P. D. (2023). *The role of small and medium UAVs in recent conflicts: Lessons and forecasting trends in the development of UAV forces in the military* [Master's dissertation]. Canadian Forces College.

North Atlantic Treaty Organization. (2020). Game of drones: How new technologies affect deterrence, defence, and security. *NATO Review*. https://www.nato.int/docu/review/articles/2020/05/05/game-of-drones-how-new-technologies-affect-deterrence-defence-and-security/index.html

O'Connor, S. (2019). *How Chinese companies facilitate technology transfer from the United States*. U.S.-China Economic and Security Review Commission. https://www.uscc.gov/sites/default/files/Research/How%20Chinese%20Companies%20Facilitate%20Tech%20Transfer%20from%20the%20US.pdf(WITA)

Olsen, J. A. (2010). *A History of Air Warfare*. University of Nebraska Press. https://muse.jhu.edu/book/28122

Outlook. (2025). Fact-check: Viral claim of Indian strike on Karachi port uses 2020 Gaza images. *Outlook India*. https://www.outlookindia.com/national/fact-check-viral-claim-of-indian-strike-on-karachi-port-uses-2020-gaza-images

Pearl, D., Intriligator, J., & Liu, W. (2025). Seamless integration: The evolution, design and future impact of wearable technology (Version 1) [Preprint]. *arXiv*. https://arxiv.org/abs/2502.05797v1

Pecht, E., & Tishler, A. (2015). The value of military intelligence. *Defence and Peace Economics*, *26*(2), 179–211. DOI: 10.1080/10242694.2014.886435

Pereira, V., & Kunkolienkar, G. R. (2013). EMP (Electro-Magnetic Pulse) weapon technology along with EMP shielding & detection methodology. In *2013 Fourth International Conference on Computing, Communications and Networking Technologies (ICCCNT)* (pp. 1-5). IEEE. DOI: 10.1109/ICCCNT.2013.6726651

Peri, D. (2025). Pakistan used 400 drones, possibly from Turkiye: India. *The Hindu*. https://www.thehindu.com/news/national/pakistan-launched-300-400-turkish-drones-across-western-border-pak-ad-radar-hit-in-indian-retaliation/article69558073.ece

Potzsch, H. (2015). The emergence of iBorder: Bordering bodies, networks, and machines. *Environment and Planning. D, Society & Space*, *33*(1), 101–118. DOI: 10.1068/d14050p

Rashi, T., & Schleifer, R. (2023). The Ethics of Psychological Warfare – Lessons from Israel. *Democracy and Security*, *19*(2), 199–210. DOI: 10.1080/17419166.2023.2210472

Regalado, A. (2024). Brainwashing, mind control, and the CIA: The history of Operation Midnight Climax. *MIT Technology Review*. https://www.technologyreview.com/2024/04/12/1090726/brainwashing-mind-control-history-operation-midnight-climax/

Roy, K. (2022). *A Global History of Warfare and Technology: From Slings to Robots*. Springer Nature. DOI: 10.1007/978-981-19-3478-0

Roy-Chaudhury, R. (2025). India–Pakistan drone and missile conflict: Differing and disputed narratives. *International Institute for Strategic Studies*. https://www.iiss.org/online-analysis/online-analysis/2025/05/indiapakistan-drone-and-missile-conflict-differing-and-disputed-narratives/

Scarpino, R., & Trainer, J. (2024). Sanctions by the numbers: 2023 year in review. *Center for a New American Security*. https://www.cnas.org/publications/reports/sanctions-by-the-numbers-2023-year-in-review

Szakacsi, I. (2016). Certain Aspects of Military Applications of Robotics in the Future Missions. *Defence Review, 1*.

Thankachan, K., Thomas, P. E., & Menon, V. A. (2025). Conflict Caught in Propaganda Nationalism: An Analysis of the India-Pakistan Newspaper Coverage of Mumbai, Uri, and Pulwama Attacks. In *Impacts of Leakage, Whistleblowing, and the Rise of Propaganda* (pp. 169-184). IGI Global Scientific Publishing.

The Hindu. (2025). Pahalgam terror attack: Operation Sindoor launch live updates. https://www.thehindu.com/news/national/pahalgam-terror-attack-operation-sindoor-launch-live-updates-may-7-2025/article69543511.ece

Times of India. (2025). Seven dead, 38 injured as Pakistan Army targets civilian areas along LoC in J&K. https://timesofindia.indiatimes.com/india/seven-dead-38-injured-as-pakistan-army-targets-civilian-areas-along-loc-in-jk/articleshow/120952157.cms

TOI News Desk. (2025). Army foils Pakistan's attempts to send swarm drones across LoC. *The Times of India*. https://timesofindia.indiatimes.com/india/army-foils-pakistan-attempts-to-send-swarm-drones-across-loc/articleshow/121014854.cms

Tripathi, S. (2024). A force for good. *Index on Censorship, 53*(4), 55–57. DOI: 10.1177/03064220241306623

Verma, V. (2022). Non-Contact Warfare: A Strategy for Future. *CLAWS Journal, 15*(2), 57–70. https://core.ac.uk/download/pdf/555351808.pdf

ADDITIONAL READING

Bhuta, N. (Ed.). (2016). *Autonomous weapons systems: law, ethics, policy*. Cambridge University Press.

Blank, L. R. (2017). Media warfare, propaganda, and the law of war. In Gross, M. L., & Meisels, T. (Eds.), *Soft war: The ethics of unarmed conflict* (pp. 88–103). Cambridge University Press., DOI: 10.1017/9781316450802.009

Gow, J., Dijxhoorn, E., Kerr, R., & Verdirame, G. (Eds.). (2019). *Routledge handbook of war, law and technology*. Routledge., https://www.routledge.com/Routledge-Handbook-of-War-Law-and-Technology/Gow-Dijxhoorn-Kerr-Verdirame/p/book/9781032092478 DOI: 10.4324/9781315111759

Gruszczak, A., & Frankowski, P. (Eds.). (2018). *Technology, ethics and the protocols of modern war*. Routledge. DOI: 10.4324/9781315410739

Javadi, M. (2025). Shaping modern warfare: Insights from conflict and innovation. *European Political Science*. Advance online publication. DOI: 10.1057/s41304-025-00515-x

Klein, J. J. (2024). *Space warfare: Strategy, principles and policy*. Routledge. DOI: 10.4324/9781003452133

Pace, P. E. (2022). *Developing digital RF memories and transceiver technologies for electromagnetic warfare*. Artech House.

Seib, P. M. (2021). *Information at war: journalism, disinformation, and modern warfare*. Polity Press.

Weissmann, M., Nilsson, N., Palmertz, B., & Thunholm, P. (2021). *Hybrid Warfare: Security and Asymmetric Conflict in International Relations* (1st ed.). Bloomsbury Publishing Plc., DOI: 10.5040/9781788317795

Whyte, C., Thrall, A. T., & Mazanec, B. M. (Eds.). (2021). *Information warfare in the age of cyber conflict*. Routledge.

KEY TERMS AND DEFINITIONS

Cyberattack: In the context of war, a cyberattack involves deliberately targeting the enemy's computer systems, networks, or digital infrastructure to disrupt communications, gather intelligence, or disable critical operations. These attacks can cripple military command, disable weapons systems, or compromise sensitive information. This is done to gain a strategic advantage. Cyberattacks often occur silently and remotely, making them difficult to detect and defend against. As warfare at present relies heavily on technology, cyberattacks have become a key tool for undermining enemy capabilities without direct physical confrontation

First-Person-View (FPV): First-Person-View (FPV) drones are small, remotely controlled aircraft equipped with cameras that stream live video to the operator's goggles or screen. This immersive setup provides a cockpit-like perspective, enhancing spatial awareness and enabling precise navigation in complex environments. FPV systems include an airborne unit (camera and transmitter) and a ground station (receiver and display). Advanced versions may feature head-tracking gimbals that move the camera with the operator's head. The Indian Army's FPV Kamikaze Anti-Tank Drone uses a live video feed to guide the drone toward armoured targets, ideal for high-risk missions like kamikaze strikes.

Line of Control (LoC): The Line of Control (LoC) is a line or military-controlled boundary that separates the Indian- and Pakistani-administered regions of the former princely state of Jammu and Kashmir. LoC was created following the 1947–48 conflict or the 1st Kashmir War and was officially recognized in the 1972 Simla Agreement. The LoC extends about 740 kilometres through difficult mountainous terrain. It is not an internationally recognized border, but functions as a heavily militarized zone marked by frequent ceasefire violations and cross-border tensions. The LoC symbolizes the enduring dispute between two nuclear-armed states, significantly impacting regional security, civilian life, and the broader dynamics of India-Pakistan relations.

Man-portable weapon systems: Man-portable weapon systems are lightweight, portable armaments designed to be carried and operated by individual soldiers or small teams. These systems typically include launchers or firing mechanisms and associated munitions, enabling rapid deployment and flexible use in various combat scenarios. Their portability allows forces to engage targets such as enemy personnel, vehicles, or fortifications with minimal logistical support. Man-portable weapon systems enhance the mobility and effectiveness of ground troops by providing them with significant firepower that can be easily carried, quickly assembled, and operated in diverse and challenging environments.

Military intelligence: Military intelligence involves collecting and making sense of information about the enemy, terrain, and potential dangers to support military decisions and operations. It covers activities like surveillance, intercepting communications, and gathering insights and data through people or technology. In modern warfare, intelligence plays a vital role in predicting enemy moves, planning strategies, and protecting soldiers and resources. It assists armed forces to maintain situational awareness, enhance their performance, and accomplish missions successfully, even in complex and demanding environments.

Pharmacological manipulation: Pharmacological manipulation in the context of war refers to the use of drugs or chemicals to influence the physical or mental state of soldiers or enemies. This can include enhancing soldier performance, such as improving alertness, endurance, or reducing fear and fatigue, to increase effectiveness on the battlefield. On the other hand, it may involve the use of chemical weapons to impair enemy forces through substances that disrupt cognition, induce confusion, or cause temporary disability. Physical disruptions from these chemicals may include loss of coordination, muscle weakness, dizziness, and impaired motor skills. In severe cases, exposure can lead to paralysis, respiratory distress, or unconsciousness, significantly hindering an individual's ability to function.

Unmanned Aerial Vehicles (UAVs): Unmanned Aerial Vehicles (UAVs) are aircraft operated remotely or autonomously, without onboard pilots. Widely used in modern military operations, UAVs perform tasks such as reconnaissance, surveillance, target acquisition, and precision strikes, often equipped with missiles or bombs. They support battlefield intelligence, logistics, communication, and electronic warfare, extending operational capabilities while minimizing risk to personnel. UAVs vary from small, hand-launched models to large, armed drones. Their incorporation into military strategy has significantly transformed warfare by enabling continuous surveillance, rapid deployment, and asymmetric advantages over traditional forces.

Unmanned Ground Vehicles (UGVs): Unmanned Ground Vehicles (UGVs) are robotic systems that operate on land without onboard humans, controlled remotely or autonomously. In military settings, UGVs handle dangerous tasks such as reconnaissance, surveillance, explosive ordnance disposal, logistics, and combat support, reducing risks to soldiers. Equipped with sensors, cameras, and AI navigation, they detect threats, transport supplies, and sometimes carry weapons. By working in hazardous or hard-to-reach areas, UGVs improve battlefield safety and effectiveness. Their growing role in modern warfare enhances operational capabilities and acts as a force multiplier, making them essential to contemporary military strategy.

Chapter 3
Dance Between Life and Death:
An Analysis of Riverdance (2021) Film

Nebahat Akgün Çomak
Galatasaray University, Turkey

Nilufer Pembecioglu
https://orcid.org/0000-0001-7510-6529
Istanbul University, Turkey

ABSTRACT

This study dwells on the observation that war is inherently within the realm of human decision-making and that people can choose to be good or bad at any moment, for any duration, or in any action, which is generally overlooked. Generally, due to the excitement of the accompanying music, colors, allure of the subject, and other stimuli, many people fail to perceive this conflict and other types of war as part of narratives. The notions of "spare time and leisure" or "entertainment" attributed to media narratives might also cause people to fail to truly perceive physical or psychological struggle. Using a mixed-method approach, this study focuses on the symbolic analysis of life and death issues in the animated film Riverdance (2021) within the framework of "death and life" and their reflections in cultures. The findings put forward that beyond simply telling a story, the narrative focuses on narrating St. Patrick's Day in a way that facilitates marketing and popularization.

DOI: 10.4018/979-8-3373-7589-2.ch003

INTRODUCTION

War and peace have been in opposition to one another from the beginning of time. Armies or armed warriors dispersed across a vast battlefield were the only things that came to mind when conflict or war was discussed in the past. However, there are other battles on the agenda now, with culture wars being the most significant. These conflicts, heavily influenced by the media, have really grown to be far larger, more widespread and have far more profound repercussions than conflicts fought with people and weapons on broad plains, sheer cliff faces, or in the ocean. Nevertheless, many individuals are unable to perceive many battles as part of narratives because of ideas like "entertainment" ascribed to the media. In the products that are depicted in several cultures, images or illustrations, the deepest human myths are included. Within the framework of "death" and "the battle of life," this study examines the 2021 animated feature Riverdance. In conclusion, this study emphasizes that Riverdance, beyond simply telling a story, focuses on narrating St. Patrick's Day in a way that facilitates marketing and popularization, much like Mother's Day, Father's Day, Valentine's Day, Mardi Gras, Halloween, Cinco de Mayo, or Thanksgiving.

Narratives usually tell us the story of being and becoming human. The public, organizational, mass, and individual perceptions of war do not stand out much among the exciting scenes presented by the media, and are even ignored, despite the fact that these wars are spreading throughout the world through increasingly digital platforms and globalized communication networks. Generally speaking, conflicts are also not highlighted in fiction since war is portrayed as an internal or individual fight - fight, a good-versus-bad conflict, or a heroic theme rather than one that occurs between people from various backgrounds, cultures, or geographies. The weakness of the challenges of mortality accompanies human beings through their transformative efforts to defend themselves, survive, and carry-on living. On the one hand, people fight to become a "self" in the face of their society today, and people in various cultures embrace the fundamental desire to live in peace and create a safe environment as long as they are loved and trusted. People can, however, develop their own identities or comprehend their own dynamics through the stories presented on media platforms, and they can generally change into whatever they observe. Humanity is currently experiencing a period in which storytelling is extremely important. In this setting, it is sometimes forgotten that, when viewed cumulatively, all of the stories of each individual, group, culture, and belief are identical.

Death is positioned as the single major threat to humanity, which is growing more powerful due to the influence of digitalization and artificial intelligence. Nowadays, transhumanism enables us to see how humanity may advance, become a better "self," and access new levels of knowledge and abilities. Today, although it is claimed that the new type of war is a war that can occur between humans and non-humans, it is considered that the real war is the internal war of man with himself. Death is always portrayed as an end, and many stories are based on metaphors pertaining to dying and death. The bravery in the face of death demonstrates the strategies by which the person fights death.

In an untenable situation, a person who is defenseless against death might sacrifice his body in defense of his culture, values, beliefs, and loved ones once needed. It is possible to witness how each hero uses his unique defense strategies and invaluable life skills to confront and defeat the other in this conflict. We are aware that everything that the media portrays alters our minds and influences how we view the world, assess it, and decide whether to give up on it. As a result, the ideas, morals, and conflict resolution techniques that are depicted in the stories and the challenges that the characters encounter are crucial. Thus, this study's primary focus is on man's struggle with life and his struggle against death rather than merely a cultural transfer illustrated by a cartoon and examples of the conflict of cultures.

Man's fundamental battle was for survival in the days of the gatherer-hunter, and this often involved a battle not against another but against nature. The ancient tribes' fight for existence evolved into "survival" for the hunters (Kelly, 2007) when they gathered together and a shared identity emerged throughout time, necessitating greater "dominance" than merely survival (Kelly, 2013). However, Ferguson claims that war in not a part of human nature (2018). Archaeologists, anthropologists, and historians are continually searching for the pieces of evidence to learn more about humanity's warring propensity. Yet, everyone is searching for archeological evidence, cave paintings, weaponry, defensive enclosures, established villages, or patterns of settlement in order to knit together the beginnings of conflict. In order to determine the causes of conflict, they also search for skeletal remains to observe projectile wounds, healed injuries, media bias, unclear interpretations, or interpretational difficulties. But, in fact, the decision-making processes were not involved in this discussion for many centuries, since we have now very recently begun to learn more about the human brain and its nature. How the decisions were made, under which conditions, etc, would vary. Given the same situation, many people might act in a different way, and the decision of war could be seen even less among most of the others. The given society, norms, decision-making styles, personalities, previously experienced occasions, history, cultivated information, etc., could act as the barriers or accelerators of the given decision.

However, when it comes to the decision-making process, the heroic tales that are presented are crucial. A person's exposure to those, since childhood may have a significant impact on their decision-making process. Every community uses myths to pass down from one generation to the next the cosmos, the natural and supernatural worlds, and the place of humans in all of these, as well as the meaning of his existence and descriptions of his past and destiny. Myths provide the foundation for discussions on the origins of human existence. Because myths are historical occurrences that may be used repeatedly under various circumstances, they have components that are similar in many communities. Bulut & Sezener (2019) argue that this is due to the fact that, with minor modifications, mythological elements can be replicated and used in various contexts and at various times. As new interpretations are ascribed to mythical events by various communities, the symbols utilized grow increasingly complex and layered. Consequently, it becomes challenging to decipher the symbols. In addition to didactic and moral components, myths include cultural, economic, and political advancements as well as various human events like wars and murders. Numerous interpretations of the Cain and Abel story could be found and analyzed with different findings, including those including drought, fertility, murder, disobedience, and urbanization, etc. Atay (2021) claims that myths are typically regarded as unique to the beliefs of small-communities (2016:50). According to this perspective, the process between living in small tribes and transforming into larger and more modern societies also brings about a war of ideas. While cities are praised in Sumerian mythology, cities are associated with immorality in Hebrew mythology. But It is possible that mythical aspects also exist in profound societies such as Jewish, Christian, and Muslim belief systems in addition to the stories of local groups It is possible to find remnants of numerous stories, including creation, the flood, sacrifice, and the Sumerian battle between Cain and Abel, in the sacred texts of monotheistic religions (Çığ, 1997:46).

Meaning Attributed to Life and Death

Reviews and critiques of biological theories of primitive warfare and/or biological explanations of the origin of warfare in human evolution are required. The major ideas of biological theories are said to come from evolutionary biology, ethology, and related fields. Van der Dennen, (1990) puts forward that Social Darwinist thought, which may be seen as a separate historical era from around the middle of the nineteenth century to the middle of the twentieth century, is the foundation of the majority of biological-evolutionary explanations of primordial war and the beginning of war. To him, the primary intellectual sources of war to be covered appear to be instinctivism (and functionalism), evolutionism/selectionism, and racism.

It's also fascinating to see how the theme of battle and heroism is delicately woven across the rich tapestry of human stories, illustrating its numerous appearances and frequently contrasting individual and communal displays of bravery and virtue. War was bound to be a persistent aspect of human existence once the means of conflict were invented. War carries with it the potential for one tribe to be conquered by another and for a powerful state to emerge in order to maintain its power. Apart from this preliminary approach, one could also see the other concepts involved in the concept of war.

Figure 1. The concept of war

(Source: https://wordvis.com/)

It is evident that the idea of war has meant different things at different times. This notion applied to conflicts in the past where armies fought one another on a single front, but more recent processes incorporating strategic structures, such the "cold war," appear to have gained prominence. Information wars, media warfare, and culture wars, particularly those using artificial intelligence, appear to be more important than conflicts fought by people against people in today's world. But we can also observe that fights over expansionist policies, famine, and the climate are still going on in some regions of the world. These new conflicts either result in the standardization of people, the stylization of their languages, customs, and ways of life, or the attempt to establish cultural groups with a common political ideology.

To Ferguson (2018), people are people and they engage in combat along with the occasional murder. Throughout history, humans have had the ability to wage war when circumstances and society so demand. However, only in the last 10,000 years—and, in most cases, considerably more recently—have such circumstances and the warlike societies they produce been widespread.

In contrast to individual bravery, collective heroism results from the coordinated efforts of communities or organizations bound together by a single cause or set of beliefs to overcome hardship and bring about significant change. That idea might also be implemented through the war depictions of the historic episodes of the past. The written works of war poets, historians, fables, myths, and fairy tales, in addition to the oral traditions of the past, had an influence on all of humanity by beckoning people to an unending struggle and bravery. Not each and every one of the existing narratives, but their cumulative effect has succeeded in transforming man into a being capable of fighting.

The Importance of Decision Making in War or Conflict Situations

Reill, (2023) points out that the ordinary adult, he notes, makes between 33,000 and 35,000 decisions every single day. The majority of these are created automatically and concurrently using the implicit understanding of what is "good" or "bad" that we have stored in our subconscious minds. Some choices, nevertheless, are more challenging and demand more deliberate consideration. It is possible for decisions to be more than one-sided; they may also involve layers upon layers. This implies that a jury, common sense, and the public conscience have to play a role in the decision-making process. Because humans have adapted to live in society rather than alone, decisions involving life and death can be regarded as significant personal and societal choices. Consequently, ministries of loneliness have emerged in England and Japan, for instance, with suicide prevention and separation units within their subunits. However, it is currently recognized that euthanasia has a legal basis in many nations, meaning that it can occur under specific circumstances in which the person gives their own consent to die. At the moment, euthonosia and /or assisted suicide currently legar in The 19 jurisdictions are: the Netherlands, Belgium, Luxembourg, Colombia, Canada, Portugal, Germany, and New Zealand; the US states of Oregon, Washington, Vermont, California, Hawaii, New Jersey, and Maine, and the District of Columbia; and the Australian states of Victoria and Western Australia. The two additional jurisdictions are Switzerland and the US state of Montana. And this means that euthanasia or assisted suicide are legal in only about 7% of jurisdictions worldwide (https://www.maxim.org.nz/article/faq6-where-is-euthanasia-legal/)

As people put the greater good ahead of their own interests, this cooperative endeavor emphasizes the value of sacrifice, solidarity, and cooperation. On the other hand, individual heroism results from the brave and resolute actions of lone heroes who face difficult obstacles with moral strength and steadfast resolve. Throughout the process of decision making during a war vs peace situation, what actually affects people is highly debatable. Because a person's internal conflict may be the true war. Due to their personal preferences, social norms, and other people's limits, each person has certain abilities and limitations. Thus, it might be argued that the greatest conflict occurs between societal rules and personal goals and wishes.

Furthermore, it is believed that distinct brain layers concentrate on various aspects of a person's life. For instance, an individual's initial 400 grams of reptilian brain weight at birth contains a trait that enables them to fight with a power that may kill anyone or anything in front of them, to survive, and to remain alert in the case that they are in danger. With various emotions and hormones, the emotional region of the brain has a characteristic that can occasionally bring forth or inhibit an individual's identity and desires. The logical, rational, and healthy structure of the individual—which also complies with societal norms—is the emphasis of the intellectual portion of the brain. The creative region of the brain can drive creativity toward the goals of the individual, which can be more social or more individual at times. Thus, a struggle for life and death is also a part of man's inner journey. Given that humans make about 35,000 decisions per day, there is a fundamental argument about how many of these decisions are focused on peace and how many are focused on conflict. An untrained mind jumps restlessly from one thought to another, hanging from branch to branch like a monkey, constantly alternating between repulsion and craving. The brain's Default Mode Network (DMN), a system linked to self-referential thought and recurrent mental loops, is neuroscientifically responsible for controlling this mental chatter. The DMN can operate as an internal taskmaster that dictates our ideas, behaviors, and emotions, trapping us in cycles of self-doubt, worry, and rumination (Raichle, 2015). Cambridge dictionary definers "war" as "armed fighting between two or more countries or groups, or a particular example of this: i.e. nuclear war, a war film, go to war, etc.". (https://dictionary.cambridge.org/dictionary/english/war). But, war in occassions, might refer to any situation in which there is strong competition between opposing sides or a great fight against something harmful. It might also include a situation in which businesses, countries, etc. compete against each other very strongly. Communities usually develope a distinct way of communication just for war occasions, and this might vary from war language, to emergencies, limitations, forbidden issues, certain bans, as well as swear words and slang.

ABOUT THE PURPOSE, METHODOLOGY AND OBJECT OF ANALYSIS

The study aims to prove that, war and conflict situations cover all steps of life and even in a leisure time activity of watching an animation it is simply there. Yet, the audience may fail to recognize the theme since it is not laid in the form of fighting heroes, armies or navies. Thus, on the one hand, this study looks at a never-ending tale about man's struggle with death. This study favors an approach that focuses on viewing the same narrative from different angles rather than methodically addressing a single approach and evaluating it. Numerous philosophical and religious viewpoints are brought to light by the examination of the significance that humans attach to the interval between life and death.

The topic is additionally evaluated in light of the idea of children and mortality since the text selected for analysis is an animated movie. The topic is linked to Celtic culture due to the Irish origins of the chosen movie, and it also explores how the topic is handled in light of the current media and culture wars.

The topic of what is done to preserve the name, deeds, and memory of the beloved gone ones is discussed separately, as is the use of the dance cult in the face of death. These concerns are among the attitudes that people adopt when confronted with death. Given this, it is also clear that a brief glance of the activities surrounding Irish dancers and the St. Patrick's celebrations is necessary. Aside from this, the film's narrative structure explores a child's perspective on death, the meanings associated with it, and the notion that one might devote themselves to the point of wishing to die in order to save others.

This research focuses on the basic conflict between life and death that man has been engaged in since the beginning of time. Based on the subjects, the findings are separated into many subcategories. At the top of them are the film's story, narrative style, and central topic. In the film, there is Irish dancing and music, traditions, magic, fantasy, legends, dark powers, prehistoric animals, as well as danger, hunting, talking animals, spirit animals, family, death, loss, feeling lost, confronting one's anxieties, pains of growing up, paying respect to the deceased, and conservation and not all theseare appropriate for all kids. Regarding the theme, Riverdance is based on a decision-making process of little boy. Mainly, after a heartfelt loss, Irish-born Keegan and his Spanish-born friend Moya learn to dance through danger and despair with a magical herd of spirit deer. Yet, the worldwide Box Office indicates 846,210$ by August 17, 2021 (https://www.the- numbers.com/movies/franchise/Riverdance#tab =summary&franchise_movies_overview=od5). Lastly, this study emphasizes that Riverdance, beyond simply telling a story, focuses on narrating St. Patrick's Day in a way that facilitates marketing and popularization, much like Mother's Day, Father's Day, Valentine's Day, Mardi Gras, Halloween, Cinco de Mayo, or Thanksgiving.

FINDINGS

The story of little Keegan does not sound a story of life and death or war or a conflict situation where strategic decision-making gains a crucial importance. Thus the findings mainly dwell on the explanations of how the actions depicted in the narrative refer to such situations. Secondly, the findings aim to find the correspondence of these actions referring to back in centuries how crucial similar narratives involved in culture and history.

Actions and Codes of the Narrative

Actually, throughout the film, the audience witness a single day of the hero Keegan. The plot actually appears to be straightforward with mini flashbacks to the childhood of Keegan. It starts with a typical day, the proprietor of his neighborhood music business selects Keegan, an Irish boy, to serve as the DJ at the St. Patrick's Day celebration. In this context, we find out that he has a lot of tracks, one of which is especially for the Riverdance, a unique dance that honors life. The audience is introduced to Keegan who lives in the lighthouse with his grandparents and has no family. We find out that among his many artistic abilities, his grandfather was able to dance well and have amazing artistic capabilities such as storytelling, singing and dancing.

Getting back to the lighthouse, Keegan hears the tale of the Huntsman and the Megaloceros Giganteus from his grandfather, who likewise loves Riverdance. There is an action scene in the movie. As part of a division of labor, the grandfather and the grandson light up their lighthouse, but Keegan's grandfather dies. Moya, a Spanish girl, sings during his funeral. In fact, this offers a crucial viewpoint on how the image of grandfather influences everyone, not just those of his own ancestry and culture. As stated by Kearney (1988), the real strength of Irish culture lies not in its uniformity but in its plurality.

As the Riverdance performance begins, Keegan receives his sailor's hat. In reality, this is a duty that has been imposed on him as well as a memory that makes him think of his grandfather. Keegan believes that he should be like him, dance like him, and see society and the future from his point of view. However, he believes that he cannot accomplish all of these things while in mourning, so he withdraws even farther. Keegan is unable to comprehend this circumstance because his grandma, who is less distressed, is more upbeat and gregarious in this setting. The following day, his music shop owner talks about how her grandfather bought her some candy when she was having a rough day. Keegan doesn't understand why he should follow the stream when they travel to the Sugar Shack.

After leaving the hamlet, Keegan rides a black sheep to outrun a flock of white ones, but not before some frogs take his gummy fish. When he encounters Moya, she forces him to go down the river. Later on, she begins dancing on the water, but Keegan falls in the water when he attempts to follow her acts.

They encounter a male and female Megaloceros Giganteus in this amazing setting. Keegan makes friends with Patrick, the king, when they participate in a hurling contest. When Keegan encounters the Huntsman, who wants to mentor him, he declines, and the party heads to the sea. Keegan discovers an ancient lighthouse full of flashing patterns. There, Keegan reunites with his grandfather; since Moya earlier disclosed that she saw her grandmother in this world, Keegan also felt that people can see their ancestors by closing their eyes.

The others pursue the Huntsman after he shoots off a deer's antlers. In return for Keegan, Huntsman grants Patrick his life and antlers. The Huntsman kills the king antler after Patrick consents. The water stops flowing just after his funeral. The Riverdance is spared when Keegan and the animals temporarily get rid of the Huntsman. After Keegan and Moya get back together and the male Megaloceros Giganteus ascends to the throne, preserving the lives and ideals and values of the past, the two kids get back to their hometown and reality.

Keegan and Moya run into the sheep once more in the field. After escaping the sheep, they head back home, where Moya informs Keegan that they were only away just a few hours. The Megaloceros Giganteus lived in harmony once more after Keegan switches on the lighthouse, serves as the DJ at the St. Patrick's Day celebration, and gives Moya a chance at the flamenco this time.

Stratificational Narrative Structure And Glimpses For Different Types Of Audiences

According to Foertsch (2022), recent cultural studies theorists have observed the rise of metamodernity as an ideal type, which is defined as a form of transcendentalism that oscillates between postmodern deconstructivism and modern idealism. Transhumanist, metamodernist, and transcendental ideas frequently reappear in important themes and narratives found in ordinary life experiences, as the new ideal known as "Transhumanism" illustrates. The movie emphasizes that everyone should do their share for an everlasting life and that certain balances should be maintained, even though it appears to be about a child's mature acceptance of losing loved ones. The main idea emphasized throughout the narrative is that death is not an end and there are several different ways to decipher how death can be commemorated together with life.

Codes of Death

The setting and narrative patterns that convey distinct concepts to various audiences are encountered. In several scenes of the movie, these techniques that offer the semiotic contradiction of continuity and closure are really powerful. For instance, everything that occurs is portrayed as though it were normal, even yet it is stressed that the day is an unremarkable one. But all of a sudden, it is evident that one particular detail is really significant. For example, the elderly man dances and suddenly we hear a crunch as he grunts and sits in a chair, complaining of back pain, warning his little grandson to keep their lighthouse lit at night or an evil hunter will destroy all the land's rivers. These scenes are the steps of sneakingly closing death, but the youngsters might fail to recognize these symptoms.

When the old man coughs for a few seconds, he goes to a chair outside and appears to fall asleep, spilling his coffee on the floor. His grandson tries to wake him up but discovers him dead and appears dejected as he recalls his final words. In another scene, the youngster dreams that he sees the old man dancing. Later, he hears his voice and gives the man a hug before he vanishes in a whirlwind of golden sparks. We anticipate that the kids will comprehend the bridge that connects life and death, as well as how the young people can feel when their appreciated seniors pass away.

All studies have demonstrated that the cult of these territorial gods existed after Buddhism. Typically, a territorial god is a mountain deity who is hard to differentiate from his "mountain dwelling." It is an earth deity that rules over a clearly defined area and is exclusive to a village or political federation. A more potent territorial deity has the ability to dominate a whole area and so dominate a number of lesser territorial deities. The Irish elk (Megaloceros giganteus) serves as the representation of the deity in this movie. It is significant to observe that they also appear to follow a hierarchical structure, with the next in line taking over as herd leader after the death of the previous one.

Bardo Pass – After Death

Many faiths emphasize that death is not the end, but the soul's eternal journey. While the exact nature of what happens to the deceased is unclear, narratives suggest a process of purification, decision-making, and acceptance of death. Depending on the good and bad deeds performed during life, a person may enter a better world and be rewarded or enter a worse world and be punished. In some faiths, this process is referred to as "staying in purgatory" or "crossing a certain bridge, a certain path." In some cultures, older people are seen going on long journeys to face their fate so that death is not visible to others and so that younger people do not have to face the concept of death at a time when they are not ready. In this film, Keegan's imaginary journey is seen as his confrontation with the concept of death. Buffetrille, (2020) highlights the relationship between pilgrimages around sacred mountains and bardo, the intermediate state between death and rebirth. As in his example of pilgrims prostrating in front of Kawakarpo Mountain in Chinese Yunnan province, Keegan in the elk-land also experiences a moment between life and death. He realizes the importance of the others' sacrifices for him, and he also sacrifices himself for the others.

In the narrative, Keegan uses his dream world to experience a similar state, while the pilgrims in real life use the landscape's features to enact the drama of the perilous way of the consciousness through the bardo, as if facing the dangers described in Tibetan texts in real life would lessen them. Buffetrille mentiones that The Tibetan Book of the Dead, a 14th-century text, and the Bardo thödröl, the "Liberation upon hearing [the instructions] during the intermediate state between death and rebirth," are frequently recited by a lama when a person is approaching death in Tibetan Buddhism. Reciting these scriptures is meant to lead the Bardo's consciousness through a purifying process until it is integrated into a new existence. Whereas in the approaching death circumstances it might be possible to cite these words of value, the unexpected situations perhaps prepare the youngsters for a different way of Bardo. In Tibetan Buddhism being liberated from the cycle of the existences (sa sāra), ordinary people know that such a goal is beyond them. Everyone is aware that crossing the bardo will be a horrifying experience in which he will be overpowered by fear and presented with terrifying visions, which are also common throughout pilgrimages, and he will be unable to discern the manifestations of genuine reality (chöni). Similarly, rebirth is viewed as a challenging journey into the unknown, beyond their control, even in Irish society. They only wanted to be sure that their ideals were understood, at least at this point, during or after the funeral. Pembecioğlu (2019) puts forward that similar customs can also be observed in Australia's aborigine culture, where the soul is freed in a river or body of water, or else rebirth is conceivable following body and soul purification. However, a condition of unconsciousness, such as not being able to completely comprehend the flow of time or experiencing the wind similarly to the pilgrims' given "wind horse-lungta" (like flying on sheep), highlights the similarities between Keegan's and the Bardo experiences. One other thing to be mentioned is the Parrott's claim (2021) referring to Bardo pass as a river to swim. In that case, the river scene that Moya dances and Keegan falls into gains more importance as Keegan fails to understand the meaning of death and life. Only after figuring out the meaning of these concepts and having his own sacrifices he could act in a similar way.

Enviromental Facts

The animation film not only refers to the mystical environments of the virtual world created with the help of AI. In the movie, the Irish elk (Megaloceros giganteus) is also animated. One of the most gigantic deer in history, also known as the enormous deer or Irish deer, is an extinct species of the genus Megaloceros. From Ireland, where it is known from numerous bogs remains, to Lake Baikal in Siberia, its range spanned northern Eurasia throughout the Pleistocene (Geist, 1998). In western Russia, the species' most recent bones were radiocarbon dated to approximately 7,700 years ago (Lister & Stuart, 2019). Referred as the biggest antlers of any deer, measuring up to 3.5 meters across, it is generally accepted that its closest living relatives are fallow deer (Dama), and it is not closely related to either of the two extant species of elk. Remains of Megaloceros giganteus have been discovered at several Middle Palaeolithic sites with cut marks that show Neanderthals butchered the animal (Blasco, et.al., 2010). Because of this, the species is treasured in many stories and is closely linked to early humans. Because of its magnitude and the abundance of its tasty products, many hunters are depicted in charts of their plans to hunt or shoot deer, despite the narratives' emphasis on the unlucky nature of doing so.

According to Ruhaak (2019), there was a great deal of respect for the animals they hunted, yet many Irish Gaelic traditions mention that hunting for food is the most glamorous way to obtain it. It is clear from Irish written sources and from the iconographic representations of the deer hunt in medieval Irish and Continental Celtic that deer hunting is an traditional component of Irish culture states Fomin (2018). The idea of seth 'deer' is recognized to have an interchangable aspect in Celtic mythology, and heroes occasionally claim to have had the experience of taking on the form of a deer (disguise as a deer). Indeed, it is known that certain stories impart morals like 'tredon seth damh', which states that a deer's lifespan is three times longer than a human's. Despite being considered a “wild” and “forest animal,” it is also said that hunting it is prohibited because it is equally valuable as all other living things, including people, trees, and nature itself (Fomin, 2018).

A Devil – Giant Hunter is depicted in the narrative as well. As Warren puts forward (2017), human presence in Ireland appears to have occurred intermittently between 12.8 and 12.5 kacal BP, regarding the archeological data. When compared to other regions of northwest Europe, such as Britain, where hominins were present between roughly 0.78 and 0.99 mya, this age is unusual. Past human activities, taphonomic conditions, and study biases are some of the explanations for this seemingly delayed invasion. The necessity of taking into account the colonization process, how hunter-gatherers came across new worlds and values, and how they ceased hunting are all heavily emphasized. This movie expresses the idea that a hunting world upset the ecological balance in a pretty unique way. So, this place was really a place where the life and death were actually dancing throughout in Irish history.

The fungus phythophtera infestans, also known as potato blight, arrived in Ireland in the fall of 1845 and was the primary cause of the Great Irish Famine (1846–52). About one-third of that year's crop and almost all of 1846's crop were killed by the fungus. It also destroyed the majority of the 1848 harvest after a season of remission. The Irish family lasted longer than most because of these frequent invasions (Ó Gráda, 2004).

All these facts, in bits and pieces were provided in the animation in a way, to raise the awareness to the fact that establishing the balance in the world is not an easy thing but It is very easy to upset the balance. Most of the children check the Megaloceros Giganteus with a kind of fascination and knowledge, admiring them.

Children's Perspective of the Narrative

The animation and adventure film is marketed to those older than seven. However, one could argue that extra caution should be used in this circumstance. We learn that the boy featured in the movie's opening is the son of deceased parents. We learn later that the girl who is with him has also experienced a loss. At this stage, the idea that boys and girls may behave differently in response to losses and have distinct perspectives on mourning is very subtly conveyed. Furthermore, directly in the start of the movie, the boy character loses his closest male relative. It seems as though the feelings of the young audience in the midst of so many losses were not taken into account. Even though the TV Parental Guidelines Monitoring Board rating states that "this program is suitable for all ages," the Kids-In-Mind.com review contains a suggestive comment, some scatological humor, a man who dies naturally, an elk that is killed and has its antlers removed, multiple dangerous situations involving children, and mild language. Even if the narrative is a bit long enough for youngsters, the message underlined seems to be clear and naive enough for the kids: We can use the arts to advocate for conservation and to honor the memory of loved ones.

Today, it is recognized that grieving as a child is a common reaction to loss and that grieving children need help and direction to deal with their loss. A variety of childhood losses, such as the passing of parents, friends, siblings, and pets, may be represented by these losses. But it was thought that children didn't grieve at all less than a century ago. Although there was little research on childhood grieving, two studies conducted by Nagy in the 1940s (Nagy, 1948), and Kübler-Ross starting by the 1970s (Kübler-Ross & Kessler, 2014) led to the creation of age-specific phases of mourning.

Dane (1994) argues that different cultures tend to face death in different ways, yet there seem to be two general cultural patterns of response to deat: death-accepting and death-denying. Greek folk tradition for example, accepts the finality of death. Acceptance is expressed in laments (Dracopoulou & Doxiadis, 1988). A child between the ages of 5 and 8 will gradually come to understand that death is final, that everyone will die at some point, and that "they may blame themselves in some way for the death and can engage in 'magical thinking'; filling the gaps when information has not been given to them," according to the Irish Childhood Bereavement Network (ICBN) (ICBN 2017:20). We can see this perspective, as typically addressed in Irish texts, as an explanation of the child's abrupt admission into an imaginative realm and his subsequent return to reality, even though the film's subject is the child's attitude toward death. Nonetheless, a child between the ages of 9 and 12 is more aware that "death is universal, irreversible, and has a cause." Physical aches and pains and difficult behavior are two ways that grief can manifest itself Clifford (2023). Even while their responses differ greatly, this does not exclude them from feeling some of the same feelings brought on by sadness. For example, the remorse and responsibility that both protagonists, Keegan and Maya have after their elders pass away are relevant aspects of this work. The telling of trauma through representations in fiction, poetry, drama, and films provides a spoken locus from which some form of psychic healing can take place, as stated by Flynn and O'Brien: "loss is only suffered by the living, and while it may put them in mind of their own death" (Flynn and O'Brien 2018: 7). Stylianou and Zembylas (2018) state that encountering new people and new places beyond consciousness can actually be seen as an intervention towards grief. states that the findings demonstrate that such interventions, called to be the good grief, positively impacted children's comprehension of grieving and sadness in two significant areas. First, children were better able to recognize their emotional responses to loss (mourning) thanks to the intervention. Second, children were able to relate pertinent personal stories and discuss about loss and mourning without feeling anxious. Gallagher on the other hand claims that it is crucial to recognize the impact of loss on kids and the kinds of losses that are illustrated in narratives for kids as well as in real life. As a "proven means of helping clients understand the issues they are experiencing, normalizes those experiences, and offers hope for positive change," bibliotherapy is a recognized technique for helping kids navigate the grieving process (Gallagher 2020: 15).

Sacrifice, Faith, Celtic Culture, and Ritualistic Practices

The Christian Celts were cautious to foster and support their children's spiritual gifts rather than prohibiting their involvement in diverse spiritual life activities and interests, Crabtree (2007) argues . Regarding children's spirituality, Celtic Christian beliefs and practices provide insightful perspectives on how to raise children today and ensure that families and communities are spiritually complete. In light of this distinctive heritage, children's spirituality has important ramifications for comprehending how today's youth develop spiritually, emotionally, and interpersonally. An approach to comprehending the human condition that places particular emphasis on the impact of media and technology on past, present, and future human cultures and civilizations is known as media ecology (ME), or the study of media as environments. As our living environments, media have an impact on how we see, think, feel, communicate, and organize as a group, to the point that it can either improve or impair our chances of survival. At least in the 21st century, we may embrace all values with a pluralistic awareness and a non-alienating understanding; nevertheless, due to media transmission allows for the sharing of all cultures, ideas, and values.

Fimi, (2017), examines the creative uses of "Celtic" myth in contemporary fantasy written for children or young adults from the 1960s to the 2000s. In the film, almost everyone makes several different sacrifices but most important them all is the one of Keegan's. Regarding the children audience, some of the sacrifices might be understoon exactly within the flow of the narrative. Even though sacrifice is regarded as the most significant human virtue, it comes at a cost to the agent. First of all, sacrifice has a price, which raises the question of how it affects wellbeing in general. After making the sacrifice, no one can be certain that they will feel better. Since sacrifice leads to more difficult value conflicts and decisions, the second handicap is seen in moral quandaries. For instance, the monarch gave his life for Keegan in the movie. On the one hand, the older and more seasoned person understood that the value of the antlers was more significant than the body. Keegan discovered while chasing it that the future is overflowing with assumptions and ambitions, and he gained a deeper understanding of the values and aspirations of previous generations. Consequentialism, which questions the sacrifice's demands on people, is another element that might be linked to the act of sacrifice. Religious sacrifice raises concerns about the sacrifice's own political, personal, and value-based connotations. The benefits of sacrifice seem to outweigh the drawbacks, which include all of them. Moral demandingness appears to be one of the most significant results of sacrifices. Debates over how much moral pressure can be imposed on humans revolve around sacrifice. However, since sacrifice is necessary to comprehend activities that transcend moral obligation, supererogation appears to be another aspect. Theories that put gratification of desires above of self-interest are contested by sacrifice. Although sacrifice's effect on one's interests is a topic of discussion in care ethics, it can also be a potent instrument of political resistance against oppression.

Being a team player, helping others, family members, accepting responsibility, taking care of the future, taking care of animals, opposing adversaries or threats, sacrificing oneself for one's loved ones, virtues and ideals, humanity, etc. are all examples of Keegan's sacrifices. It's critical to comprehend the idea of death and to embrace life after realizing its significance. Dancing, talking, interacting, and appreciating other people are the finest ways to do it.

Media and Culture Wars: St. Patrick's Celebrations

According to Blyn (1995), children's play in Celtic-speaking cultures has some distinctively Celtic elements, such as the focus on complex verbal texts, the element of chance, and indoor, fireside settings that create a children's version of the "ceili," an adult form of social or communal entertainment. However, in an effort to portray the Celts as distinct from other ethnic groups, the role of these fireside games in the overall corpus of children's play activities in the Celtic-speaking territories has been over-represented in print. The games that have been marketed as "Celtic" nowadays, usually fit within this framework, which also includes games that are played in Celtic-speaking regions but aren't thought of as specifically Celtic.

The Wearing of the Green is the first to document the entire history of St. Patrick's Day. Cronin and Adair (2004) examine the history of this widely celebrated event, analyzing how the day has been used for political and commercial purposes. They also look at more general Irish customs and Irish communities around the world, and they examine the Irish people's shared heritage through the creation of this distinctive patriotic holiday.

Nagle (2005) describes the first significant St. Patrick's Day parade in London, emphasizing how the Irish community used the parade to raise their profile and visibility in the city by articulating an inclusive experience of Irishness and fostering a positive Irish identity. Many members of the Irish community view this high level of visibility as a formal acknowledgement of the contribution that the Irish make to multicultural London, a place where the Irish have long been marginalized due to their representation as a pariah group. Nagle asserts that instead of seeing these spectacles as either co-opted by the "dominant power" or radically liminal and progressive, the organizers and sponsors are perceived as a "new social movement," with the belief that their goal is to pluralize society and offer fresh examples of cross-cultural communication.

Recently, McIvor (2012) explores the performative "essences" from different minority groups by following City Fusion's 2009 process from conception to performance. McIvor also argues that these dramatic decisions expose City Fusion's interpretation of interculturalism, which is largely negotiated by Irish-born artists and designers. As a result, to her, the histories, concerns, and performance styles of the participants are not placed in their proper perspective.

It was frequently asserted that St. Patrick's Day celebrations were significantly more prominent among the Irish diaspora than in Ireland until the recent revitalization of a St. Patrick's Day festival in Dublin. Instead of just adopting the national customs of the "homeland," the festivals have their own uniqueness and are significant among the diaspora states Hickman (2015). The great majority of St. Patrick's Day celebrants do it on their own initiative. In their extensive research of St. Patrick's Day across the diaspora and Cronin and Adair (2004) discovered that many people are conscious of their family history on the day, and for many of their older correspondents, the church service was an important aspect of commemorating the day.

Different Angles of Death and Philosophical Insights

There are numerous stories about this throughout history and one of them, the Epic of Gilgamesh is among the earliest and best-known instances. In addition to highlighting how people can adopt various attitudes and behaviors toward death, the Gilgamesh story highlights how helpless and powerless a person is in the face of death, regardless of how strong they are. According to Kovacs (1989), The Epic of Gilgamesh has been the primary source of information ever since a collection of Mesopotamian poetry preserved on clay tablets was discovered more than a century ago. Among the earliest known works in the world is the Epic of Gilgamesh and Heidel, (1949) argues how Gilgamesh Epic and Old Testament Parallels each other. Although it is an old narrative, whose discovery dates back centuries, interpretation and decipherment of the Gilgamesh story, goes back over about two thousand years. It's been accepted as an important narrative affecting the human beings struggle of life and death and having this main theme, it's been one of the most cited narratives. Dalley, (1991). On the other hand puts out tah Gilgamesh in the Arabian Nights seem to be similar to each other even centuries later.

The dancing show Riverdance served as the inspiration for the 2021 animated adventure movie Riverdance: The Animated Adventure. Even if the production of the film began in February 2020, with an estimated budget of 35 million Euros, Cinesite produced the movie for River Productions and Aniventure (https://en.wikipedia.org/wiki/Riverdance:_The_Animated_Adventure). The film, which has an IMDb score of 5.6 out of 10, is classified as Aniventure (Animation Adventure) (https://www.imdb.com/title/tt10841088/).A Spanish girl Moya and an Irish lad Keegan explore the legendary realm of Megaloceros Giganteus, who teaches them to value Riverdance as a celebration of life. This is an animated journey of sight, sound, and movement like nothing one could ever seen or heard before, inspired by the worldwide stage sensation and featuring brand-new music by Grammy winner Bill Whelan.

This movie emphasizes that man is accountable for his own moral deeds and that he is not alone in his battle between life and death. Networked actors bearing cumulative responsibility throughout the network is also important. In addition to them all, a social community's accountability area is another layer where each person is accountable for their influence on society. This study helps to understand the individual's stance against death, his/her struggle with death, and to analyze the semiotic representations of our individual and social main unit that is attributed with the character of death.

The animation film mainly concentrates on the Irish Dances. According to Sherlock (1999), Riverdance is a means of opposing and fighting Western culture and globalization. Masero (2010) describes how the popularity of Riverdance themes led to modifications in Irish dance. Brennan (2001) claims that early English traveler recordings of Irish dance date back to the late 17th century. Dancing's presence may only be presumed, even if it is mentioned in English poetry from the 14th century. Nothing is known about the history of dance prior to the 16th century because of Queen Elizabeth I's restrictions on traditional Irish dance and music. According to Irish dance historian Cullinane (2003), there were three occasions that changed Irish dancing. These include the founding of the Gaelic League in 1893, the introduction of traveling dance masters in the early eighteenth century, and, just one hundred years later, Riverdance's performance during the Eurovision Song Contest's interval act in 1994, which really helped the dance gain popularity. With its regular steps and movements, the actual standardization of the dances amounted to a near-reinvention of traditional culture, even though there were multiple artistic variations of dancers or dance outfits. The Riverdance team became well-known and inspired several traveling groups while nurturing the love of dancing in all Irish-originated people worldwide. The following succinctly describes the cause of their success. The reason of their success has many aspects. It doesn't focus solely on dancers' physical prowess, virtuosity, or modernity in performance. Simply said, it raises awareness of cultural resonance, sentimental recollections, and a greater sense of cultural connection. Through the institutionalization of dance, it also advances the notion of the commodification of values by blending them with tools of capitalist profit-making. Nevertheless, it also highlights Ireland's socioeconomic realities, paradoxes, and portrayals of women.

Figure 2. Analyzing the Success of Riverdance

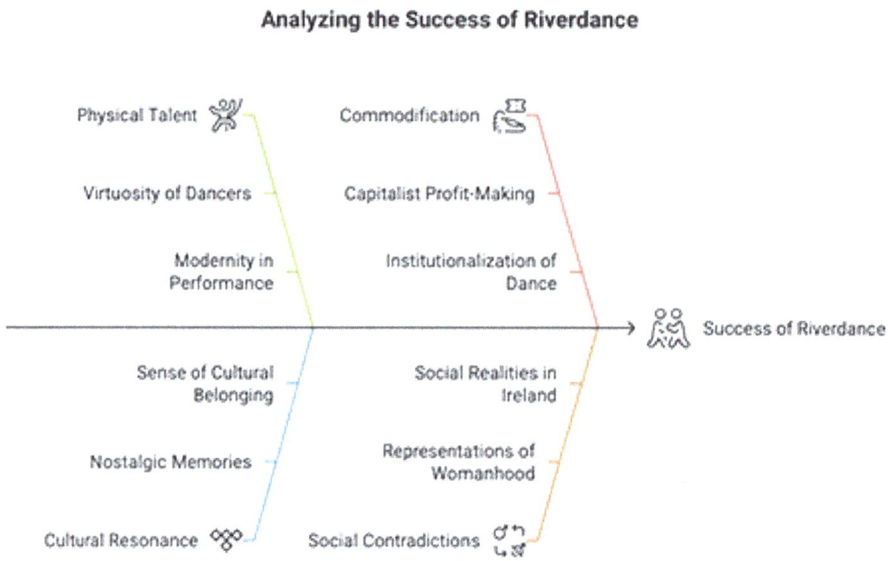

CONCLUSION

"Diachronic analysis, which is defined as the change in the meaning of words over time, can be used to study language as a system of signs as a complete system at any given point," according to Saussure (1964). By focusing on the relationship between language and historical transformations, the diachronic approach—also known as historical linguistics—is the study of language as it manifestly evolves in usage. To begin with language philosophy, it is important to understand that the "mother tongue" of any individual is intimately associated with individual's culture (Pieris, 1951). Once the language philosophers have revealed the general principle of an idea, they begin to wonder about how certain characteristics of a certain language are related to the characteristics of a particular society. It is also something that changes or develops the thoughts at the same time (Altınörs, 2012). The foundation of Wilhelm von Humboldt's philosophy of language is the idea that every language carries a unique worldview. Since the language that a person speaks and writes is a sign of who they are, the language of that country is a sign of that person's identity. These days, when the issue of identity is also explored, the problem of the language-culture connection is becoming more significant (Akarsu, 1998).

"All the signs confirm that a language expresses the character of those people speaking that language," wrote French philosopher Etienne de Condillac. "Every language reflects the understanding and character of those who speak it," thought Condillac's young German colleague Johans Gottfired Herder (Deutscher, 2013). Akarsu quotes Humboldt as declaring: "Being constantly influenced by the world-view of one's own language allows one to fully immerse oneself in the worldview of a foreign language." Thus, people are dominated by the language's worldview, which provides them a specific path in life (Akarsu 1998).

The mythical perspective holds that an object's name conceals its significance. The ruler role over items is awarded to the one who can control the names. Naming is the term used to describe a man's capacity for intellectual thought in connection to either concrete or abstract information. Since the emergence of sacred texts, the capacity for thought and naming has been seen as equal. (Aydın, 2011:318). One of the most essential features of contemporary Western thought is the study of nominalism, which first appeared in scholastic philosophy in the 19th century. Nominalism's central tenet is that concepts are acquired through abstraction that does not encounter an object. For instance, the notion of "justice" has no object to which it may be referred. (Aydın, 2011:317).

To Heraclitus, every word limits its object, and this limitation may result in misunderstandings. When an entity or object is reduced to a "word," its content is constrained by concentrating on a particular facet. This might lead to a loss of perspective on the object's or entity's greater existence in its life stream, which could not be fixed. According to Heraclitus, language is not constrained by things or entities since it is composed of fluid dynamics (Aydın, 2011:318).

The naturalization of nominalism in all spheres of thought in contemporary civilizations encourages narrow-minded and biased thinking. Even though this problem doesn't seem to be recognized adequately in daily lives, yet it is causing misunderstandings, prejudices, because everything depends upon the perception of the different values.

Children and the values attributed to them are carried in their names, in their style of bringing up. With their child names, people who have fought death for thousands of years and yearned for a longer, healthier, and better life hope to make an impression on future generations. For instance, a 2019 study on Turkish children's names found that 82% of them emphasized the ideas of respectability and power, 3% were religious names, 5% were historical names, 7% had environmental connotations, 2% were positive, and 1% were negative. Nowadays, it is evident that celebrities in particular choose names with more unusual and mathematical combinations in an effort to sever connections to the past and create a new, futuristic future. For instance, Musk's child's name deviates from conventional child-naming customs in terms of coding. The Elvish spelling of AI, or artificial intelligence, is represented by the letter "Æ" in the name AE A-XII, which also means "love" in other languages. While the "A-12" is a reference to the SR-71 aircraft's ancestor, which they respect for its non-violent combat capabilities, the letter A represents the initials of their favorite song, Archangel (Gencer, 2023).

Even if sometimes it seems to be all clear, the language has some tricks evolving many different perceptions. Dascal, (2008) for example, states that dichotomies are ubiquitous in deliberative thinking, in decision-making and in arguing in all spheres of life. Thus, sticking uncompromisingly to a dichotomy may lead to sharp disagreement. Yet, there are some dicohtomies that we always know what to choose, in case of life and death for example. While Callahan discusses on defining a 'natural death', Brueckner and Fischer discus "Why is Death Bad?". There are more to be discussed in the 21st century since nowadays "brain death" is also discussed in its various ways (Brueckner & Fischer, 2019). Some other researchers give importance to the quality of dying and death (Hales, Zimmerman & Rodin, 2008). In short, the meaning attributed to death varies across cultures (Palgi & Abramovitch, 1984).

Almost everyone is aware that of all the film genres, animation is the most relatable to young audiences. Additionally, it can be easier, more convincing, and more durable to deliver all types of information to everyone—adult or child—with this sort of film. It is evident from the study's framework that both the youngster who plays the role of the film's hero and the young viewers are exposed to visuals that contain adult perception. It is also evident that children are unable to judge their lives as "positive" or "negative" at this stage and cannot comprehend the difference between life and death. Children who are unable to discern between reality and dreams in the real world may be able to recall certain things by being captivated by cartoons, instead of confronting death or making sacrifices. They may think that it is a game with deer, watch sheep being blown around like balls in the wind, and laugh, among other things. Even many adults may find it challenging to understand a time when people have varied perspectives on their religion, fortitude, sins, and good actions. One example of this is the "Bardo Pass," which is viewed differently by many different faiths worldwide. Beyond the religious themes, the primary lesson that kids can learn from this film is that St. Patrick's Day is a day to celebrate, and people should dress in green and dance. This supports the idea that only those who create them can prevail in the culture wars of today.

In conclusion, it would be overly optimistic to assume that other children would be affected in the same way by Keegan's half-realistic, half-imaginary tale of loneliness, significant losses, growing pains, and mythological metaphors, as well as the necessity of reiterating the struggle between life and death for both him and others in order to keep the world going. Children's conclusions from this logical and spiritual conflict, on the other hand, will typically live on in their memories as a joyful entertainment custom in which everyone dresses in green, has a good time, eats, and dances. This implies that Irish and Riverdance cultures will prevail in the current worldwide cultural conflict. Despite being rated as appropriate for kids of all ages, it is evident that many of the words and scenes are inappropriate for young audiences. In addition, it is believed that many meanings and stratificational structures that are given semiologically outside of context will be difficult for youngsters to understand. It appears that youngsters from all cultures cannot comprehend the metaphors used to explain death and life, which are only metaphorically expressed through deer and hunters and connected to rivers and lighthouses. The prevalent Irish traditional culture is highlighted in this movie. Beyond all of this, though, the culture's components and specifics are not completely depicted; instead, everything is boiled down to the greatest St. Patrick's Day celebration, which also happens to be the grandfather's death day. To put it another way, there is a global viewpoint that encourages people to go out, dress in green, dance, and enjoy life regardless of how much suffering, grief, or despair they are experiencing. However, this fashion may not be appropriate for everyone, particularly children.

FUTURE RESEARCH DIRECTIONS

The purpose of this study is to distinguish between the representations in the movie, the beliefs associated with the images, and the time and war between life and death. It is believed that additional multifaceted research on the topic is necessary to preserve and spread traditions, cultural history, values, and artistic and dancing practices, as well as to comprehend the significance of other public holidays like St. Patrick's Day. Irish or comparable movies and animations must be examined for this purpose, and the key distinctions and parallels must be examined.

REFERENCES

Akarsu, B. (1998). *Bedia, Wilhelm von Humboldt'da Dil-Kültür Bağlantısı*. İnkılâp Yayınları.

Atay, T. (2021). Din Hayattan Çıkar. Ankara: İletişim Yayınları. 6th publication.

Aydın, M. (2011). *Güncel kültürde temel kavramlar*. Açılımkitap.

Blasco, R., Peris, J. F., & Rosell, J. (2010). Several different strategies for obtaining animal resources in the late Middle Pleistocene: The case of level XII at Bolomor Cave (Valencia, Spain). *Comptes Rendus. Palévol*, *9*(4), 171–184. DOI: 10.1016/j.crpv.2010.05.004

Brennan, H. (2001). *The story of Irish dance*. Roberts Rinehart.

Buffetrille, K. (2020). The Pilgrimage to Mount Kawakarpo: A Metaphor for Bardo?. , (67), 99-122.

Bulut, M., & Sezener, A. (2019). Dumuzi-Enkimdu'dan, Kabil-Habil'e Yerleşik Hayatın Serüveni. *Turkish Studies*, *14*(6), 3071–3081. DOI: 10.29228/TurkishStudies.23495

Çığ, M. İ. (1997). Kur'an İncil ve Tevrat'ın Sümer'deki Kökeni. İstanbul: Kaynak Yayınları. 9. Basım.

Clifford, R. (2023). Good grief: changing attitudes to childhood grief in children's literature.

Crabtree, M. L. (2007). Forbid Not the Little Ones: The Spirituality of Children in the Celtic Christian Tradition. *Christian Education Journal*, *4*(2), 238–251. DOI: 10.1177/073989130700400204

Cronin, M., & Adair, D. (2004). *The wearing of the green: a history of St Patrick's Day*. Routledge. DOI: 10.4324/9780203007143

Cullinane, J. P. (2003). *An Coimisiún le Rincí Gaelacha (Irish Dancing Commission)*. Its Origins and Evolution. John Cullinane.

Dalley, S. (1991). Gilgamesh in the Arabian Nights. *Journal of the Royal Asiatic Society*, *1*(1), 1–17. DOI: 10.1017/S1356186300000031

Dane, B. O. (1994). Overview Of Cultural Attitudes Toward Death And Grief. AIDS and the New Orphans: Coping with Death, 13.

Dascal, M. (2008). Dichotomies and types of debate. In *Controversy and confrontation: Relating controversy analysis with argumentation theory* (pp. 27–49). John Benjamins Publishing Company. DOI: 10.1075/cvs.6.03das

Deutscher, G. (2013). *Dilin Aynasından Kelimeler Dünyamızı Nasıl Renklendirir, Çev: Cemal Yardımcı*. Metis Bilim.

Dracopoulou, S., & Doxiadis, S. (1988). In Greece, lament for the dead, denial for the dying. *The Hastings Center Report, 18*(4), 15. DOI: 10.2307/3563249 PMID: 3220738

Ferguson, R. B. (2018). War is not part of human nature. *Scientific American, 319*(3).

Fimi, D. (2017). *Celtic myth in contemporary children's fantasy: Idealization, identity, ideology*. Springer. DOI: 10.1057/978-1-137-55282-2

Flynn, D., & O'Brien, E. (Eds.). (2018). *Representations of loss in Irish literature*. Springer. DOI: 10.1007/978-3-319-78550-9

Foertsch, S. (2022). Metamodernity, American transcendentalism and transhumanism in Japanese Anime.

Fomin, M. (2018). Hunting the deer in Celtic and Indo-European mythological contexts. Celtic myth in the, 21, 73-87.

Gallagher, M. (2020). Bibliotherapy as a Counselling Intervention for Bereavement, Loss and Grief. Irish Association for Counselling and Psychotherapy, 20(1).

Geist, V. (1998). Deer of the world: their evolution, behaviour, and ecology (Vol. 272). Mechanicsburg: Stackpole books.

Gencer, B. (July 26, 2023) Elon Musk ile Grimes oğlu X Æ A-12'nin adını değiştirdi, NTV Haber, https://www.ntv.com.tr/galeri/n-life/magazin/elon-musk-ile-grimes-oglu-x-a-12nin-adini-degistirdi,8zWSDE6xW0iR5wsw_IIrhw/4wVqA5KjvE2jcQEJ2V1NVg

Heidel, A. (1949). *Gilgamesh Epic and Old Testament Parallels* (Vol. 136). University of Chicago Press.

Hickman, M. J. (2015). *Diaspora space: celebrating St. Patrick's Day in New York and London. Consuming St Patrick's Day*. Cambridge Scholars Publishing., https://dictionary.cambridge.org/dictionary/english/warhttps://en.wikipedia.org/wiki/Riverdance:_The_Animated_Adventurehttps://kids-in-mind.com/r/riverdance-the-animated-adventure-parents-guide-movie-review-rating.htm#:~:texthttps://www.imdb.com/title/tt10841088/ https://www.maxim.org.nz/article/faq6-where-is-euthanasia-legal/ https://www.thenumbers.com/movies/franchise/Riverdance#tab=summary&franchise_movies_overview=od5https://wordvis.com/

Irish Childhood Bereavement Network. (2017). *Standards for Supporting Bereaved Children & Young People – A Framework for Development*. ICBN.

Kearney, R. (1988). *Transitions: Narratives in modern Irish culture*. Manchester University Press.

Kelly, R. J. (2007). *The foraging spectrum: Diversity in hunter-gatherer lifeways*. ISD. DOI: 10.2307/j.ctv2sx9gc9

Kelly, R. L. (2013). *The lifeways of hunter-gatherers: the foraging spectrum*. Cambridge University Press. DOI: 10.1017/CBO9781139176132

Kovacs, M. G. (1989). *The epic of Gilgamesh*. Stanford University Press.

Kübler-Ross, E., & Kessler, D. (2014). *On grief and grieving: Finding the meaning of grief through the five stages of loss*. Simon and Schuster.

Lister, A. M., & Stuart, A. J. (2019). The extinction of the giant deer Megaloceros giganteus (Blumenbach): New radiocarbon evidence. *Quaternary International*, *500*, 185–203. DOI: 10.1016/j.quaint.2019.03.025

Masero, A. (2010). The Changes in Irish Dance Since Riverdance.

McIvor, C. (2012). Essences of social change: City Fusion, interculturalism and the Dublin St. Patrick's Day festival in post-Celtic Tiger Ireland. *Public*, *23*(45), 180–191. DOI: 10.1386/public.23.45.180_1

Nagle, J. (2005). "Everybody is Irish on St. Paddy's": Ambivalence and Alterity at London's St. Patrick's Day 2002. *Identities (Yverdon)*, *12*(4), 563–583. DOI: 10.1080/10702890500332733

Nagy, M. (1948). The child's theories concerning death. *The Pedagogical Seminary and Journal of Genetic Psychology*, *73*(1), 3–27. DOI: 10.1080/08856559.1948.10533458 PMID: 18893204

Ó Gráda, C. (2004). Ireland's Great Famine: an overview.

Parrott, J. (2021). *Swimming through the Bardo: Death as a Means to Self-Discovery.* Pacifica Graduate Institute.

Pembecioğlu, N. (2019). *Duyarsızlaşma, Kültürlerarası Anlayış ve Ölüm Üstüne: Bir Jindabyne Çözümlemesi. Sinema Kültür İletişim Okumaları.* Parşömen.

Pembecioğlu, N. (2019). *Policy Makers.* Educators & Parents Up For Media Literacy At The Crossroads Of Rights, Values And Identity, CiCea Conference.

Pieris, R. (1951). Speech and Society: A sociological approach to language. *American Sociological Review, 16*(4), 499–505. DOI: 10.2307/2088282

Raichle, M. E. (2015). The brain's default mode network. *Annual Review of Neuroscience, 38*(1), 433–447. DOI: 10.1146/annurev-neuro-071013-014030 PMID: 25938726

Reill, A. (2023, December 5). A Simple Way to Make Better Decisions. *Harvard Business Review.*

Ruhaak, R. (2019). Towards an Alternative Black Death Narrative for Ireland: Ecological and Socio-Economic Divides on the Medieval European Frontier. *Journal of the North Atlantic, 2019*(39), 1–16. DOI: 10.3721/037.006.3901

Sherlock, J. I. (1999). Globalisation, western culture and Riverdance. In *Thinking identities: Ethnicity, racism and culture* (pp. 205–218). Palgrave Macmillan UK. DOI: 10.1057/9780230375963_10

Stylianou, P., & Zembylas, M. (2018). Dealing with the concepts of "grief" and "grieving" in the classroom: Children's perceptions, emotions, and behavior. *Omega, 77*(3), 240–266. DOI: 10.1177/0030222815626717 PMID: 29940830

Tigay, J. H. (2002). *The evolution of the Gilgamesh epic.* Bolchazy-Carducci Publishers.

Van der Dennen, J. M. (1990). Origin and evolution of 'primitive'warfare. In *Sociobiology and Conflict: Evolutionary perspectives on competition, cooperation, violence and warfare* (pp. 149–188). Springer Netherlands. DOI: 10.1007/978-94-009-1830-6_9

Warren, G. M. (2017). The human colonisation of Ireland in Northwest European context. Advances in Irish Quaternary Studies, 293-316.

ADDITIONAL READINGS

Grof, S. (2013). *Books of the dead*. WW Norton & Company Incorporated.

Masero, A. (2010). The Changes in Irish Dance Since Riverdance.

Oruç, C. (2021). Çocuk medyalarında dinin sunuluşu. Çocuk ve Medeniyet, 6(12), 147-171.

KEY TERMS AND DEFINITIONS

Hero: A person who is admired for having done something very brave or having achieved something great

Myth: These are the stories that human societies tell in order to interpret the universe, the world and nature.

Narration: A way of telling a series of events in literary genres.

Persona: The particular type of character that a person seems to have and that is often different from their real or private character

Chapter 4
Exploring Music, Art, and Literature in Conflict Narratives Using Fantasy Fiction:
War Through a Cultural Lens

Anushka Malik
https://orcid.org/0009-0008-0288-159X
Christ University, Bangalore, India

Bidyut Bhusan Jena
https://orcid.org/0009-0002-1285-8744
Christ University, Bangalore, India

ABSTRACT

The study of war has long been confined to disciplines such as history, political science, and journalism. However, with the evolution of modern conflicts and the shifting global landscape, it is imperative that scholars also explore more unconventional or often overlooked mediums, including literature, art, and music—not just to study how civilians respond to war and the trauma it leads to, but also the intricacies of war itself. This study examines war and conflict narratives through a cultural lens, using Critical Discourse Analysis of Letters of Enchantment (2023), a fantasy duology by Rebecca Ross. Fantasy fiction, once perceived as an escapist genre, has recently developed into a medium capable of exploring serious socio-political themes, including warfare, propaganda, and trauma. This study explores the

DOI: 10.4018/979-8-3373-7589-2.ch004

ways in which the duology portrays war through themes of media influence, artistic expression, censorship, and music suppression, drawing parallels with real-world historical and contemporary conflicts.

WAR THROUGH A CULTURAL LENS: EXPLORING MUSIC, ART, AND LITERATURE IN CONFLICT NARRATIVES USING FANTASY FICTION

The fictional representation of war has traditionally been analyzed through the genre of non-fiction and historical fiction–the more preferable genre in this context being the former, i.e., nonfiction journalistic accounts, and political discourse. However, modern literature, especially modern fantasy fiction, has begun to tackle complex issues and themes in a way that it had not in the past. Fantasy fiction, once dismissed as a mere literary diversion for children and middle schoolers (mainly boys) alongside their comic books (Reinke, 2001; Traw, 1993), has over time evolved into a rich and layered genre that provides nuanced explorations of contemporary issues, including but not limited to warfare and its psychological, sociocultural, and international political repercussions. And yet, it is usually the sub-genre, Historical fantasy fiction, that is considered preferable, given that the historical facts and instances cited in the fiction can still be verified to an extent. Researchers in the field of fantasy fiction have been dealing with issues such as the lack of acceptance extended to fantasy novels as subject matter for serious study and unclear classification of sub-genres to be included and clearly defined for some time (Angelskår, 2005; Fredericks, 1978; Ivanova & Ryzhchenko, 2019; Marizka, 2024; Nasriddinov, 2023; Petzold, 1986; Swinfen, 2019) and contributing to the ongoing effort of garnering the genre a respectable seat at the larger proverbial table of serious and valuable literature. While researchers agree that fantasy fiction benefits children (Jones, 2008), there is still a need for further research on how Fantasy fiction benefits the universal readers at large. Fortunately, one of the greatest successes in the field of literary studies has to be the recent shift where no realm or genre of literature is now entirely considered to be either too simple or too erudite to pull apart and observe when it comes to the portrayal of sociocultural, psychological, or international political themes. War, as everyone can come to a common agreement, fits all of the themes mentioned above and therefore impacts every aspect of life and a nation and its people.

In the present study, the researchers, using the *Letters of Enchantment* (2023) duology, explore themes of war propaganda, trauma, and public perception, offering a fresh perspective on conflict by blending imaginative storytelling with real-world themes. *Letters of Enchantment* is a fantasy duology (a series of two books) by

Rebecca Ross, where the first book is titled *Divine Rivals* (2023a) and the second one, *Ruthless Vows* (2023b). Both the fantasy fiction works engage deeply with the depiction of war through media, primarily newspapers, and the inescapable presence of propaganda. Moreover, Ross brilliantly portrays the complexities of war reporting, capturing the traumatic experiences of not only those on the battlefield but also the civilians who live in constant fear far from the front lines.

CONFLICT NARRATIVES AND PUBLIC PERCEPTION OF WAR

Fantasy Fiction's Narrative: Rebecca Ross' **Letters of Enchantment** *(2023)*

The rationale behind choosing this duology for the present topic of analysis includes not just the war themes that it focuses on – mainly how media represents war, deals with propaganda, ultimately tampering with the public's perception of the war; but also how, owing to the very nature of the genre, this fantasy duology will undoubtedly lead the researchers to a very different outlook than if one was dealing with journalism and real world repercussions of war alone. After all, modern-day literature that deals with grave issues such as armed conflict, war, propaganda, and more, takes on a position that no researcher, in good conscience, can afford to ignore. Where media has been acknowledged as possessing the power to shape and mold the public at large's views on a particular international or domestic political issue or conflict (McCombs, 2002; McCombs & Valenzuela, 2020), the power of art, music, and literature needs to be constantly studied and revisited as well.

Studying Conflict Narratives through a Cultural Lens

The *Letters of Enchantment* (2023) duology by Rebecca Ross deals with all three of the elements that make up a cultural lens. While the role of media in shaping war narratives is a key theme in the duology, the narrative also explores how propaganda influences public perception, a phenomenon historically observed in real-world conflicts. In a similar vein, the study of traditional war narratives—such as memoirs like *Goodbye to All That* (1929) by Robert Graves; *Memoirs of an Infantry Officer* (1930) by Siegfried Sassoon; and *The Forever War* (2008) by Dexter Filkins, to name a few, offer a first-hand insight into the authors and journalists in war time. Barring the first two, *The Forever War* is a journalist's memoir from Iraq and Afghanistan, reflecting the disorientation of modern conflict. Admittedly, the influence of propaganda through media in these cases would be decidedly less, given that people would understandably be more prone to trusting their experiences over journalistic fodder and hence, independent research on such memoirs has been carried out over the years reflecting all aspects with or without propaganda (Mac Ginty, 2021; Thomas, 1930; Winter, 2018). However, what could be more beneficial in changing times is comparative research of such memoirs and more traditional war narratives with modern/contemporary fictional adaptations of war. Rebecca Ross kept this avenue of research open in her duology when the ending showcased one of the protagonists (Roman Kitt) changing their profession from Journalist to full-time author. The reader would take a moment to appreciate that a character who served for years as a war correspondent would do justice even to fictional war narratives. The role of journalism in wartime can in no way be overstated, as observed from historical events where media was used to manipulate public opinion, such as during World War I and II (Cozma, 2014; Socolow, 2007), Israel-Palestine (Damanhoury & Saleh, 2024; Neureiter, 2016), Russia-Ukraine (Gerber & Zavisca, 2016; Oleinik, 2023; Young & Omosun, 2024), more recently, India-Pakistan (Saud & Kazim, 2022; Thankachan, Thomas & Menon, 2025), and every contemporary conflict where misinformation and media bias play a crucial role is evident in Ross' (2023) fantasy world at war. Ross' protagonists, Roman Kitt and Iris Winnow, serve as war correspondents, and the fantasy duology, *Letters of Enchantment* thus provides an alternative perspective on war reporting, examining the responsibilities and ethical dilemmas journalists face. The protagonists, both relatively young journalists, experience firsthand how war narratives are constructed and weaponized, raising questions about truth, bias, and the role of storytelling in shaping history.

Portrayal of Censorship Faced by Journalists

The first novel in the duology opens up with the main characters, Roman Kitt and Iris Winnow, working for the same newspaper in a safe, if stressful and competent, environment. After a few chapters and a tragic personal loss, Iris Winnow leaves her desk job and interviews for another position (as a war correspondent working and reporting straight from the border front) at a rival news agency. Her decision was in no way driven by monetary benefits but rather by her personal belief that her fellow citizens needed to know the realities of how the war was progressing. Iris' brother, Forest Winnow, had chosen to fight in the war and had been amongst the first ones to enlist, owing to which a certain sense of patriotism could be felt in the scenes showcasing the Winnow household. When her then boss, Zeb Autry, the editor of *Oath Gazette,* refused to publish her work with its true authenticity, Iris knew not to waste the editor's or her own time, and she quit. The editor had ordered:

"Don't waste my paper and ink ribbons on a war that is never going to reach us here in Oath. It's a western problem and we should carry on as normal. Find something good to write about, and I *might* consider publishing it in the column next week." (Ross, 2023, p. 26).

And in the interview for her new job in the capacity of a war correspondent, when asked why she wanted to be one, she did not give any stock answers. She did not say she wanted her articles to enrage the public about the brutality of the war, nor call to their emotions, and put them in agony over the conditions their soldiers were braving. She instead gave a very honest and pragmatic response – knowledge and information needed to trump anger and sadness in times of war. An informed citizen would make informed choices. Iris' response to her new editor felt, at the same time, like a much-needed response to her previous editor when she said:

"People in Oath think they're safe. They think that because the war is far away, it will never reach us here. But I believe it will come to the city one day, sooner than later, and when it does… there will be a lot of people unprepared. Your choice to report the news on the war front is going to help change that." (Ross, 2023, p. 124).

Propaganda and Personal Bias

Upon being hired as a war correspondent and reaching the front lines, Iris is faced with another glaring reality, however, that makes her (and the readers) introspect about what it means to be an unbiased reporter even when you want your readers to have all the facts. There is a scene where she is conversing with an army captain of Enva (the patron goddess of music) and the latter says:

"I hear that you're to be neutral reporters, but I also don't think that's quite possible, if I'm frank. I highly doubt you'll be welcomed over to Dacre's side, let alone return whole from it. I think the best piece of advice, Miss Winnow, is to write what you see happening and what you feel and who we are and why it's vital that the people in Oath and the cities beyond join our effort." (Ross, 2023, p. 234).

The dilemma here is one that war journalists have faced since the beginning. War journalists must carefully navigate the delicate balance between the presentation of facts and their ethical possibilities. If their tone becomes too forceful, they risk being perceived as biased or even as propagandists—jeopardizing their credibility entirely. The role of a journalist in conflict zones is both problematic and vital; they must persist in their duty, even as the weight of truth, danger, and responsibility presses heavily on their shoulders. Moreover, after walking the fine lines and sidestepping all the potholes, there remains the issue of possibly facing flak from powerful people in positions of power who might be displeased by the headlines being reported. While Iris' previous boss shied away from engaging the displeasure of bureaucrats and businessmen by stating his position in terms of war, her new boss, Helena Hammond, cowered from no one. The newspaper *Inkridden Tribune,* where Hammond is the Chief Editor, received serious flak for their honest reporting, resulting in the Chancellor himself resorting to threatening them. This is revealed in the second book of the duology, *Ruthless Vows,* by Helena to her employees, "He sees it as fearmongering and propaganda. He thinks I'm trying to drive up my sales with such headlines." (Ross, 2023, 21). And it did not matter to the Chancellor that the reporters were, in fact, eyewitnesses to the attack they had written about; simply by publishing anti-Dacre news, news that did not align with the Chancellor's personal ideology, Helena was now facing trouble from high-ranking Dacre sympathizers.

THE PSYCHOLOGICAL AND SOCIOCULTURAL IMPACT OF WAR IN FANTASY FICTION

Civilian Anxiety and Survivor's Guilt

One of the most profound ways in which war manifests in literature is through its psychological impact on characters. The fear of a parent with a child on the battlefield, the confusion of a child wondering why their parent has been gone for so long, the anxiety of a sibling, and the anguish of a spouse are merely some of the deeply personal emotions that fiction captures to convey the true horrors of war. There is a scene where Iris is telling Roman where he, as a journalist, went wrong while interviewing the family of a soldier recently declared MIA (missing in action). While Roman Kitt asked all the stock questions, the tone was off.

"There's no emotion, Kitt," Iris said. "You asked the parents things like 'When was the last time you heard from your daughter?' 'How old is she?' 'Why did she want to fight for Enva?' And you have the facts, but you didn't ask them how they're doing or what advice they would give for someone experiencing a similar nightmare. Or even if there's something the paper or community can do for them." She handed him his notepad. "I think for this particular article, your words should be sharp as knives. You want the readers to feel this wound in their chest, even though they've never experienced a missing loved one." (Ross, 2023, 83).

Through a seemingly everyday conversation between colleagues and journalists, Ross has portrayed here that asking the right questions matters when the subject matter is as delicate as war. Having a daughter that is an MIA soldier is its own kind of 'nightmare' as Iris correctly describes it, quite possibly having that clarity and empathy due to the fact that her own elder brother, Forest Winnow, had not responded to any of her letters for months and she did not know if he was safe or even alive. When Roman urges her to describe what that felt like, she responds as follows:

"It feels like wearing shoes that are too small," she whispered. "With every step, you notice it. It feels like blisters on your heels. It feels like a lump of ice in your chest that never melts, and you can only sleep a few hours at a time, because you're always wondering where they are and those worries seep into your dreams. If they're alive, or wounded, or sick. Some days you wish that you could take their place, no matter the cost. Just so you can have the peace of knowing their fate." (Ross, 2023, 83).

A close reading of Iris' response would require breaking it down into several chunks – the first few lines are more of a physical description (too small shoes, blisters on your heels, lump of ice in your chest) and evoke a physical response from the listener/reader that could be a certain tingling in the feet, a sharp pang or a fleeting shiver in one's chest. The lines following that are more psychological in nature (broken sleep, anxiety about someone's whereabouts and well-being, worrying dreams/nightmares) that suggest what prolonged anxiety over a loved one's health and safety can do to the mental health of someone. The closing lines suggest severe survivor's guilt – a phenomenon that occurs in the family members of a lot of people deeply affected by war, violent conflicts, death, and loss (Berman, 1978; Field & Bonanno, 2001; Kubany & Manke, 1995). One of the established methods that promote gradual healing from sibling bereavement and letting go of the survivor's guilt is to talk about it –to try to communicate one's feelings in a safe environment to people one loves and trusts, such as family or close friends (Gibbons, 1992). While there are studies that suggest survivor's guilt might not be as high in cases of loss due to terrorism, as seen in the World Trade Centre tragedy, for example (Flomenhaft, 2007), there may be differences observed if the factor of chance were to be removed from the case. Ongoing research in the field still states that not enough is known about the phenomenon of Survivor's guilt despite it being a pretty common experience following traumatic events (Murray, Pethania & Medin, 2021) and is as yet, poorly and insufficiently described (Hutson, Hall & Pack, 2015). In such a scenario, a lot of the testing for symptoms and treatment would be guesswork. The death of a civilian sibling, while still a tragedy, might not always result in severe survivor's guilt for the surviving sibling because society would repeatedly tell them that it could have happened to anyone. That fact alone removes certain responsibility that one might otherwise be burdened with. On the other hand, when one's sibling is in the army, or on the frontline of any kind of savior professions (firefighters, policemen, etc.), there is a certain shame and self-doubt that accompanies the pride. While the sibling will be proud of having a brother/sister who died or was injured, trying to save lives, there will also be an underlying sentiment that would suggest they themselves had somehow failed their sibling. Maybe if they had been in the same profession, they could have aided their siblings in their jobs and kept them safe. Maybe it could have been them instead of the one that was lost. It is these unrealised possibilities and some others that lead to the surviving sibling or surviving individual forming a survivor's guilt –a phenomenon that is more prevalent in peers, siblings, and people who either were in the same precarious situation (the holocaust: Garwood, 1996; Leys, 2007) or could have been.

War Veteran Rehabilitation Difficulties

Apart from merely provoking a conversation on Survivor's guilt, grief, and mental health problems (in terms of the family members), Ross' fantasy duology also sparks the conversation from the perspective of the war veteran himself. The changes Iris noticed in her brother upon his return were not inconsiderable. He is a changed man in more ways than one. He survived not just active service in a war zone but also faced captivity by Dacre's army, survived physical and mental torture, escaped, and became a deserter from his own army because his spirit was not the same, and he could not continue serving. And after all of that, he came home to a society that looked at him with either disdain or pity, neither emotion a veteran could possibly want aimed at them. The reader gets to see glimpses of Forest's (and later, Roman Kitt's) traumatic experiences. Forest suffers from considerable survivor's guilt for having survived where a lot of others in his unit did not –a common psychological reaction of grief and shame in veterans (Glover, 1984; Okulate & Jones, 2006; Singer, 2004). Another aspect of guilt and shame that veterans suffer from is what is called the "bad penny syndrome" at the Post-Trauma Treatment Center. Veterans who cannot let go of the guilt of committing certain acts or atrocities that their conscience does not clear them of are more prone to face problems in the recovery process (Williams, 1988). The violence that veterans sustain, commit, and witness over the period of their service will be considerable, and resolving all those feelings can quite understandably be complicated.

Post-War Healing and Rebuilding

Ross' fantasy world brings to the reader's attention not just matters of individual healing or the healing of a family or a community, but it also focuses on the individual journey of various characters that turns into a collective redemptive experience. It sheds light on the nature of post-war healing and rebuilding. Rebuilding a nation, a people, and a world after any war, conflict, or armed violence takes a great deal of time and effort, and this slow, arduous process can be studied through various lenses. There has been research on patients of chronic illnesses and bereavement, healing through art creation, storytelling, and coming together as a support group (Heiney et al., 2017), and the calming effects of activities like pottery, where you create things with your hands (O'Brien, Gallagher, & Malafouris, 2025). War leads to destruction on a large scale, and therefore, it should be no surprise that to come back from that, people need to involve themselves in some form of creation.

Healing through Creation

Creation, whether pragmatic in the form of great architectural pursuits such as building new centers of healing, education, peace, or creation in the form of more aesthetically inclined ventures of art, literature and music, are equally important when it comes to bringing a sense of calm and order to a land and people shaken from war and violence. War brings in its wake widespread devastation. While the loss of human life is often seen as the most tragic consequence, the destruction of places and objects integral to human existence should not be overlooked. The obliteration of a cherished place of worship or a museum, carries its own profound sorrow. War not only claims lives but also erases physical memory—places and structures tied to childhoods, communities, and significant personal or cultural moments. Genocide stands among the gravest of human crimes, but alongside the human toll lies the systematic erasure of the sites of national importance. Such loss is irreparable, as though the history and identity they once embodied never existed at all (Bevan, 2007). The post-war period, therefore, turned towards reconstruction—not only of lives but also of monuments (Barakat, 2007; van Schaik, 2023; Verpoest et al., 2020). The path of healing leads through creation. The study of architecture, then, of what a land looks like a decade after a great war is no less important than the study of how the themes of poetry, or the styles and the hues of painting, have changed. These are all forms of recovering and rebuilding after loss and trauma, and are crucial and intriguing processes to study from a research perspective as well. And one of the ways that most, if not all, of the aforementioned themes can be analyzed all at once is through fantasy fiction. Fantasy fiction often employs a stream-of-consciousness narrative technique, allowing readers to delve into the inner turmoil of war victims, soldiers, and civilians. Traversing along said stream of consciousness, one may find solace or sympathy, as required, and be able to embark on their own journey of healing after conflict, trauma, and loss.

Role of Employment in Healing

When Iris' brother, Forest Winnow, returned from war, his journey to healing had been a difficult one, even if not a lot of scenes portrayed his struggle of trying to re-assimilate himself in a city and neighbourhood he had grown up in. He felt like a stranger in his own skin, his own home, and his own city –something that researchers say a lot of war veterans are suffering from and can relate to (Crowe et al., 2017; Wilson, Hill & Kiernan, 2018). One of the things that Ross highlighted in her novel about Forest Winnow's return was how difficult he was finding it to get a job. Lack of steady and fulfilling employment has been cited as a key factor when it comes to problems with transitioning from military to civilian life (Keeling et al., 2018). This aspect, although being researched thoroughly, has not reached the masses due to a lack of media adaptation in the form of books and movies that build their place in pop culture and eventually become the talk of a regular household. Ross' portrayal of this aspect in her fantasy novel highlights this side of necessary rehabilitation for veterans to her readers, and that is how fiction contributes to social change.

MUSIC AS RESISTANCE AND CULTURAL IDENTITY

Censorship of Music

Letters of Enchantment duology focuses on the power of music in a land riddled with war and violence in a roundabout way, when the readers are forced to dwell on the fact that an entire city has been prohibited from either creating or listening to any form of music, vocal or instrumental. Being in possession of a musical instrument has been declared a criminal offense. It brings to mind Plato's ideal *Republic* and how it had no room for poets because their creations might end up jeopardising the stability of a nation. Music can be provocative, there is no denying that, and researchers have studied the power of music through a political lens, garnering interesting results on the power it holds (Allen, 2004; Gilbert, 2007; le Roux-Kemp, 2014). However, the risk of rebellion should not be enough to steal away an entire people's freedom of expression, be that expression through any mode. In the context of the fantasy duology, *Letters of Enchantment,* the reasoning behind this ban on music is due to the fact that the ongoing war is between the mortal armies of two immortal divinities – Dacre and Enva. Where Dacre is a God of the Underworld and healing, Enva is a Goddess who is the patron of music. The fact that these two divinities waging war against each other are actually patrons of healing and music is in itself an irony, an idea that the author, Rebecca Ross, surely made a brilliant use of. The present focus however needs to be the Governor of Oath, a fictional city in the fantasy duology that has yet to get actively involved in the war, favors Dacre in a not very subtle way throughout the duology and has found it imperative to issue a ban on music within the borders of Oath, supposing that the power of music empowers Enva and her armies. This fantasy element possibly draws its ideation from how soldiers across history had found peace and solace and even strength in poems, ballads, and lyrics whenever the blood and death and violence got to be too much for their souls and sanity. Many soldiers during the Great Wars had turned to song and poetry because that is the part of themselves –their legacy –that they wanted to leave behind, not the number of enemy soldiers they killed in battle. Wilfred Owen and Siegfried Sassoon were undoubtedly two names that became synonymous with war poetry after World War I. Both of them served in the Great War, but only one came back from it. There have been studies that have researched music concerning war, the various ways it keeps up the morale of soldiers, and the ways in which it instills fear in their enemies on occasion (Botstein, 2022; McWhirter, 2012). However, there are also studies on music as a form of resistance to war.

Resistance Through Music

The prohibition of music in *Letters of Enchantment* (2023) is, therefore, a powerful allegory for the suppression of culture in times of war. Music has historically been used as a form of resistance and solace during conflict, from the songs of enslaved African Americans during the Civil War (Ellison, 1994; Garofalo, 1982; Wilson, 1990) to protest music during the Vietnam War (Auslander, 1981; Deaville, 2018; Lee, 2009). The need to explore the historical significance of music in wartime and how its suppression in the *Letters of Enchantment* duology reflects real-world instances where oppressive regimes such as Algeria, Afghanistan, Iran, South Africa, Nazi Germany to name a few, have banned artistic expression to control public sentiment is crucial (Aghamohseni, 2024; Kirkegaard, 2017; Korpe, Reitov, & Cloonan, 2006; Moody, 2021; Rastovac, 2009). Moreover, the study would be incomplete if it did not also address the psychological effects of music deprivation and its role in cultural resilience and identity (Simmons, 1992). Censorship of music and curbing the freedom of cultural expression can go two ways. For the people who need music for its healing properties and to be able to express a crucial part of themselves, censorship may either lead to psychological issues and breakdown, or it may push the individual to rebel and break the silence. While *Letters of Enchantment* (2023) does not portray either of the extreme reactions on the censorship of music, one of the major characters, Thea Attwood, or more commonly known as Attie, who is a fellow war correspondent with the protagonist Iris Winnow, serves as a heartbreaking example of how war and totalitarian regimes shatter innocent dreams of innocent people and force them into professions and life choices that they, in more peaceful circumstances, would never have otherwise made. Attie's father was a musician, and growing up, he had taught her how to play the violin –an instrument she grew to love back when their world was not turned upside down by war. The following lines from Book One of Ross' duology, *Divine Rivals* (2023a), Chapter Twenty-four: *Dangerous Instruments,* portray how dire the situation for musicians became right before the war:

"I once thought I could earn a place with the symphony," Attie began. I practiced hours a day, sometimes until my fingertips were bloody. I wanted it more than anything. But of course, things changed last year, when the war broke out. When suddenly everyone was afraid of falling prey to Enva's songs, and Oath began to shed its musicians like we were a sickness. The constable actually came to our house, to confiscate anything with strings. You can imagine how many of them we had in our house." (Ross, 2023, p. 190).

Attie's pain as she recounts her experience with law enforcement officers and a violation of her home, her safe space, as the constable invades her home and takes away the musical instruments that symbolized her dreams of a peaceful and happy future is the pain of a lot of people who had to make hard choices that completely reshaped their lives. Artists living under censorship and their families deserve this sort of representation in popular fiction so that the issues do not die away in silent oblivion.

WAR AND ARTISTIC EXPRESSION

Graffiti: The Art of Rebels

Art has often flourished in response to conflict, serving both as a means of protest and as a reflection of societal upheaval (Teune, 2005). Therefore, when it comes to artists, conflicts have been known to be the inadvertent precursors to some of the greatest art movements in our world. Many art movements have stemmed from wars, revolts, and rebellions. Although *Letters of Enchantment* features graffiti –one of the most rudimentary forms of angry art that crops up around conflict– it is fair to say that any form and style of art is a symbolic nod to how the general public and amongst them the more emotionally attuned individuals (artists), feel about the mindless war that is causing widespread deaths and destruction, creating an atmosphere of dread that no one can ignore. In the duology, the graffiti is portrayed as a form of resistance against the leaders and the regime in power as well as the Gods Dacre and Enva, in the names of whom the wars are being waged in the first place. There is a scene in *Ruthless Vows* (2023b) where the protagonist witnesses the displeasure of the people regarding their current circumstances, expressed in the form of angry graffiti:

Iris paused when she saw words painted in bold red paint on a white brick wall.

Gods belong in their graves.

It wasn't the first time she had come across that statement. Last week, she had seen it painted on the side of a cathedral and on the library doors. The words were always in red, bright as blood, and often followed by a single name: *Enva.* (Ross, 2023, p. 15).

This particular graffiti requires being analysed in three stages.

Stage 1. The surface analysis is, of course, the anger of the graffiti artist on the war, and the blood red color paint being used for it was self-explanatory.

Stage 2. The second, deeper analysis led to the element of faith and religion, two things that frequently cropped up in war-torn nations. When you lose a loved one to war or witness large-scale destruction, it is bound to shake one's faith in God. The graffiti artist using cathedral walls to write such an inflammatory statement about the Gods wanted to make as bold a statement as one could, and this hints towards great personal loss or rage stemming from helplessness.

Stage 3. The third, an even deeper insight into the art, focuses on the name the artist targets: *Enva.* Even though it was the feud between Dacre and Enva that had resulted in the war, the artist had only named one of the parties. That led to the question as to whether the artist was against war or merely expressing his anger towards whom he/she believed their enemy to be. The artist could be a Dacre sympathiser, then. The line of questioning that follows any piece of art, be it through a well-read and informed critic or a layman, always aims to unearth the artist's intent. The art that crops up surrounding an armed conflict or war can either be polarized in favor of one side of the conflict (usually their own) or can be anti-war entirely. Ross' work urges the readers and the literary community (including critics) to understand the motive, politics, and sheer rage behind the creation of art.

Major Historical Art Movements

This theme and provocation for art as a way to fight back mirrors historical instances where art has emerged from war-torn societies in the past and the way it will continue to in the future. Some of the significant art movements that arose from conflict include:

Dadaism after World War I, which was a reaction to the absurdity of war and political chaos (Buelens, 2006; Jones, 2002; Mowery, 2021);

Expressionism in the pre-World War II era, depicting the emotional and psychological turmoil of war (Dzhakupova, 2024; Gilbert, 2004; Jachec, 1991; Sarsılmaz & Yiğitler, 2012), and

Street Art and Graffiti (Awad, Wagoner, & Glaveanu, 2017; Benavides-Vanegas, 2005; Marche, 2012; Miladi, 2018) that seem to thrive in modern day conflicts, big and small as seen in regions affected by political turmoil, such as the murals of Belfast during the Northern Ireland conflict (Migeon, 2024; Hill & White, 2011) or Banksy's works in Palestine that target the wall that does not simply separate Israel and Palestine but also traps the people of Palestine and restrict movement and agency (DeTurk, 2015; Hazou, 2022).

All these art forms and movements affected the mood and environment of their nation and their time in a big way. The feeling of dread and shock that pervaded the people of Oath when they saw the graffiti was palpable in Ross' description. Hence, by comparing the aforementioned art movements to the symbolic use of graffiti in the fantasy world of *Letters of Enchantment,* specifically, the study aims to highlight how artistic expression remains a crucial aspect of cultural identity and resistance in times of war, whether the war is real or an imagined fantasy creation.

Graffiti in Ross' Fantasy Duology

Interestingly, in the context of Ross' fantasy duology, the reader is only given to see one single art form in its pages –the art of graffiti, which often appears far removed from traditional notions of art. At first glance, graffiti may seem like nothing more than scribbled words on a wall and essentially something anyone could do. It is, therefore, an art reduced to something that requires no actual training, talent, or even forethought. There's no immediate sense of artistry, only what seems like the defacement of public spaces, buildings, or walls. However, what is essential here is to understand that this so-called defacement is often an act of protest and a visual outcry from someone who feels silenced or wronged. Some people perceive graffiti among the more angry art forms, the one that inspires extremely passionate responses from its audience (Myllylä, 2020; Nordmarker, 2016), even if in certain instances, this uncommissioned art form is also viewed and decreed unlawful (McAuliffe & Iveson, 2011; Young, 2012). The strong emotions that graffiti inspires can possibly be one of the reasons why Ross' chose this art form to feature in her fantasy novel rather than some other. Whether or not it resembles conventional art is beside the point. The graffiti artist doesn't need formal training or recognized talent to express themselves through paint as custom demands. Expression is a fundamental human right, and we must respect each individual's choice to voice their truth, however unconventional their medium may be.

THE EVOLUTION OF GOOD AND EVIL IN FANTASY WAR NARRATIVES

Moral Ambiguity in Pop-Culture

Unlike earlier fantasy literature, which often depicted war in black-and-white terms, modern fantasy explores the moral ambiguity of war. The *Letters of Enchantment* duology reflects how modern fantasy literature deals with serious warfare issues and the politics behind it better than the black and white manner in which fantasy dealt with good and bad centuries ago. Modern fantasy literature does not deny the existence of the gray areas that the political leaders and capitalist businessmen dwell in on either side of the border, and the conflict on said border.

When pop culture first brought in the idea of "origin stories", it took the media and the audience by storm (McNeal, 2023). Although the idea has existed for quite some time now, focusing on how people who make the wrong choices also have sound reasoning for it still intrigues researchers and readers alike. The reasoning, of course, does not excuse the culprit or remove the culpability factor in any way, but it does humanize them to some extent. Every villain, the closer you get to reality, retains parts of their humanity and every hero, as long as they are portrayed with some semblance of truth, have the potential to tap into their more villainous side when pushed too far and this is what aids their survival in the long term as they cross one hurdle after another (Jonason et al., 2012). Fantasy fiction, intriguingly, focuses on all these elements of heroes and villains, peace that is always on the verge of slipping into violence, wars that are waged in the name of lasting peace, murdered children who provoke their parents to be the reason hundreds more are orphaned, children who are saved and rehabilitated but who deal with lasting trauma not just for the rest of their lives but something they pass down as intergenerational trauma that never really goes away. The stream of consciousness element in fantasy fiction is a goldmine to study the psychological impact of war and trauma, and how it affects its readers who may or may not have had first-hand or second-hand experiences with some form of armed conflict.

Traditional Ideas of Heroism and Villainy in Fantasy and Mythology

The *Letters of Enchantment* duology challenges traditional notions of heroism and villainy, portraying war as a complex web of political maneuvering, personal vendettas, and ethical dilemmas. The objective of this study, then, is to analyze how modern fantasy narratives reflect real-world complexities, comparing the duology's approach to the moral ambiguity that existed in earlier works. While depicting the "origin story" of her villain, God Dacre, Ross does not portray him as all bad and with no clear motive to wreak havoc on the world at large. Dacre had been beguiled by the Goddess Enva, her voice and musical talent, and proposed to her to come live in the underworld with him. The beginning of the love story of Dacre and Enva had clear parallels with that of Hades and Persephone, but the ending was nowhere near the same. When Enva agreed to go with Dacre initially, it was only to keep the peace on Earth because she had seen the reaction her rejection had provoked in Dacre. Therefore, it was no surprise that the first chance Enva got, she tried to escape the Underworld and Dacre and return to her Skyward life. Through her magic of music, Enva had sung Dacre and his people to slumber for centuries, and during that time, there had been peace. Ross provided her readers the room to decide the severity or magnitude of villainy they wanted to associate with Dacre. For some readers who enjoyed the Greek mythology folklore of Hades-Persephone and felt pity for the loneliness Dacre felt, his crimes would not be considered unforgivable. However, for other readers, especially ones who did a feminist reading of this part of Dacre-Enva's back story, Dacre would be perceived as nothing short of a boor and brute who bullied Enva into submission and forced her into a life of captivity for his selfish reasons. For them, Enva's cleverness in putting Dacre and his family into a comatose sleep would be an act of reclaiming her agency and would be celebrated as such. While for the first category, this act by Enva could be perceived as breaking her marriage vows and cheating her husband into a false sense of security, only to cast a spell on him and his family that robbed them of centuries of their life. To reclaim her agency, she stole his. There was no conversation that she tried to have with Dacre, to see if he would be amenable to her going away for brief periods from the Underworld, a place of darkness and stagnancy, and visit her Skyward family above, the place of light and life and sun and air that she had grown up and felt deeply nostalgic for. When Enva had agreed to Dacre's proposal, she had only made two demands:

"I will follow you into the realm below," she said. "I will live with you in the shadows on two conditions: you will uphold peace and you will permit me to sing and play my instrument whenever I desire."

Dacre, who was enchanted by her, readily agreed. (Ross, 2023, p. 74).

Going by the agreement between them, Dacre had done no wrong to justify Enva's musical-magical attack on his person. This entire debacle and debate is Rebecca Ross' clever creation to show her readers that good and evil, right and wrong are never as black and white and obvious as earlier fantasy stories and pop-culture media had us believe. Through the war narrative in *Letters of Enchantment,* Ross showcases how the representation of Good and Evil has changed and evolved over time, at least in fantasy fiction and future reading of the genre cannot be basic, for there are complexities in plot, world building and character development that call for a more nuanced understanding and close observation of how each strand progresses.

Fantasy Writers' Personal Opinions Portrayed in Literary Devices

While Ross wrote about both Dacre and Enva's actions in a gray light, rather than letting her readers see which of the two she condemned, she did drop specific hints across the novel that a keen observer (reader) can pick out and understand the author's stance on the matter. Fantasy fiction is best known for its magical world and symbolism, and the answers to most of the questions and riddles can usually be found in the symbolism. There is a scene in the second book of the duology, *Ruthless Vows,* where Iris is listening to her friend, Marisol, talk about birds. The conversation might feel vague and out of place if the reader does not try to decipher what is being alluded to rather than what is simply being said. Marisol said:

"Years ago, a study was done on nightingales, and quite a number were caught and put into captivity."

"Why?" Iris asked.

"They wanted to trade the birds, as well as study their songs. Most of the nightingales died, but the ones who lived until autumn...they eventually killed themselves trying to escape, bashing their wings and their bodies against the cages that held them. They felt the need to migrate, and they couldn't." (Ross, 2023, p. 119).

While the above conversation neither followed nor preceded any mention of Dacre and Enva, it felt like Ross was trying to allude that the captive bird was how Enva had felt before fleeing the Underworld and pushing Dacre to the limits of waging a war that was now causing so much destruction and death. While a lot of civilians condemned Enva for singing the songs that called their sons and husbands to battle, and a lot of graffiti art painted her as the culpable party in this war, this tiny mention of how a caged bird might feel is what Ross wants her readers to focus on. Enva could not go back into that life of captivity, and if that called for a war where innocent lives must be lost, then that was the price of her freedom. The irony here is that Enva is a goddess in the fantasy world. And if the freedom of Gods and Goddesses can be stolen, mere mortals need to be twice as vigilant with the freedom their ancestors had fought so hard for them to enjoy. Hence, while war is an evil that should be avoided at all costs, freedom might be the only thing that could justify the cost.

CONCLUSION AND FUTURE SCOPE

Role of Fantasy Fiction in War Studies

This study analyzed Ross' duology's portrayal of trauma, intergenerational grief, and the lasting psychological scars left by conflict. Compared with elements of historical and contemporary literary works that explore post-war trauma, the analysis brought new and seemingly unexplored findings to light that would function as excellent points to study in further detail by future researchers in the field. *The Things They Carried* (1990) by Tim O'Brien and *Slaughterhouse-Five* (1969) by Kurt Vonnegut are just some fictional texts stemming from war that would function as excellent counterparts for a comparative study between literature of the previous century and literary texts today. By drawing parallels in this regard, one may aim to demonstrate how modern fantasy literature serves as a conduit for exploring the human cost of war beyond mere battlefield strategies and political machinations.

As the nature of war evolves, so must the methods through which we study its impact. With the changing times, modern-day conflicts look nothing like the Great Wars of the previous century, and the study of their portrayal needs to change accordingly. Where war could only be studied through the disciplines of war history, political science, journalism, trauma psychology, and social work, now that approach needs to be reevaluated. Even when delving into fiction books, the more obvious choices for studying war remained limited to historical fiction and memoirs. The scholars of literature and cultural studies, even upon acknowledging how different modern-day conflict looks in comparison to the conflict of decades and centuries past, rarely turn to other genres to study it from a different lens. War has shaped and reshaped civilizations since the beginning of time, and studying war through every lens at our disposal is imperative. For the authors of this chapter, this included researching the portrayal of war through a cultural lens, using the depictions of music, art, and literature in the primary texts, among other things. In this study, the choice of using fantasy fiction as the genre and primary text to study war and conflict narratives was well-informed and ideated.

Fantasy fiction, with its ability to weave intricate narratives that mirror real-world struggles, offers an invaluable perspective on war and conflict that also has the potential to study all the previously mentioned themes in one single text. This study, hence, proposed that cultural studies scholars should broaden their scope to include fantasy literature, more specifically, Modern fantasy fiction, in the study of war, as it provides a unique and necessary lens through which to examine the psychological, social, and political ramifications of conflict. Through the exploration of media, music, art, and the moral complexities of war, *Letters of Enchantment* exemplifies how fantasy fiction serves as both a reflection of and a commentary on the realities of warfare. This study is meant to contribute to the growing discourse on literature's role in shaping war narratives and its broader implications for cultural studies. By analyzing *Letters of Enchantment* through this multifaceted lens, the study underscores the importance of re-evaluating how we perceive war literature and the genres that house it.

REFERENCES

Aghamohseni, K. (2024). Forbidden Music in Iran: The Tension between Communism and Imperialism. *Interdisciplinary Researches of Art*, *2*(1), 81–102. DOI: 10.22124/ira.2024.28037.1024

Allen, L. (2004). Music and Politics in Africa. *Social Dynamics*, *30*(2), 1–19. DOI: 10.1080/02533950408628682

Angelskår, S. (2005). *Policing Fantasy: problems of genre in fantasy literature* (Master's thesis). http://urn.nb.no/URN:NBN:no-12249

Auslander, H. B. (1981). "If Ya Wanna End War and Stuff, You Gotta Sing Loud"—A Survey of Vietnam-Related Protest Music. *Journal of American Culture*, *4*(2), 108–113. DOI: 10.1111/j.1542-734X.1981.0402_108.x

Awad, S. H., Wagoner, B., & Glaveanu, V. (2017). *The street art of resistance*. Springer Singapore., DOI: 10.1007/978-3-319-63330-5

Barakat, S. (2007). Postwar reconstruction and the recovery of cultural heritage: critical lessons from the last fifteen years. *Cultural heritage in postwar recovery*, 2(1), 26-37.

Benavides-Vanegas, F. S. (2005). From Santander to Camilo and Ché: Graffiti and Resistance in Contemporary Colombia. *Social Justice (San Francisco, Calif.)*, *32*(1 (99)), 53–61. https://www.jstor.org/stable/29768290

Berman, L. E. A. (1978). Sibling Loss as an Organizer of Unconscious Guilt: A Case Study. *The Psychoanalytic Quarterly*, *47*(4), 568–587. DOI: 10.1080/21674086.1978.11926858 PMID: 715121

Bevan, R. (2007). *The destruction of memory: Architecture at war*. Reaktion Books.

Botstein, L. (2022). Music in Time of War. *The Musical Quarterly*, *105*(3-4), 235–244. DOI: 10.1093/musqtl/gdac011

Buelens, G. (2006). Reciting Shells. Dada *and* Dada *in* & Dadaists *on* the First World War. *arcadia*, 41(2), 275-295. https://doi.org/DOI: 10.1515/ARCA.2006.019

Cozma, R. (2014). Were the Murrow Boys Warmongers? The relationship between sourcing, framing, and propaganda in war journalism. *Journalism Studies*, *16*(3), 433–448. DOI: 10.1080/1461670X.2014.882098

Crowe, T. K., Sánchez, V., Howard, A., Western, B., & Barger, S. (2017). Veterans transitioning from isolation to integration: A look at veteran/service dog partnerships. *Disability and Rehabilitation*, *40*(24), 2953–2961. DOI: 10.1080/09638288.2017.1363301 PMID: 28805082

Damanhoury, K. E., & Saleh, F. (2024). Mediated Clash of Civilizations: Examining the Proximity-Visual Framing Nexus in Al Jazeera Arabic and Fox News' Coverage of the 2021 Gaza War. *Digital Journalism (Abingdon, England)*, *13*(1), 136–158. DOI: 10.1080/21670811.2024.2332614

Deaville, J. (2018). Vietnam: Music of Patriotism and Protest. In *Music and War in the United States* (211-227). Routledge.

DeTurk, S. (2015). The "Banksy effect" and street art in the Middle East. *Street Art & Urban Creativity,* 1(2), 22-30. https://doi.org/DOI: 10.25765/sauc.v1i2.25

Dzhakupova, I. (2024). Greenberg, The Cold War, and French Existentialism: How Abstract Expressionism became Constructed as Inherently Masculine. *Groundings Undergraduate Journal*, *15*, 175–191. DOI: 10.36399/GroundingsUG.15.139

Ellison, M. (1994). African-American Music and Muskets in Civil War New Orleans. *Louisiana History: The Journal of the Louisiana Historical Association*, 35(3), 285-319. https://www.jstor.org/stable/4233124

Field, N. P., & Bonanno, G. A. (2001). The Role of Blame in Adaptation in the First 5 Years Following the Death of a Spouse. *American Behavioral Scientist,* 44(5), 764-781. https://doi.org/ (Original work published 2001)DOI: 10.1177/00027640121956485

Filkins, D. (2008). *The Forever War: NATIONAL BOOK CRITICS CIRCLE AWARD WINNER*. Vintage.

Flomenhaft, D. (2007). *The forgotten ones: The grief experience of adult siblings of World Trade Center victims* (Doctoral dissertation, New York University, School of Social Work).

Fredericks, S. C. (1978). Problems of fantasy. *Science Fiction Studies*, *5*(Part 1), 33–44. https://www.jstor.org/stable/4239155. DOI: 10.1525/sfs.5.1.0033

Garofalo, R. (1982). Music and Musket: Bands and Bandsmen of the American Civil War: By Kenneth E. Olson. Westport, Connecticut: Greenwood Press, 1981. 299 pages. Illustrations, bibliography. Hard cover, $27.50. *Music Educators Journal*, *68*(5), 23–25. DOI: 10.2307/3397736

Garwood, A. (1996). The Holocaust and the power of powerlessness: Survivor Guilt an Unhealed Wound. *British Journal of Psychotherapy*, *13*(2), 243–258. DOI: 10.1111/j.1752-0118.1996.tb00880.x

Gerber, T. P., & Zavisca, J. (2016). Does Russian Propaganda Work? *The Washington Quarterly*, *39*(2), 79–98. DOI: 10.1080/0163660X.2016.1204398

Gibbons, M. B. (1992). A child dies, a child survives: The impact of sibling loss. *Journal of Pediatric Health Care*, *6*(2), 65–72. DOI: 10.1016/0891-5245(92)90123-L PMID: 1573530

Gilbert, G. (2004). Robert Motherwell's World War Two collages: Signifying war as topical spectacle in abstract expressionist art. *Oxford Art Journal*, *27*(3), 311–337. DOI: 10.1093/oaj/27.3.311

Gilbert, S. (2007). Singing Against Apartheid: ANC Cultural Groups and the International Anti-Apartheid Struggle*. *Journal of Southern African Studies*, *33*(2), 421–441. DOI: 10.1080/03057070701292848

Glover, H. (1984). Survival guilt and the Vietnam veteran. *The Journal of Nervous and Mental Disease*, *172*(7), 393–397. DOI: 10.1097/00005053-198407000-00003 PMID: 6726209

Graves, R. (2025). *Goodbye to All That*. Aegitas.

Hazou, R. (2022). Arts and conflict: Banksy, art and Palestinian solidarity. In *Tu Rangaranga Rights, responsibilities and global citizenship in Aotearoa New Zealand* (pp. 223–236). Massey University Press.

Heiney, S. P., Darr-Hope, H., Meriwether, M. P., & Adams, S. A. (2017). Healing by Creating: Patient Evaluations of Art-Making Program. *The Journal of Creative Behavior*, *51*(1), 35–44. DOI: 10.1002/jocb.84

Hill, A., & White, A. (2011). Painting Peace? Murals and the Northern Ireland Peace Process. *Irish Political Studies*, *27*(1), 71–88. DOI: 10.1080/07907184.2012.636184

Hutson, S. P., Hall, J. M., & Pack, F. L. (2015). Survivor guilt: Analyzing the concept and its contexts. *Advances in Nursing Science*, *38*(1), 20–33. DOI: 10.1097/ANS.0000000000000058 PMID: 25635503

Ivanova, N., & Ryzhchenko, O. (2019). Meta-Genre Of Fantasy In The Context Of Modern Science. *Research Trends In Modern Linguistics And Literature*, *2*, 23–37. DOI: 10.29038/2617-6696-2019.2.23.37

Jachec, N. (1991). “The Space between Art and Political Action”: Abstract Expressionism and Ethical Choice in Postwar America, 1945-1950. *Oxford Art Journal*, *14*(2), 18–29. https://www.jstor.org/stable/1360522. DOI: 10.1093/oxartj/14.2.18

Jonason, P. K., Webster, G. D., Schmitt, D. P., Li, N. P., & Crysel, L. (2012). The Antihero in Popular Culture: Life History Theory and the Dark Triad Personality Traits. *Review of General Psychology*, *16*(2), 192–199. DOI: 10.1037/a0027914

Jones, A. (2002). Equivocal Masculinity: New York Dada in the Context of World War I. *Art History*, *25*(2), 162–205. DOI: 10.1111/1467-8365.00312

Jones, G. (2008). *Killing Monsters: Our Children's Need For Fantasy, Heroism, and Make-Believe Violence*. Basic Books.

Keeling, M. E., Ozuna, S. M., Kintzle, S., & Castro, C. A. (2018). Veterans' Civilian Employment Experiences: Lessons Learnt From Focus Groups. *Journal of Career Development,* 46(6), 692-705. https://doi.org/ (Original work published 2019)DOI: 10.1177/0894845318776785

Kirkegaard, A. (2017). Silencing artists: Reflections on music censorship in the case of Mahsa Vahdat. *Researching music censorship*, 28-51.

Korpe, M., Reitov, O., & Cloonan, M. (2006). Music censorship from Plato to the present. *Music and manipulation: On the social uses and social control of music*, 1, 239-263.

Kubany, E. S., & Manke, F. P. (1995). Cognitive therapy for trauma-related guilt: Conceptual bases and treatment outlines. *Cognitive and Behavioral Practice*, *2*(1), 27–61. DOI: 10.1016/S1077-7229(05)80004-5

le Roux-Kemp, A. (2014). Struggle Music: South African Politics in Song. *Law and Humanities*, *8*(2), 247–268. DOI: 10.5235/17521483.8.2.247

Lee, R. A. (2009). Protest Music as Alternative Media During the Vietnam War Era. *War and the media: Essays on news reporting, propaganda and popular culture,* 24-40.

Leys, R. (2007). *From guilt to shame: Auschwitz and after* (Vol. 8). Princeton University Press.

Mac Ginty, R. (2021). Temporality and Contextualisation in Peace and Conflict Studies: The Forgotten Value Of War Memoirs And Personal Diaries. *Cooperation and Conflict*, *57*(2), 191–209. DOI: 10.1177/00108367211027605

Marche, G. (2012). Expressivism and resistance: Graffiti as an infrapolitical form of protest against the war on terror. *Revue Française d'Études Américaines, 131*(1), 78–96. DOI: 10.3917/rfea.131.0078

Marizka, A. (2024). *Retracing The Past Through Historical Fantasy Novel: The Poppy War* (Doctoral dissertation, UNIMED).

McAuliffe, C., & Iveson, K. (2011). Art and crime (and other things besides...): Conceptualising graffiti in the city. *Geography Compass*, *5*(3), 128–143. DOI: 10.1111/j.1749-8198.2011.00414.x

McCombs, M. (2002, June). The agenda-setting role of the mass media in the shaping of public opinion. In *Mass Media Economics 2002 Conference, London School of Economics*: https://sticerd.lse.ac.uk/dps/extra/McCombs.pdf

McCombs, M., & Valenzuela, S. (2020). *Setting the agenda: Mass media and public opinion*. John Wiley & Sons.

McNeal, J. (2023). *Ambiguous Morality in the Human Condition as Reflected by Modern Villains from Popular Culture* (Doctoral dissertation). https://hdl.handle.net/2104/12591

McWhirter, C. (2012). Christian, *Battle Hymns: The Power and Popularity of Music in the Civil War* (Chapel Hill, NC, 2012; online edn. *North Carolina Scholarship Online, 24*. Advance online publication. Retrieved May 15, 2025, from. DOI: 10.5149/9780807882627_mcwhirter

Migeon, M. (2024). Peace and Conflict in Public Space: Gendered Murals Shaping Belfast. *Geopolitics*, *29*(5), 1604–1634. DOI: 10.1080/14650045.2024.2321164

Miladi, N. (2018). Urban Graffiti, Political Activism and Resistance. In *The Routledge Companion to Media and Activism* (pp. 241–249). Routledge. DOI: 10.4324/9781315475059-26

Moody, C. (2021). Iranian music censorship & international human rights law. *Brooklyn Journal of International Law*, *47*(1), 232–259.

Mowery, N. (2021). *Even Objects Revolt Against Regimentation: Art and Censorship in Post-WWI Germany and the Rise of Dadaism. A Film and Thesis* (Doctoral dissertation, University of Oregon).

Murray, H., Pethania, Y., & Medin, E. (2021). Survivor guilt: A cognitive approach. *Cognitive Behaviour Therapist*, *14*, e28. DOI: 10.1017/S1754470X21000246 PMID: 34557258

Myllylä, M. (2020). The Good, The Bad, and The Ugly Graffiti. Emotions In Technology Design: From Experience To Ethics, 87-104. https://doi.org/DOI: 10.1007/978-3-030-53483-7_6

Nasriddinov, D. (2023). Fantasy Genre and Its Scientific Interpretation in Theoretical Views. https://doi.org/DOI: 10.47689/2181-3701-vol1-iss2-pp56-62

Neureiter, M. (2016). Sources of media bias in coverage of the Israeli–Palestinian conflict: The 2010 Gaza flotilla raid in German, British, and US newspapers. *Israel Affairs*, *23*(1), 66–86. DOI: 10.1080/13537121.2016.1244381

Nordmarker, A. (2016). Graffiti–For Joy and Confirmation. Motivational Aspects, Triggering and Inhibiting Factors, and Emotional Satisfactions In Graffiti: The Creative-Interactive Dimension Of Vandalism.

O'Brien, C., Gallagher, S., & Malafouris, L. (2025). Handmade Therapy: The Hedonic Impacts of Engaging in Pottery Making. *Topoi*, *44*(4), 1–14. DOI: 10.1007/s11245-025-10181-9

O'Brien, T. (1990). *The Things They Carried*. Houghton Mifflin.

Okulate, G. T., & Jones, O. B. E. (2006). Post-Traumatic Stress Disorder, Survivor Guilt and Substance Use–A Study Of Hospitalised Nigerian Army Veterans. *South African Medical Journal*, *96*(2), 144–146. PMID: 16532084

Oleinik, A. (2023). War Propaganda Effectiveness: A Comparative Content-Analysis Of Media Coverage of the Two First Months of Russia's Invasion of Ukraine. *Atlantic Journal of Communication*, *32*(4), 526–544. DOI: 10.1080/15456870.2023.2187801

Petzold, D. (1986). Fantasy Fiction and Related Genres. *Modern Fiction Studies*, *32*(1), 11–20. DOI: 10.1353/mfs.0.1205

Rastovac, H. (2009). Contending with Censorship: The Underground Music Scene in Urban Iran. *The McNair Scholars Journal*, *273*, 59–82.

Reinke, J. A. (2001). What Are My Middle School Students Reading? *Michigan Reading Journal*, *33*(4), 6.

Ross, R. (2023a). *Divine Rivals*. Macmillan Publishers.

Ross, R. (2023b). *Ruthless Vows*. Macmillan Publishers.

Sarsılmaz, N., & Yiğitler, K. (2012). How World War II Shaped the Art and Literature? *Anadolu Bil Meslek Yüksekokulu Dergisi*, (25), 50–58.

Sassoon, S. (1930). *Memoirs of an Infantry Officer*. Faber & Faber.

Saud, A., & Kazim, N. (2022). Disinformation and Propaganda Tactics: Impacts of Indian Information Warfare on Pakistan. *Journal of Indian Studies, 8*(02), 335–354.

Simmons, B. K. (1992). The effect of censorship on attitudes toward popular music. *Popular Music and Society, 16*(4), 61–67. DOI: 10.1080/03007769208591497

Singer, M. (2004). Shame, Guilt, Self-Hatred and Remorse in the Psychotherapy of Vietnam Combat Veterans Who Committed Atrocities. *American Journal of Psychotherapy, 58*(4), 377–385. DOI: 10.1176/appi.psychotherapy.2004.58.4.377 PMID: 15807083

Socolow, M. J. (2007). "News is a Weapon": Domestic Radio Propaganda and Broadcast Journalism in America, 1939–1944. *American Journalism, 24*(3), 109–131. DOI: 10.1080/08821127.2007.10678081

Swinfen, A. (2019). *In defence of fantasy: A study of the genre in English and American literature since 1945*. Routledge., DOI: 10.4324/9780429321221

Teune, S. (2005, September). Art and the re-invention of political protest. In Paper presented at the *3rd ECPR Conference,* 8, 10.

Thankachan, K., Thomas, P. E., & Menon, V. A. (2025). Conflict Caught in Propaganda Nationalism: An Analysis of the India-Pakistan Newspaper Coverage of Mumbai, Uri, and Pulwama Attacks. In *Impacts of Leakage, Whistleblowing, and the Rise of Propaganda* (pp. 169-184). IGI Global Scientific Publishing.

Thomas, T. H. (1930). Some War Memoirs. *The Journal of Modern History, 2*(4), 629–640. DOI: 10.1086/235644

Traw, R. (1993). *Nothing in the Middle: What Middle Schoolers Are Reading.*

van Schaik, M. (2023, July). Place-Identity and Post-war Rebuilding. In *World Congress of Architects* (95-105). Design for Resilient Communities. UIA 2023. Sustainable Development Goals Series. Cham: Springer International Publishing. https://doi.org/DOI: 10.1007/978-3-031-36640-6_7

Verpoest, L., Engelen, L., Heynickx, R., Schmidt, J., Uyttenhove, P., & Verstraete, P. (Eds.). (2020). *Revival after the Great War: Rebuild, remember, repair, reform.* Leuven University Press. DOI: 10.2307/j.ctv1b0fvbf

Vonnegut, K. (1969). *Slaughterhouse-Five.* Delacorte Press.

Williams, T. (1988). Diagnosis and treatment of survivor guilt: The bad penny syndrome. In *Human adaptation to extreme stress: From the Holocaust to Vietnam* (pp. 319–336). Springer US., DOI: 10.1007/978-1-4899-0786-8_14

Wilson, G., Hill, M., & Kiernan, M. D. (2018). Loneliness and social isolation of military veterans: Systematic narrative review. *Occupational Medicine*, *68*(9), 600–609. DOI: 10.1093/occmed/kqy160 PMID: 30551163

Wilson, K. (1990). Black Bands and Black Culture: A Study of Black Military Bands in the Union Army During the Civil War. *Australasian Journal of American Studies*, *9*(1), 31–37. https://www.jstor.org/stable/41054165

Winter, J. (2018). War Memoirs, Witnessing and Silence. *War Stories: The War Memoir in History and Literature*, Berghahn Books, 27-47.

Young, A. (2012). Criminal Images: The Affective Judgment of Graffiti and Street Art. *Crime, Media, Culture*, *8*(3), 297–314. DOI: 10.1177/1741659012443232

Young, A., & Omosun, F. (2024). How Does It Look from Where You Are? A Visual Media Framing Analysis of the 2022 War in Ukraine. *Digital Journalism (Abingdon, England)*, *13*(1), 18–36. DOI: 10.1080/21670811.2023.2296560

KEY TERMS AND DEFINITIONS

1. **Censorship:** The suppression, alteration, or prohibition of speech or ideas that might be deemed objectionable, harmful, or sensitive by authorities such as governments, institutions, or media platforms.
2. **Cultural Lens:** Viewing of any research material through a perspective shaped by the customs, beliefs, values, and experiences of a particular culture, in the case of the present study, that lens comprises art, music, and literature.
3. **Fantasy fiction:** A literary genre that uses magical, supernatural, or imaginary elements as key components or key supporting elements of the plot, setting, or characters set in fictional worlds. This is also viewed as an escapist genre, seeing as it allows the reader to briefly escape reality.
4. **Flak:** Originally a term for anti-aircraft fire, it is now also used figuratively to describe strong criticism or opposition in news and journalistic settings.
5. **Graffiti:** Unauthorised visual artwork or markings, created in public spaces, usually using spray paint or markers—sometimes viewed as art, sometimes as an act of vandalism.
6. **Intergenerational Trauma:** Psychological effects of trauma that are passed down from one generation to the next, often through behavioral patterns, cultural narratives, or unresolved grief.

7. **MIA:** A military designation for a soldier whose whereabouts are unknown following combat, and who has not been confirmed dead or captured. The families of MIA soldiers, due to a lack of information and closure, suffer immense trauma the longer their family member remains missing-in-action.

8. **Origin Story:** A narrative that explains how a character, usually an anti-hero or villain, came to be. This story or narrative highlights formative events—especially common in mythology, comics, and pop culture.

9. **Pop-Culture:** Cultural products—such as music, movies, fashion, memes, and trends—that are widely consumed and influential within mainstream society. It may start off as one thing but get unrecognizably transformed into another through a number of retellings, the longer a fandom experiments with it.

10. **Stream of Consciousness Narrative Technique:** A literary style that attempts to depict the flow of thoughts, feelings, and perceptions in a character's mind, often in a continuous, unstructured manner. This narrative has been known to lean more towards the inner psyche than the external circumstances of the character concerned.

11. **Survivor's Guilt:** A psychological condition where a person feels guilt or remorse for having survived a traumatic event when others did not, common among survivors of war, disasters, or accidents.

Chapter 5
Intersections of Gender, Nationalism and Identity in Select Indian Films:
Sita Ramam and Amaran

Deepthi Kirshna Thota
https://orcid.org/0000-0002-0171-9418
Mahindra University, India

ABSTRACT

This chapter examines the representation of gender and identity in Indian war films, focusing on nationalism and militarism. Films Amaran (Rajkumar Periasamy, 2024) and Sita Ramam (Hanu Raghavapudi, 2022) place women at the centre of military narratives shaping historical understanding and soldiers' experiences. This chapter analyses the impact of gender relations on nationalism and patriotism, linking personal and national identity during conflict. Cross-cultural romantic relationships between Hindu soldiers and women from minority religions Christianity and Islam challenge and reinforce societal norms within tense militaristic contexts. Through Benedict Anderson's framework of nationalism, the research emphasises women's roles in military narratives, going beyond traditional romantic portrayals. This analysis is framed within popular media discourse on war and selected films and supplemented by secondary sources on media research to capture public perceptions. It also aims to deepen the understanding of how gender dynamics in Indian war cinema impact contemporary national identity.

DOI: 10.4018/979-8-3373-7589-2.ch005

INTRODUCTION

Many Indian films made post-independence depict military realities. This chapter will refer to them interchangeably as war films or military films.

In Hindi cinema, war films typically focus on India's geopolitical struggles with its neighbours. Films like *Samadhi* (Grave) (Saigal, 1950), *Lalkar* (Challenge) (Sagar, 1972), and *Haqeeqat* (Fact/Reality) (Anand, 1964) highlight various conflicts, including World War II and the Sino-Indian war. Famous ones like *Hindustan ki Kasam* (I swear on India) (Anand, 1973), *Vijeta* (The Victor) (Nihlani, 1982), *Border* (Dutta, 1997), *Uri: The Surgical Strike* (Dhar, 2019), and *Shershaah* (Lion King) (Kulasekaran, 2021), to name a few, feature war scenes and soldiers in action. Additionally, some Indian films focus on Pakistani-sponsored terrorism in India, addressing the issue without emphasising military involvement. However, war films are a recent phenomenon in films in other Indian regional languages.

India has a long and gory history with Pakistan, fraught with communal violence and heart-wrenching loss on both sides. The trauma of the India-Pakistan partition runs deep. Wars and constant skirmishes over the border keep their enmity alive. India versus Pakistan cricket matches turn the ground into a symbolic battlefield of patriotism and national honour, which war films also depict.

Indian Cinema has many Hindi war films evoking the patriotism and honour of Indian soldiers. Like popular cinema worldwide, mainstream Indian war cinema fuels patriotism and nationalism through images of military discipline and the sacrifice and gallantry of soldiers. Common themes are soldiers as heroes and military life as a source of pride.

Indian films are "an integral unit of India's cultural heritage and nationalistic discourse" (Raj & Suresh, 2022, p. 172). Military films enhance a sense of belonging to the imagined community of India's nation, leading to an idealised vision of nationalism. Recent hits like *Uri: Surgical Strike* and *Shershaah* depict war and battle scenes. *Uri: Surgical Strike* is a semi-fictional film based on the 2016 surgical strikes and features strong nationalistic themes (Iyer & Das, 2025).

War films focus on soldiers in homosocial contexts, featuring masculine symbols of artillery and war machines in hostile terrain. The feminine presence emerges primarily through *Bharatmata* (Mother India), anthropomorphising the nation as requiring protection. Meanwhile, women appear primarily as love interests or mothers, confined to limited roles as 'heroic mother, chaste wife and celibate warrior' (Banerjee, 2006, p. 63).

FILMS IN FOCUS: *AMARAN* AND *SITA RAMAM*

The war films in focus in this chapter, *Amaran* (Immortal) (Periasamy, 2024), and *Sita Ramam* (Raghavapudi, 2022) are grounded in passionate romantic narratives.

Amaran is a Tamil film based on the true story of Major Mukund Varadarajan (played by Siva Kartikeyan), narrated by his wife Indhu Rebecca Varghese (played by Sai Pallavi). It follows Major Mukund Varadarajan's military service journey through the ranks from Lieutenant in 2006 to Captain and then Major in 2012, the position he held till his death.

Sita Ramam is a fictional Telugu-language film set in 1964 against the backdrop of Pakistan's Operation Gibraltar[1]. It follows the life of Lieutenant Ram (Dulquer Salmaan) and his romantic interest, Sitamahalakshmi (Mrunal Thakur). A third character, Afreen (Rashmika Mandanna), is a young Pakistani girl who burns with hatred for India. She is searching for Sita, to whom she must deliver a letter from Ram. Sita is later revealed to be a Muslim princess.

Both films explore India's military history during times of conflict—*Amaran* in the 2010s, and *Sita Ramam* between the 1960s and 1980s. As central figures driving military narratives, women illuminate perspectives long overlooked in war cinema: those of families shaped by service life.

This chapter analyses both films through the lens of gender relations within nationalistic and militaristic frameworks. It examines communal dynamics during war and cross-border conflict, particularly how Hindu soldiers' romantic relationships with Christian and Muslim women symbolise patriotism and the nation. Benedict Anderson's (2006) frameworks of nationalism illuminate these relationships through an intersectional gendered perspective.

RESEARCH OBJECTIVES

This chapter examines gender in military and war narratives.

First, we investigate how female characters—mothers and protagonists—and male military figures shape narrative and gender dynamics. We explore their representation, familial military ties, and influence on nationalism and patriotism.

The chapter then examines how romantic partnerships between Hindu soldiers and women from minority religions (Christian and Muslim) affect national identity and communal relations during conflicts. This exploration reveals how these relationships reflect or challenge societal norms about gender, love, and loyalty within militaristic and cross-border tensions.

These objectives explore gender dynamics within military narratives, revealing women's critical influence beyond romantic roles—particularly at the intersection of nationalism and identity in contemporary film discourse.

The analysis draws primarily from two conceptual frameworks: Anderson's theory of nationalism and reception studies. Building on existing literature, I first conceptualise Indian nationalism, then trace its historical connection to communalism. Then, I examine how these elements manifest in Indian war films. Finally, I segue into academic work on war films in India.

Another object of inquiry follows: the depiction of gender, particularly in war films.

Here, I draw from popular discourse through primary and secondary sources: academic research on Indian war media, online news reports, and social media discussions from Instagram and Reddit about *Amaran* and *Sita Ramam*. The chapter concludes with a summary of key findings and suggested directions for future research.

CONCEPTUAL FRAMEWORK

Nationalism and Benedict Anderson

The concept of nationalism and nation-states gained significant momentum after the end of the Second World War (Anderson, 2006; Chatterjee, 1996; Sherfudeen et al., 2024).

Nation versus The Nation State

Before exploring the literature on nationalism, let us clarify the distinction between nation and nation-state:

> *The nation-state is a politically constructed category, which holds within it the judicial, legislative, and constitutional domains. It is the nation-state that has proved to be inadequate in granting those who owe their allegiance to it, all that it set out to grant in terms of rights, equality, choices and space. (Viswanath, 2014, p. 32)*

Meanwhile, the "nation…will be understood as a culturally determined subjective, affective category, and state as its political expression" (Viswanath, 2014, p. 32).

In his work *Imagined Communities*, Anderson (2006) posits that nationalism is a modern phenomenon (Shelef & Koo, 2022) arising from specific historical and cultural conditions. He contends with the paradox of its universality and modernity—however, Sherfudeen et al. (2024) note that a precise definition of "nation" or "nationalism" is still debated.

One definition of the nation is "an imagined political community – and imagined as both inherently limited and sovereign" (Anderson, 2006, p. 6).

Apart from Anderson's definition, we can define a nation as a group of people living in a specific territory and wishing to create their own state (Sherfudeen et al., 2024). Iyer and Das (2025) add that among its various definitions, nationalism commonly includes a call for loyalty and affiliation to a specific nation. Hence, conceptualising a nation rather than defining it appears more suitable. Viswanath (2014) finds the post-structuralist and postmodernist discursive nature relevant to conceptualising the nation.

In addition, nationalism has the core element of connection to the land in the form of homelands, cultural regions, and shared identity (Anderson, 2006; Musolff, 2021; Shelef & Koo, 2022). It is shaped through power relations and contested through specific practices. It is not a static, pre-constituted ideal (Shelef & Koo, 2022). While established as a universal concept, nationalism is inherently limited and defined by borders, as no nationalist advocates for worldwide unity (Iyer & Das, 2025).

Borders, Territoriality and Home(lands)

A nation's finite boundaries limit the people connected with it, and national identity transcends actual inequalities among people. A common language, race, religion, or ethnicity is not strictly necessary to form a nation. Instead, a shared culture and consciousness often connect people residing in a particular place they call home. (Sherfudeen et al., 2024)

On the implication of homelands in nationalism, Shelef and Koo (2022) opine that it occurs on two fronts. One, it is in international conflicts regarding secession, partition, and restoration of territory; two, it is borne by intangibility, materiality, and demographics of the territories.

Borders territorialise the national identity and provide a space to produce and construct *national consciousness* (Athique, 2012). The national consciousness is embedded in narrative to drive nationalism as an ideology (Iyer & Das, 2025). In this context, "nation-ness" and "nationalism" are constructs which Anderson (2006) calls *cultural artefacts*.

Religion, Symbols and Codes in Nation Building

In his discussion, the religious community is where one "can detect the seeds of a territorialization of faiths which foreshadows the language of many nationalists ('our' nation is 'the best' - in a competitive, *comparative field*)" (Anderson, 2006, p. 17). Most of the time, the symbols and values of nationalism are derived from religious mythologies and inadvertently filled with religious signifiers. Hence, homelands possess *quasi-religious power* and are considered *sacred ground* imbued with nationalism (Shelef & Koo, 2022).

There needs to be a tangible space for the production and representation of culture in everyday living. For refugees, the displaced, and the diaspora, the materiality of a land to imagine their roots is necessary. Even if it is a long-lost land that no longer exists in contemporary geography.

Homelands are fertile for the production of national symbols and other values and signifiers within nationalism's cultural systems. Communication systems—from pre-technological forms like folk culture, myths, theatre arts, and oral histories to the revolutionary printing press—enable these symbols and codes to circulate through mass media. Print capitalism, vernacular languages, elite-driven official nationalism, historical narratives, and political discourse all fuel nationalism (Anderson, 2006; Musolff, 2021; Shelef & Koo, 2022).

Media and Nation

Media plays a key role in forming national identity. Cinema has become an important form of sustaining nationalism as a cultural unit that disseminates stories of the nation, including metaphors and historical narratives.

The relationship between cinema and nation-building has long been a central topic within film studies, drawing on theoretical frameworks from scholars such as Anderson (2006) and Hall (1993). These discussions interrogate how films construct, reinforce, or contest ideas of national identity and collective belonging.

In *Films and Nationalism,* Williams (2002) examines cinema's broader roles in nation-forming processes. He engages with questions of national propaganda, the globalisation of cinematic culture, and the role of cinema in defining national character. He also states that films help shape national sensibilities across diverse socio-political landscapes. Cinema reflects the ongoing process of nationalism, revealing subtle interpretations in everyday practices, known as *banal nationalism* and discursive practices (Musolff, 2021; Shelef & Koo, 2022).

Violence plays a key role in producing nations and nationalism, manifesting through territorial disputes and conflicts between in-groups and out-groups (Shelef & Koo, 2022). Viswanath (2014) cites Ernest Renan's theorisation of nation-building on cultural memories, where grief plays a role. He then brings in Shahid Aman's emphasis on subaltern histories as well. War films serve as a direct expression of violent nationalism, highlighting self-sacrificing hero soldiers.

Anderson's (2006) framework of nationalism combines nationalism and patriotism theories with reception studies, integrating an intersectional gendered perspective. This approach analyses films portraying gendered military narratives and audience perceptions.

Through Anderson's (2006) imagined communities theory, I examine how military-associated women's experiences shape nationalism and patriotism. This intersectional lens crystallises women's nuanced religious experiences, showing both empowering and restrictive dynamics.

Building upon this religious aspect, the framework examines gender's role in cross-border conflicts, particularly through romantic relationships between Hindu soldiers and their Christian or Muslim love interests—relationships that mirror broader societal tensions and the complexities of national identity.

LITERATURE REVIEW

Conceptualising Indian Nationalism

Anderson's Imagined States

According to Anderson (2006), nationalism is a modular concept, where the idea and practice of nationalism can be copied, adapted, and applied in different global contexts. Indian nationalism at its inception can be termed as anti-colonial and anti-imperial nationalism.

He adds that the colonial administrative units nurtured arbitrary imagined communities of a nation of diverse linguistic and multi-cultural communities. Censuses, maps, museums, visual studies, and shared culture played a significant role in creating an imagined community of the nation in colonial states (Anderson, 2006).

Chatterjee (1996) questions the contradictory ideas of the nation as an imagined community and the modular nation put forward by Benedict Anderson. He accuses Anderson of regarding the post-colonial states as passive consumers of modernity and Western ideas of the nation. If indeed there were a modular template for nationalism, Chatterjee (1996) asks, then why "imagine" a nation and who is doing the imagining? Chatterjee later very deftly conceptualises Indian nationalism, which I will discuss later.

I understand that the nation's replicable modular nature finds itself in apparently universal values because of their idealistic appeal, such as democracy, loyalty, patriotism, and collective belonging.

A utopia is an ideal unachievable state that remains an enduring vision. This vision manifests in practical efforts to realise universal national values, always shaped by each nation's unique sociocultural and geopolitical context.

India's Variegated Nationalism

Indian nationalism encompasses diverse forms. There is communal Hindu nationalism, communal Muslim nationalism, secular nationalism, progressive and inclusive nationalism (Bhutia, 2017), and cultural nationalism (Sherfudeen et al., 2024). The former two nationalisms are from the colonial era.

Indian nationalism diverges from the Western/British constructs of nations (Chatterjee, 1996; Daiya, 2008). In postcolonial societies, "The study of nationalism ... must begin with the idea of nationalism as anti-colonial struggle" (Viswanath, 2014, p. 13). During this struggle, progressive, secular and inclusive nationalism were the dominant features of Indian nationalism (Bhutia, 2017; Chatterjee, 2012) and the constitutive principles of post-independence India (Gopinath & Deshmukh, 2023). However, these very traits created contests between multiple internal nationalisms or sub-nationalisms (Iyer & Das, 2025).

It is commonly accepted that the concepts of the Indian nation and Indian nationalism originated with the establishment of the Indian National Congress (INC) in 1885 (Chatterjee, 1996). However, that idea is contested (Chatterjee, 1996). According to Daiya (2008), equating Indian nationalism with anti-colonial nationalism is an oversimplification because it cannot be solely understood within the context of the *modern state system* (See Breuilly, 1985). Instead, consider nationalism "...a forceful, enduring process" which "cannot be subjected to closure" (Daiya, 2008, p. 40).

Indian Nationalism According to Partha Chatterjee

The deservedly oft-cited exploration of Indian nationalism by Chatterjee (1996) discusses Indian nationalism operating in two registers or domains during colonial rule: outside/material and inside/spiritual.

Outside includes economy, statecraft, science, and technology. The West has clearly dominated and progressed in this domain, which can be *carefully studied and replicated.*

The inner domain comprises "'essential' marks of cultural identity" (Chatterjee, 1996, p. 217). The inner/spiritual domain is impervious to Western materialism and immateriality, which is crucial for Indian nationalism.

Chatterjee (1996) argues that the inner domain emerges during the latter phase of social reform, when societies recognised the need for change but rejected colonial powers as reformers—unlike the first phase, when colonial powers dominated reform efforts.

The second phase of social reform points to the nationalism that internally emerged thanks to reformers like Raja Ram Mohan Roy, Ishwar Chandra Vidyasagar, Swami Dayanand Saraswati and Swami Vivekananda. Unlike B.R. Ambedkar, Savitri Bhai

Phule and Jyotirao Phule, these reformers harked back to Vedic Hinduism to fight social evils and advocate for progressive reforms.

The evolution of Indian nationalism during colonial rule, rooted in all these factors, distinguishes it from Western nationalism. The postcolonial states' nationalistic fervour comes from the trauma of colonial oppression and anti-colonial struggle; thus, the nation is an affective unit (Fanon, 2000).

Nationalism underwent a postcolonial process involving a return to the past regarding culture, codes, dress, ethnicity, language, and values in colonies freed as independent nations (Sherfudeen et al., 2024).

The push for a unified "national culture" faced challenges due to the privileging of an "essential tradition," which excluded many (Chatterjee, 1989, p. 632). Exclusionary nationalism grew roots during the anti-colonial struggle.

Evidenced by the discarding of reform movements and the rise of communal politics within the anti-colonial struggle, a Hinduised image of the nation-state evolved to equate Indian nationalism with Hindu nationalism. On paper, it is secular and modern, continuing through the post-colonial era (Banerjee, 2006; Chatterjee, 1989; Daiya, 2008; Gabriel, 1998; Iyer & Das, 2025; Raj & Suresh, 2022). However, despite the Constitution affirming that the state has no religion, administrative practices reflect Hindu nationalism. For example, missiles are named after Hindu gods (Upadhyay & Robinson, 2012). Concepts of freedom, equality, and cultural refinement are not practised universally.

In this configuration of nationalism, women have become important symbols of the nation and its nationalism. Nationalism addressed the "women's question" as a problem of Indian tradition, resolved by demarcating a cultural essence distinct from the West and the mass of the people.[2] (Chatterjee, 1989).

Indian Nationalism and Communalisation

Having examined how communalisation permeated Indian nationalism, this section explores how communalism infuses religious ideals into the construction of national identity (Daiya, 2008).

Communalism is often defined as a conflict between two religious groups (Hindu and Muslim) with a *history of togetherness* over *secular issues* to seek or resist a power hierarchy in asserting their identities (Upadhyay & Robinson, 2012). Communalism is often associated with religious communities but can be at microlevels of cults and sects (Upadhyay & Robinson, 2012). This chapter focuses specifically on religious communalism between Hindu communities and Islam, and to a lesser extent, Christianity as a minority religion.

Jones (1968) defines "communalism as consciously-shared religious heritage which becomes the dominant form of identity for a given segment of society" (p. 39 as cited in Upadhyay & Robinson, 2012, p. 36).

Daiya (2008) narrates how British colonial policy fostered religious division: first by encouraging Muslim elites to form the Muslim League in 1906, then by establishing separate Muslim electorates through the Indian Council Act in 1909—all based on their assertion that Hindu and Muslim interests were incompatible.

Colonial policies of *separate electorates, census enumeration*[3], and British practices of *written accounts of colonial knowledge of India*[4] (Upadhyay & Robinson, 2012) fomented further communal tensions.

Hence, the period from 1905 to 1917 saw a rise in Hindu-Muslim conflicts over *cow protection, music near mosques* and the *Urdu language* (Daiya, 2008, p.34). The communal tensions did not remain consistent with the return and rise of MK Gandhi and the non-cooperation movement. The Hindu-Muslim unity (alliance as Daiya (2008) rightly terms it) collapsed in the 1922 communal riots.

The partition of India was the consequence of growing communalism in the imagination of Indian nationalism. At the time of the declaration of independence, communal riots broke out in India, including the princely state of Hyderabad.[5]

Media and society's dominant narratives about Hindu-Muslim sectarian and communal divisions emerged from the India-Pakistan partition and the collective trauma it created (Raj & Suresh, 2022).

I refer to Shelef & Koo's (2022) pertinent scholarship, which finds that those who promoted the partition as a solution to communal conflict often downplayed the attachment to homelands. In this case, "territory is fungible; any territory will do for the location of the national state as long as it has relatively defensible borders and is reasonably ethnically homogenous" (Chapman and Roeder, 2007; Downes 2001; Johnson 2008; Kaufmann 1996, 1998; Mearsheimer 2000; Saideman and Ayers 2000; Tir 2005; Van Evera 1994 as cited in Shelef & Koo, 2022, p. 419).

In independent India, many more riots further cut deep into the severed communal ties and displayed open hostility towards minorities as the *Other*. Vote bank politics and the aggressive stance of political parties and socio-religious organisations in the Sangh Parivar[6] did not help (Upadhyay & Robinson, 2012). Within the nation, Muslims remain strangers, facing suspicion about their loyalty. Although they are citizens, their acceptance is grudging and conditional. According to Viswanath (2014), the nation's idealistic vision of creating citizens in a complementary relationship inadvertently created *unredeemed minorities.*

The popular discourse characterises Christian converts as betrayers of religious nationalism, viewing them as victims of missionary manipulation (Misra, 2011). National identity, particularly in its prominent Hindu/tva nationalist form, fundamentally depends on the concept of the *Other* (Athique, 2012; Iyer & Das, 2025).

The borders provide the spatial and material framework for distinguishing between the *Other* and the self. This includes imaginative geographies of land alienated by irredentism, secessionism, and partition. However, these geographical claims may not align with the collective sentiments of populations living in those territories.

In India, hyper-nationalism often emerges through the othering and antagonising of outsiders, particularly Pakistanis (Iyer & Das, 2025). This antagonistic *Other* inevitably takes the form of a terrorist—either a Pakistani or an Indian Muslim man sympathising with Pakistan's cause.

Contemporary Formulation of Indian Nationalism

There is no single idea of a national culture in India. Today's nationalism is not the revival of a pre-existing nation but the development of a new concept of nation (Iyer & Das, 2025; Sherfudeen et al., 2024). This is the shift from political nationalism to cultural nationalism that occurred in the 1990s (Sherfudeen et al., 2024; Viswanath, 2014).

Cultural nationalism, a concept proposed by Friedrich Meinecke, is expressed through fine art and literature (cited in Sherfudeen et al., 2024). Cultural nationalism finds roots in religious nationalism and cultural identity, which, in India, is *Vedic Brahmanism*[7] (Daiya, 2008; Viswanath, 2014). Cultural nationalism, hyper-nationalism, and jingoism—along with their inherent othering—become obstacles to a multicultural ethos, leading to a homogenised national culture that promotes a single, dominant version of nationalism (Iyer & Das, 2025; Sherfudeen et al., 2024).

Nationalism in the public sphere demands that its citizens rise above the racial, ethnic, religious, class, caste and gender differences and go through the process of *cultural normalisation*[8] (Chatterjee, 1996). This normalisation occurs in a close nexus between nationalism and populism, deeply embedded in the Indian socio-political system.

Normalisation is achieved by pushing for a one-nation and one-culture policy, propagating myths of oneness and employing strategies of assimilation and exclusion (Iyer & Das, 2025; Sherfudeen et al., 2024). Any constructive critique of the cultural norm is labelled harmful and intolerable. The trend toward hyper-nationalism, cultural normalisation, and populism dominated the arena. In time, "the domain of subaltern politics has increasingly become familiar with, and even adapted itself to, the institutional forms characteristic of the elite domain" and construction of the "hegemonic project of nationalist modernity" (Chatterjee, 1996, p. 224). Cultural nationalism is a process of regenerating opinions by expanding art, music, theatre, and thought (Sherfudeen et al., 2024, p. 125).

Indian War Films

Cinema has emerged as a powerful force in shaping and fostering nationalist ideologies. Sherfudeen et al. (2024) demonstrate how films influence nationalism through multiple channels: shaping public opinion, constructing national narratives, reinforcing cultural values, and creating metanarratives ranging from melodrama and fantasy to divisive social commentary.

Indian nationalism encompasses different interpretations and ideological perspectives. Cinema shapes and reflects these ideas through powerful narratives, visual symbolism, and the manipulation of historical events. It has been used both to promote inclusivity and unity and to incite exclusion and division within India. "Nation in war cinema is an object of fetish" (Viswanath, 2014, p. 92).

More so than in other genres, war films frequently depict the Indian soldier as an icon of national masculinity, highlighting themes of patriotic sacrifice, loyalty, and xenophobia (Gopinath & Deshmukh, 2023). War films consolidate all the influences on nation and nationalism (Viswanath, 2014), including hyper-nationalism. Sherfudeen et al. (2024) note that Indian popular cinema only recently began exploring hyper-nationalism.

Military films in India have become more prominent due to rising nationalistic sentiments, jingoistic ideologies from the 1990s (Gopinath & Deshmukh, 2023), and the development of a national identity (Banga & Kapoor, 2020).

The genre's popularity initially focused on romance and family, but later shifted to more unrestrained representations of war (Gopinath & Deshmukh, 2023). War films often use religious and militaristic nationalism. These "… specific nationalisms have also been present in Hindi films over the years." (Iyer & Das, 2025, p. 161). We can categorise war films based on what they represent: 1) War as an immediate threat, 2) War as a source of historical perspectives, 3) A glorification of heroism, 4) The impact of war (Viswanath, 2014, p. 23).

Indian war cinema's thematic evolution spans from the late 1990s to the 2020s. Two films particularly illuminate this shift: the critically-studied box office hit *Border* (Dutta, 1997) and the contemporary war narrative *Uri: The Surgical Strike* (Dhar, 2019).

Aside from firing up nationalist sentiment, *Uri* creates a positive image of the central government ruling at that time and reinforces the concept of the ultimate enemy: Pakistan (Iyer & Das, 2025).

Border established military personnel as authentic representatives of the nation and patriotism—figures who, as Gabriel (1998) reasons, "stake their lives on their 'national' identity" (p.839). While Uri demonstrates two nationalist themes: militarised nationalism and patriotic familial bonds to the nation (Iyer

& Das, 2025), Border employed two overlapping registers: 1. discursive (nation and religious identity) and 2. affective (personal lives) (Gabriel, 1998, p. 830).

Over decades, the depiction of familial bonds through soldiers' personal lives endures as they relate to the nation. In both *Border* and *Uri,* there are loud expressions of nationalism, which also reflect in contemporary Indian politics where emphatic nationalistic sloganeering proves one's belonging to the nation (Athique, 2012; Gabriel, 1998; Iyer & Das, 2025).

> *Uri articulates the nation in two modes: membership and othering; us versus them; self versus other; and inside versus outside (Higson, 2000, cited in Raj & Suresh, 2022). Uri indicates the shift in close relationship with the populist government, heightening their nationalist appeal. The film's popular dialogue "How's the Josh?" was even used by BJP workers in election rallies (Iyer & Das, 2025). In Border, despite the constant othering, the enemy is humanised in the film. Our glimpses into their private lives reveal they are filled with love and family (Athique, 2012). The film acknowledges that the Pakistanis across the border and the terrorists penetrating it look similar to the Indians, which is not the case in Uri.*

Within these narratives, the Indian Muslims are portrayed as a deceiving anti-national group who find brotherhood with the enemy for no other reason than their shared religion (Athique, 2012). "India-Pakistan border...emerges not as a physical location, but as a psychological condition underpinned by a regime of violence" (Athique, 2012, p. 39).

> *...the terrorist is, therefore, an antagonistic Other who is always presented in a shaggy appearance; armed with heavy weaponry, they work in an underground bunker and are associated with the iconography of the Muslim religion. (Raj & Suresh, 2022, p. 174)*

The academic literature reveals a decisive shift toward populist nationalist narratives that increasingly embrace hyper-nationalism. This shift manifests through eroding trust in minorities and the aggressive rise of cultural nationalism.

However, scholarship on Tamil and Telugu war films remains limited, primarily because these regional film industries have produced very few war narratives and only started doing so recently. This chapter's multidisciplinary analysis of *Amaran* and *Sita Ramam* addresses this gap.

Gender in Indian War Films

War narratives are essentially masculinist discourses, foregrounding male history, valorising male deeds, and portraying war as a ground for proving one's masculinity (Viswanath, 2014). Banerjee (2006) adds that *armed masculinity* and *nation* at a portentous juncture express themselves forcefully to defend the *national community* (p.63).

In media discourse, nationalism, with its violent machismo, is of the masculine domain. Hard physical labour is highlighted as a key aspect of masculinity, with the Indian soldier's uniform celebrated as a signifier of power and patriotism. The military's construction of masculinity centres on the male body's strength, endurance, and toughness. However, Aravind (2023) reveals a paradox: the true source of this lionised masculinity lies not in its physical attributes, but in the symbolic power of the soldier's uniform. War films' battlefield narratives not only contoured national identity and gender roles, but also served as an arena for proving and recovering masculinity (Gabriel, 1998).

Military films evoke a certain sense of national consciousness. They call upon the *microculture of everyday life masculinity* folded into the national culture and serving its militaristic functions (Sharma & Pal, 2023).

However, there is a notable absence of non-upper caste Hindu protagonists (Gopinath & Deshmukh, 2023). Aravind's (2023) analysis of 300 films released from 2013 to 2014 reveals that, except for six lead characters, all other soldier-protagonists ostensibly belong to the upper caste. This also alludes to the nation the soldier carries, which is Hindu and of the Upper Caste.

Hypermasculinity in war films panders to currently prevalent national aspirations and majoritarian attitudes (Raj & Suresh, 2022). Moreover, there is "the circulation of a particular imagination of warfare that complements the 'muscular nationalism' of the contemporary moment" (Aravind, 2023, p. 88; Viswanath, 2014).

> *"Nationalism has typically sprung from masculinised memory, masculinised humiliation, and masculinised hope" (Enloe, 2014, p. 93). Normalising masculinised war imagery renders war as ungendered, embedded in the process of producing invisible masculinities (Aravind, 2023) and pushing women to the margins.*

In war narratives, women frequently inhabit liminal spaces, portrayed as family members celebrating men's exploits, or as victims, widows and orphans (Gopinath & Deshmukh, 2023; Viswanath, 2014). They remain subservient to national objectives. Ironically, this diminutive construction of womanhood coexists with the ideal of the nation as a mother (Bharatmata).

In war films, women are also expected to transcend familial ties to demonstrate loyalty to the nation. A *nationalist love story* of the trio, Bharatmata, and an ideal male and female citizen, compels the collective imagination (Viswanath, 2014). Bharatmata is often called a hallowed figure (Viswanath, 2014), which I would like to contest, as it is overflowing with the symbolism of a mother goddess.

This portrayal reinforces traditional patriarchal gender roles, suggesting that women, sometimes equated with Mother Earth (Viswanath, 2014), need protection within male-dominated environments like militaristic institutions (Gopinath & Deshmukh, 2023).

Hence, according to Hutchinson and Smith, the war films reflect the status of women in a nation-state as: "a) biological reproducers b) reproducers of national/ethnic/racial boundaries c) transmitters of culture d) signifiers of ethnic/national difference e) participants in national/political/military struggles" (as cited in Viswanath, 2014, p.33).

The research on Indian war films pertains to exploring nationalism, masculinity, and femininity on the fringes. The films I examine place women at the centre of their narratives and provide unique insights into the complex relationships between gender, nation and identities.

The scarcity of literature on war films in Telugu and Tamil can be attributed to the fact that films depicting modern military warfare are a recent emergence. This work will add to the research on Indian war films with conceptualisations drawn from a multidisciplinary approach and provide diversity in research on Indian war films from South Indian language film industries.

METHOD

Reception Studies

The chapter uses the cultural studies approach in reception studies to understand the perceptions of the films selected for analysis.

The media plays a crucial role in amplifying dominant public discourse. Media research gave birth to reception studies, which offer the methodological framework for analysing these narratives.

Media texts come alive through audience engagement, as readers and viewers actively construct meaning from their encounters. Livingstone and Das (2013) emphasise their capacity for multiple interpretations rather than fixed meanings. These interpretations always occur within power structures and social dynamics. Livingstone (2019) describes this meaning-making process as 'decoding', shaped by both specific and dominant social contexts.

To elaborate on the importance of contextualisation in reception studies, Staiger (2005) uses arguments of three scholars: Terry Eagleton, Jonathan Culler, and Wolfgang Iser.

Staiger (2005) says that Terry Eagleton describes reception studies as a way to understand how meaning is created based on social and historical contexts. She adds Jonathan Culler's argument that it is not just about interpreting texts. It is also about understanding how people's interpretations change over time based on different cultural codes and assumptions. Finally, she brings in Wolfgang Iser's metaphor, that while the main elements (the "stars") in a literary work remain the same, their connections or interpretations between them can vary.

Reception studies from a structuralist approach moved onto poststructuralism, and Umberto Eco's journey between these two is articulated in recognising the fluidity of meaning of a media text. However, the reader is "implied" or assumed to be "ideal" (Livingstone & Das, 2013). Livingstone and Das (2013) distinguish between interpretation and reception, or rather call for a bridging of interpretation with reception.

It is important to converge empirical audience research with theoretical methodological approaches put forward by scholars such as Umberto Eco and Stuart Hall, as well as approaches by the Frankfurt School and British Cultural studies.

For this chapter, I focus on the latter approach for reception studies. I focus my interpretations within the media's sociocultural contexts and public discourses on social media.

To analyse these films, I draw upon two key approaches in media reception studies: 'text-activated research' as put forward by Barthes, Eco, Genette, and others, and 'context-activated research' developed by Jauss and Leenhardt.

Unlike structuralist research methods like Semiotics and Marxism that treat the text as a closed and whole unit, the poststructuralist methods I have chosen consider texts as living in constant interaction with outside discursive processes.

In the following paragraphs, I will present literature that emphasises theoretical approaches in reception studies, showing how these frameworks can deepen our understanding of audience interpretations within social and cultural media and public discourse contexts.

John Fiske, as cited in Livingstone and Das (2013), explores how texts generate multiple meanings through two interconnected processes: textual polysemy and *audiencing*. These interpretative activities, deeply embedded in cultural contexts, create ongoing cycles of meaning-making and cultural exchange.

Empirical audience research is important, but let us bear in mind John Hartley's critique of the excessive celebration of "audience activity" where it is easy to lose sight of the operation of power (as cited in Livingstone and Das, 2013, p.13).

In the production and reproduction of meanings via media, it is important to understand who owns the cultural capital, who speaks for whom, and why. Post-World War II media research has found that the media has at best a peripheral effect. Audiences consume media to reinforce or facilitate the ideas and predilections they already possess.

To answer the question of whether the media texts alone provide insights into audience perceptions, I will discuss relevant literature.

Metz (1991) opines that "narrative..." in film "...represents one of the great anthropological forms of *perception*" (emphasis added by the author) (pp. 27–28). The perceptual codes in films are derived from existing cultural signs related to filmic objects, spaces and characters. He maintains that films replicate the existing dominant cultural codes into the filmic narratives that enable viewers to read the film.

Hence, *bricolage* has emerged to comprehend the individual reading of the filmic text by filling in the gaps with their connotations with an ad hoc combination of signs (Staiger, 2005). This approach has assumed that spectators have fairly similar cultural experiences.

Poststructuralists counter this by asserting that social institutions (family, religion, educational systems, etc) reproduce particular power structures and fold the individual within them (see Althusser, 1971). This opened space for negotiated and resistant readings of media.

My argument is that the media is intertwined with various social institutional processes. While mainstream cinema mostly reflects dominant discourses and ideologies, it is important to recognise that cinema is inherently polysemous as an art form. This means it can provide alternative spaces for resistance and transgressions.

Of the many approaches Staiger (2005) outlines in reception studies, the sociological approach of conflict theory assists with understanding how privileged and well-resourced groups manufacture certain ideologies and consent through "representational systems."

The audiences, whether "active or passive," and "normative or resistant," depend on the questions of individual or social agency (Staiger, 2005). Who holds power, and who is in control? This chapter focuses on the more decisive influence of the text, emphasising the need to explore the text itself and the potential responses it enables in the viewer.

Other developments in reception studies are also relevant to this chapter. Firstly, the theorisations of the unconscious as laid out in Mulvey's (1989) seminal essay "Visual Pleasure and Narrative Cinema." She argues that the film form is structured by the *unconscious of patriarchal society* (page 6).

Secondly, the ritual function of communication, as proposed by Carey (1989), is helpful to understand the maintenance of the nation's imaginary. The ritual model shows how national belonging operates: it maintains social cohesion, fosters collective identity, and negotiates shared cultural understanding.

Lastly and most importantly, according to Staiger (2005), Stuart Hall revolutionised reception studies by introducing agency, consciousness and pleasure to media research. This, alongside his emphasis on the socio-cultural contexts of decoding in encoding/decoding (Staiger, 2005). Though media texts invite multiple interpretations, there are limits to the pluralities of their meaning.

War cinema's portrayal of national enemies offers rich material for studying the "reception and globalisation studies, pinpointing the gender and cultural politics inscribed into popular culture texts" (Livingstone & Das, 2013, p. 10).

ANALYSIS

Gender in the Military

Sita Ramam

Afreen

Sita Ramam centres on Waheeda, sister to Rasool—a reformed child militant. When Major Abu Tariq, a high-ranking Pakistani military officer, adopts her, she becomes Afreen.

The Indian cinematic imagination often depicts Pakistani men as aggressors and terrorists. Pakistani women, on the other hand, are depicted as kind, as in the film *Roja* (Maniratnam, 1992) and as love interests, as in the film *Veer-Zara* (Chopra, 2004) and *Agent Vinod* (Raghavan, 2012).

In contrast, Afreen imbibes the masculine nationalism of violence and hatred for India and Hinduism. Her militancy first surfaces when she torches a wealthy Indian philanthropist's car in the UK. She embodies traditionally masculine traits—belligerence, pride, aggression—inverting Indian cinema's typical portrayal of the "aggressive Pakistani male."

Her relations with the army stem from her adoptive grandfather, Maj Abu Tariq. She stresses her own otherness through open hostility towards Hindus and India as she searches for Sita to deliver Ram's letter to Sita, which carries a great symbolic value of prominent Hindu Gods venerated across India as the ideal couple.

Afreen is required to deliver a letter in order to pay a sum of 10 lakhs (in Pakistani currency) as an apology for being reinstated as a student at Oxford after burning the car. According to her late grandfather Maj Abu Tariq's will, she can only successfully inherit his wealth if she delivers Ram's letter to Sita. She is metaphorically placed in a war by her grandfather who stresses that the delivery of the letter written by Ram to Sita is crucial, and says, "In this war, Ram and Sita have to win."

The employment of mythological Gods and symbols deliberately points towards Hindu nationalism. Afreen, who, unlike others before her, has loose parallels with the character Trijata in Valmiki's Ramayana. Trijata is a demoness who serves as a companion to Goddess Sita in Ravana's garden, delivering news about Lord Rama.

The film producers compare Afreen to Lord Hanuman,[9] who is legendary for his devotion to Ram. This comparison overlooks the fact that Hanuman is inherently a devotee, while Trijata is a demoness reflecting Afreen's animosity towards Hindus and India.

The media often emphasises Hanuman in cultural discussions because his image is more accessible as a revered God with many famous temples dedicated to his worship. But Trijata remains less represented. This dynamic suggests that stories about characters like Afreen resonate more with established cultural myths rather than the obscure one of Trijata, which seems particularly relevant given her Pakistani Muslim heritage.

Afreen's Muslimness is intermittently stressed with her donning a headscarf. This is a positive representation of Pakistani Muslim women departing from the image of the burkha and its associated stereotypes of illiteracy and oppression. She wears her headscarf by choice.

In one of the final scenes, Afreen realises her loyalty to Lt Ram when she discovers he had saved her twenty years ago when she was called Waheeda. And she runs from the military interrogation dungeon into the light towards Sita, who has read Ram's letter. As she does so, her scarf slips off in the wind, and sunlight hits her face. As the scarf falls away while she hugs Sita, the transformation of a Pakistani Muslim woman is complete.

Symbolically, she has also freed herself from the prevailing global perception of the Hijab as an oppressive Islamic practice. Ironically, many Muslim-majority countries have banned the Hijab.[10]

The Muslim woman's progressive stereotype and its subterfuge in the conclusion return to familiar Indian patriarchal nationalism. Her femininity is restored by discarding the scarf at the military award ceremony in honour of Lt Ram.

Sita/Noorjehan

Sitamahalakshmi, the love interest of Lt Ram, is a disservice to the progressive representation of Afreen just before the climax. Sita pretends to be the wife of Lt Ram in a series of letters after he reveals that he is an orphan during an All India Radio program.

Mythological metaphors provide a sacred nature to Lt Ram and Sitamahalakshmi, who are named after the deified characters of noble Prince Ram and his romantic partner, the chaste Sita. In the film, Sita describes her first encounter with Ram in a letter: "Ravana's[11] annihilation in the *Kurukshetra*[12] war and in the shimmer of the battle, Sita's *Swayamvaram*[13]."

The *Kurukshetram* is the attack of Kashmiri Muslims on Kashmiri Pandits, and when Kashmiri Muslims come across Sita alone, they try to assault her. The attackers are Ravana, whom Lt. Ram defeats, and the *Swayamvaram* is when she chooses Ram to be her husband. The quote from the film is filled with metaphors alluding to the actions of a soldier performing his duty, protecting civilians, equated to the Gods in Hindu epics.

Comparisons to Hindu Gods belie the secular nature of the Indian army soldier, which is firmly planted in Hindu nationalism.

The use of Hindu imagery in cinema can be traced back to Dada Saheb Phalke's films, which created a unique Indian identity by invoking Hindu gods and their mythologies. This differs from colonial movies inspired by films from Western countries (Dennis, 2020; Shoesmith, 1988).

I further discuss the public perception and audiences' meanings woven into the representation of Ram and Sita in the films in the section on public perception/reception.

Sita's screen appearance comes in the 48th minute of the film. The delayed reveal gives her an enigmatic aura. This is a narrative device frequently used to elicit a sense of mystery. Sita's unattainability and ethereality are further enhanced at the interval point when Sita is revealed to be Noorjehan, a Muslim Princess from Hyderabad. The unattainable princess falls only for the unimpeachable keeper of *dharma* (righteousness), *desam* (land) and fulfiller of *dharmam* (duty).

Sita represents the ideal Hindu woman, cultivating the inner spirituality that Chatterjee (1989) identifies as central to modern patriarchal expectations. She embodies the feminine virtues of "chastity, self-sacrifice, submission, devotion, kindness, patience and the labours of love," which are also reflected in "her dress, her eating habits, her social demeanor, her religiosity," leading the woman of the nation to a "legitimate subordination" (Chatterjee, 1989, p. 629).

Sita/Noorjehan dresses in a Hindu-coded appearance of sarees, half-sarees, and skirts with modest, collared blouses, and a *bindi*[14] on the forehead, which is the mark of a Hindu woman.

She fashions a *bindi* out of red vermilion that falls on her while reminding herself of Ram's marriage proposal. Ram comments on her lack of *bindi*, saying that she looks beautiful even without it.

She practices Bharatanatyam, a classical dance form with roots in Hinduism. The art is steeped in Hindu symbolism, and its performances are mostly narratives of Hindu religious stories, mythologies and Gods. She even reads the *Natyashastra*, the ancient book on Bharatanatyam.

In the aspirations of Indian nationalism, the ideal Muslim woman is a Hinduised woman. To complete her Indianisation (Hinduisation), the Muslim leading lady defiantly demands that her brother call her Sita.

In discussing the Indian Muslim woman Noorjehan (Sita) and the Pakistani Muslim woman Afreen, I bring in Chatterjee's (1989) relevant point of how a new woman in the nationalist discourse excluded Muslims. In extension to his argument, there were no interventions or changes allowed in the Muslim family and home terrains. Muslim women will get space in Indian (Hindu) nationalism only if they shed all their Muslimness.

Amaran

Indhu

Indhu Rebecca Varghese (from here on referred to as Indhu), in *Amaran*, is the keeper of her husband Major Mukund Varadarajan's (from here on referred to as Mukund) heroics as an Indian soldier.

The film opens with Indhu's solipsism. She describes her emotions as deep and vast as the ocean, her memories are ripples, and the distance between her and her husband, Major Mukund, like sea and sky. The opening places the female lead at the centre of the narrative, promising a compelling story of war and its impact on a soldier's wife and daughter.

Indhu is demure, while Mukund is assertive. Her initial promise as the locus of the war film is denied by her portrayal as a passive biographical narrator of Mukund's life as an army officer.

Indhu offers no nuance in her feelings towards her life as an army wife. Her emotions oscillate in reaction to Mukund's army life, swinging between fear (of his death) and sadness (from the threat of his death). She is not depicted as an individual separate from her husband. Her commentary and insights as a woman are not offered. Her existence in the film comes into being from her meeting with Mukund. While Mukund's dreams of being an army officer and his mission are constantly repeated and reinforced throughout the film, her individual aspirations are lacking, only to merge into Mukund's.

Indhu calls him her "baby despite his big post as Commander Mukund," fitting the role of chaste nurturer. Mukund's masculinity complements her femininity. To refine her femininity a bit more, Mukund helps her transform from a tomboy dressed in a man's clothes and dirty shoes into a saree-clad woman.

Christian Indhu is visiting a temple to pray to his favourite Hindu god during his interview for selection into the army. Christian identity is to be sacrificed at the altar of the nation for the soldier coded in Hindu markers. She finds a job to help him pay for the house that he wants to buy for his parents.

A light-hearted discussion about films and actors between Mukund and his fellow soldiers is interrupted by an ambush. Indhu, on the phone with Mukund, listens to guns blazing and bombs blasting. She remains a helpless and passive witness to her husband's life in danger, and begins to pray. Indhu's silence and passivity are necessary for aspirational masculine nation-building and Hindu-coded nationalist discourse.

Mothers and Sisters in Sita Ramam and Amaran

The role of the mother, Geetha Vardarajan, in *Amaran,* is a departure from the usual depictions of the sacrificial and patient mother sending her hero son off to lay his life for the nation. She adds much-needed diversity to the representation of mothers in war films.

Her character provides a refreshing response to Mukund's appeal to her nationalism and patriotism. He cries, "Mother, please think of the mothers of 11 lakh Indian soldiers of the country!" She retorts, "Why should I? I will think of other crores of mothers of sons who do 9-5 jobs?" It is from Geetha's role that we hear more about the complex emotional turmoil of a soldier's mother. Her character successfully rejects the hegemonic nationalism that dictates the virtues of a warrior's mother. She provides an alternative nationalistic discourse for women and mothers.

The sisters' roles in *Amaran* are peripheral and have minimal dialogue, whereas in *Sita Ramam* the brief appearance of an adoptive sister, Radhika, is more substantial.

She writes a letter to Lt. Ram pretending to be her sister, and when he visits her, he finds that she is a sex worker. Her unwilling entry into sex work—forced by traffickers—and her shame about her profession make her worthy of Ram's rescue. These qualities mark her for restoration within nationalism's new patriarchal framework.

Modern patriarchy constructs an ideal woman in contrast to her 'common' counterpart, marked by a lower caste and class. Chatterjee (1989) lists some examples of common women: "maidservants, washing women, barbers, peddlers, procuresses, prostitutes" (p.627).

The discussion of women in gender in war films is comprehensive when the men and their masculinities are discussed. I discuss them in later sections.

Public Perception and Reception of Women

Nationalist discourse relies heavily on women's symbolism casting them as embodiments of nation because women have become the synecdoche for the nation (Daiya, 2008), especially its cultural and religious spirit.

Despite their significance in national imagination, "...women do not speak here. It is a discourse which assigns to women a place, a sign, an objectified value; women here are not subjects with a will and a consciousness" (Chatterjee, 1989, p. 632).

Afreen, Sita/Noorjehan and Indhu are reduced to being carriers and upholders of this nation and nationalism via the affective relations with the army men intersecting with their lives.

The portrayals of femininity often serve to reinforce and reflect masculine identities. Women's representations are shaped by and contribute to the understanding of masculinity, rather than existing independently. As Gabriel notes,

The semiotic burden of the affective is borne primarily by figures of femininity; the discourses they generate congeal around figures of masculinity. Figures of femininity are thus emptied into vehicular signs that record a history of masculinities, their construction and significance. (Gabriel, 1998, p. 831)

This portrayal of Hinduised minority women in war films becomes a fundamental cultural reference point feeding into cultural normalisation in the nationalism discourse.

The discourse on social media (Instagram and YouTube) supports this sentiment. Many Instagram reels and YouTube videos discussing *Sita Ramam* refer to the female lead as Sita and not Noorjehan, which is her real name.

Of all the reviews, discussions and appreciation posts, one Instagram post by a user *@telugintiammayi* (Telugu girl) sums up the perception of gender in the film *Sita Ramam.*[15]

In her edit, she compares Lt Ram's devotion to India to Lord Rama's devotion to Ayodhya, for which he sacrifices his love for Sita. She ends with the question of who is the greatest? Was it Lt Ram who sacrificed his love for the country, or Sita who sacrificed her identity and wealth for Ram? She thanks the director Hanu Raghavapudi for creating a film with the Gods that *we* worship (italics added by me for emphasis). This indicates that the leads in the film are another iteration and story of the couple and Gods, Sita and Ram, in the Hindu mythology of Ramayana.

Like TV news journalists (Pandit & Chattopadhyay, 2017), the creator's use of 'we' assumes a Hindu national identity that idealises its military and sees soldiers as heroes. The Indian Army is widely perceived as a "symbol of state power, a flag-bearer of national resolve, and an enforcer of national will" (Parashar, 2018, p. 128)

On YouTube, I analyse the reviews on *Sita Ramam* with the top views. The second-highest-viewed review by user *Pranavreacts*[16] observes that the theatre only had couples in the corner seats, alluding to their sexual intimacy. He ends the review by saying that the couples seated in corner seats later moved to the front seats, and that, as portrayed in the film, love is not just about sexual intimacy.

Echoing his sentiments, many comments on YouTube and Instagram were iterations of the phrase, "no nudity, no nudity, no vulgarity…… true/real/natural love" as seen in the screen grab (Figure 1) from the top-viewed video on YouTube.

Figure 1. Sita Ramam review comment section

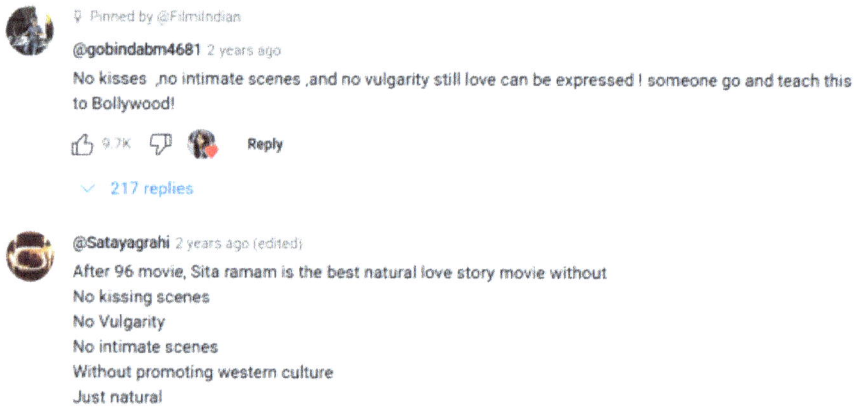

Source: YouTube

The viewers praise Sita's chastity and sacrifice. Reflecting the findings in literature review, the public discourse cements the notion of femininity as the vehicle of masculinity in the military. Similarly, Indhu's understanding and devotion to her husband, and her stoicism upon learning about his martyrdom, are praised, with appreciative posts receiving high levels of engagement.

Society demands that women embody sacrifice in their roles as wives and lovers. This ideal found its perfect symbol in Himanshi Narwal—her stoic vigil beside her slain husband, a naval officer killed in Pahalgam's terror attack, captured the nation's imagination as her image spread across social media.

This didn't last long. Her pleas to refrain from spreading hatred and targeting Muslims and Kashmiris evoked extreme vitriol from right-wing groups and individuals.[17] She had deviated from society's expectations of a grieving army widow. She was vilified, and her chastity was immediately questioned. Her secularism was equated to illicit relations with Muslims.

Military, Muslim, and Terrorist Masculinities

British colonisers deliberately feminised Indian men. Yet their imperial aggression ironically sparked the rise of a distinctly Hindu masculinity—an ideal that emerged as anti-imperial resistance (Aravind, 2023; Banerjee, 2006). This benevolent Hindu warrior archetype prominently shapes both *Sita Ramam* and *Amaran* (Aravind, 2023; Banerjee, 2006). I will begin with Hindu military masculinities.

Sita Ramam: Ram

Lt Ram in *Sita Ramam* despises communal discord and strives to win over antagonised Muslims who were subjected to manipulation. When his superior suspects five persons with mules to be terrorists, he orders Lt Ram to shoot, but Lt Ram defies his superior's order, justifying that he felt in his gut that those Muslims were not terrorists. He states that he would rather face punishment than kill an innocent.

The benevolent and noble Hindu military masculinity can differentiate between good and bad Muslims. This is possible as the national imagination and hegemonic nationalism discourse put a Hindu man, especially a military Hindu man, in an ideal place. As the product of the codes and socio-cultural values embedded in a nation and nationalism, he is familiar with the terms of exclusions and inclusions into nationalism. Hence, the most privileged militarised Hindu masculinities, compared to other masculinities (Hindu civilian and then minorities), can authoritatively recognise the belongingness of good Muslims.

On the ground, Muslims can only aspire to be considered fully Indian, patriotic and nationalist. Their loyalty to the nation is constantly questioned. After the Pahalgam attack, there were reports that the Muslim community was targeted, harassed and assaulted.[18] Many Kashmiri students at various educational institutes had to flee,[19] and a 15-year-old Muslim schoolboy was assaulted and forced to urinate on the Pakistan flag.[20]

Unlike the Muslim soldiers, Ram is the ideal national masculinity. His idealism is reinforced with his statement, "I am an orphan, but I never got angry at my mother", alluding to Bharatmata. He says this while languishing in a Pakistani prison, where a Pakistani army official asks him whether he is angry that India has abandoned him.

Aravind (2023) rightly points out that the military masculine has overthrown other forms of non-state and civilian masculinity in the Indian cinema landscape. It is reflected in the polarising discourse since the BJP's win in 2014. A study of 779 news articles on demonetisation found that those in favour of demonetisation had the highest mention of values of bravery and correlated it with being an Indian soldier (Prateek et al., 2017).

Social media weaponised military sacrifice—contrasting soldiers standing eight-hour border watches against civilians who complained about demonetization queues. These posts shamed dissenters, suggesting that enduring bank lines pales against defending the nation. These posts undermined any public criticism of the government's mishandling of demonetisation.[21]

Amaran: Maj Mukund

In *Amaran,* 44 Rashtriya Rifles, the Cheetah Company base is shot in a sequence to depict the diversity and various backgrounds the soldiers come from. All soldiers are shown doing everyday activities of talking to their mothers and wives, playing carrom, bathing and other mundane routines. Then, the camera pans briefly to show two Muslim soldiers offering prayers, with the *azaan* playing in the background. They are removed from the humdrum activities of their mates. Everyone else is perceived to be Hindu. Only *namaaz* marks the distinct identity of a Muslim man.

In the film, a fundamental divide shapes Indian religious identity: Muslims remain perpetually marked by their faith, while Hinduism weaves invisibly through everyday life. This contrast emerges clearly in military narratives, where Hindu soldiers move fluidly between civilian and military worlds while Muslims remain sequestered in religious otherness.

The performance of a benevolent Hindu military requires good Muslims, such as local Kashmiris Waheed and Saifuddin in *Amaran*, who share their knowledge of the harsh terrains of Shopian, Pulwama, and Tral with Major Mukund.

Good Muslims' loyalty to the nation is rewarded. When an informer passes away, his family is financially compensated for their loss. Waheed is rescued when he is captured by terrorists.

A Kashmiri soldier, Saif, accidentally fires the gun that hits Mukund, and he, as a true leader, insists upon shifting to a Kerala hospital to keep Saif safe from inquiries. He wants more men from Kashmir to join the Indian Army, and he wants Saif to set a good example for Kashmiris, thus providing an idea of nationalism inclusive of Muslim masculinities.

Mukund's battalion's battle cries are "Bolo Bajrang Bali Ki Jai[22]" and occasionally, "Har Har Mahadev[23]", which the Muslim soldiers are seen repeating.

Consider Subedar and Maulvi Ishrat Ali of Rajputana Rifles, who, for 22 years, chanted "Jai Hind." He was censured in 2014 for not chanting "Ram Ram" and "Jai Mata Di" (Ohri & Sharma, 2014). This produces subjective and biased binaries regarding Muslim soldiers. The Muslim soldiers in the film are good as they have risen above "the narrow-mindedness and religious fundamentalism"[24] that Ishrat Ali allegedly displayed, according to his superiors.

Indian Muslim identity is articulated through "*ghar* (home/family), *qaum* (community/faith) and *mulk* (homeland/nation)" (Vazira Fazila-Yacoobali, 2002, pp. 183-198, as cited in Athique, 2012, p. 36). Good Muslims are obedient and prove loyalty by constantly and repeatedly placing themselves in danger and sometimes going against their Islamic tenets by choosing *mulk* above *ghar* and *quam*.

The Army

The opening shot of *Sita Ramam* establishes the spatial and temporal context, indicating that the year is 1964. The location is abbreviated as POK, which, for an Indian audience, clearly refers to Pakistan-Occupied Kashmir.

Before the words "Terrorist Camp" appear on the screen, Indian citizens familiar with Indian nationalism can anticipate that the characters depicted will be terrorists. Displaying "Terrorist Camp" is unnecessary for the seasoned Indian audience — they recognise the dressing, names, manners and markers of Muslim men and terrorists.

Terrorism is conflated with Muslims and Pakistan, and the recent military films deal with fighting terrorism rather than with other nations. This shift started from the 1990s in war and military action cinema, where the terrorist is the villain, and terrorism is a grand system of evil (Ganti, 2004). Contemporary war films present a "more violent version of nationalism" (Raj & Suresh, 2022, p. 182), battling the evil of terrorism.

This violent version demands nation's sons and idealised masculinities (Mukund and Ram) to sacrifice their love and lay down their lives for the nation. In cinematic images, "death is eroticized in the discourse of war as the dead are invariably young men" (Viswanath, 2014, p. 21). The aspirational nation with these aspirational masculinities becomes hegemonic, feeding into the national imagination.

Sharma & Pal (2023) contend that this nationalistic culture's hegemonic masculinity produces the other and manifests the nation in the man himself. This is contradictory to the notion of the nation as Bharatmata. The nation is constructed by the collective microculture of hegemonic masculinity imbued with symbolic value. Its populist performativity leads to cultural normalisation.

The symbolism and rhetoric in mediating and legitimising war are also well-documented in the research by Devakumar (2025), Lakho et al. (2025), Musarat (2024), Pandit and Chattopadhyay (2017), and Sharma and Pegu (2023).

Pandit and Chattopadhyay (2017) commented on TV news journalists' symbolic and metaphorical descriptions of Diwali (as part of commentary over a successful surgical strike). The Indian defence minister called the Indian Army Hanuman, and there were election banners depicting Modi as Ram and the Prime Minister of Pakistan as Ravana.

The name *Operation Sindoor*[25], in retaliation for the Pahalgam terrorist attack, is heavy with symbolism. The attackers selectively killed Hindu men, leaving the women and children alive. It alludes to the *sindoor* that was wiped off the foreheads of the surviving wives of the terrorist attack.

Similarly, the affective appeals of the Indian government and the Indian Military to gather support for war were witnessed in the aftermath of the Pulwama-Balakot terrorist attack.

The media rallied in support of retaliation, using emotionally-charged rhetoric. There were titles in the news coverage of the surgical strike like "Modi-ka Masterstroke" ("The Masterstroke of Modi"), "Modi Ka Muhthod Jawab" ("The Ruthless Reply of Modi") and "Modi ka Badlapur" ("The Revenge of Modi") (Pandit & Chattopadhyay, 2017, pp. 168–169).

Besides the surgical strike, in the Pulwama-Balakot conflict as well, the use of hashtags (#SurgicalStrike2, #IndiaStrikesBack), reliance on military visuals converting TV studios into war rooms, and media is in direct alignment with the ruling government (Devakumar, 2025; Pandit & Chattopadhyay, 2017; Sharma & Pegu, 2023). The news coverage also highlights the underlying phallocentric logic present in the dominant nationalist discourse.

Media (including news channels, social media and films), particularly during periods of national conflict, is characterised by an *assertive nationalistic rhetoric* (Pandit & Chattopadhyay, 2017) and frequently celebrated and associated with national pride (Parashar, 2018). It frequently includes a blend of militant nationalism along with the ideological labelling of the discourses from particular communities (Banga & Kapoor, 2020; Pandit & Chattopadhyay, 2017).

With the elevation of Prime Minister Modi as the justice-bringer and the equivalent of Ram commanding the army, the nation's cultural nationalism and masculine-imbued values also impose a performativity on other genders, castes and religious groups.

With 'Bharatmata', the nation is also coded in this familial relation to a female figure. Likewise, in *Amaran,* Mukund's training officer likens the homeland and the Line of Control to the soldiers' mothers' *saree pallu*[26]. In both films, they prove their physical prowess on the battlefield and in conflict zones by their victory over the terrorists.

Both Mukund and Ram take pride in the uniforms they wear. Mukund's ego is wounded when Indhu's father rejects him as a suitor because of his military background. Meanwhile, Ram frequently refers to himself as a soldier, conveying that nothing can deter him and that he can overcome any challenge. He holds on to this thought even when he faces silence from Sita regarding their uncertain romantic future. The characters use that military masculinity prowess to win over good Muslims and provide retribution to evil Muslims who are terrorists.

For a Hindu military masculinity to perform, the *Other* – Muslim masculinities, both good and evil, need to exist, so that they can be outperformed.

Public Perception of the War and the Army in Amaran and Sita Ramam

In social media, the public discourse favours the Indian Army. Surveys consistently show that Indians have more faith in the Indian military than any other institution, including traditional media (Seth et al., 2022).

It is a fact that the military demands discipline and sacrifice, and a soldier's life is a tale of heroism, suffering, victories and losses (including loss of life). The respect and reverence soldiers command are not unique to India. The comments engaging with the top-viewed film reviews on YouTube and Instagram reels of *Sita Ramam* and *Amaran* deeply empathise with the sacrifices of the soldiers and the grief of army widows.

To capture the different interpretations of the film, I analysed Reddit discussion threads. I chose Reddit because it is non-mainstream, user-moderated, anonymous and enjoys reduced censorship, with a low risk of doxing[27] (only in extreme cases). This allows alternate thoughts and ideas to be expressed freely. The popularity of an opinion is reflected in the number of upvotes and downvotes a user receives on their comment or a post.

Sita Ramam and Amaran Discussion on Reddit

A user's comedic take on the film *Sita Ramam* in Figure 2 was appreciated with the most upvotes.[28]

Figure 2. Reddit post on Sita Ramam; Source: Reddit

There is consensus and tacit agreement on Indian subreddits to support the Indian Army and back any Indian soldier in almost all the posts on the film. Contradictory comments are downvoted heavily and criticised by other users.

This post has a different take on how *buhkts*[29] rejoice that the soldier is a Hindu named Ram, but are disappointed that Rashmika (playing Afreen) is a Pakistani, and Sita in the film is a Muslim princess. Reactionary anti-Islamic rhetoric is rife among reactions to this post, as shown in the figure 3 below.

Figure 3. Comment on a Reddit post on Sita Ramam; Source: Reddit

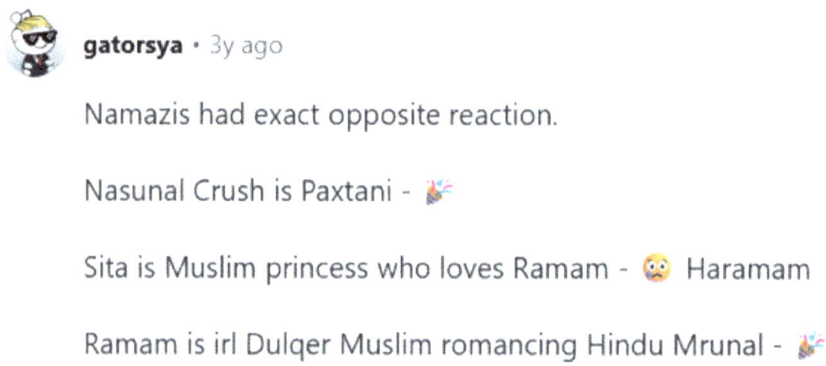

The disparaging word Namazis is used for Muslims. Other discussions on the film's potential to be a pan-Indian film are also met with the same Islamophobic rhetoric.

In their objections to the positive depiction of the India-Muslim brotherhood in the film, they called it *pseudosecular* and *Hinduphobic antinational Urduwood influence* in South Indian films. This tracks with the prevailing dominant discourse that alienates the *Other* based on religious identities. This line of discourse follows an aggressively polarised stance towards "militant aggressive nationalism" centring on "fraught identity politics" (Pandit & Chattopadhyay, 2017, p. 174)

The film discussion threads on *Amaran* were extensive and more nuanced, with many opinions offered by users in Tamil film subreddits. The consensus here was also the unconditional support for the Indian Army and empathy for the actual martyr Major Mukund Varadarajan. There were users objecting to the focus on Indhu, which took away from the storytelling about Maj Mukund's army life, and its potential to be an action-packed war film glorifying the Indian soldier.

This near-unanimous support for the Indian military during ongoing conflicts and wars show citizens fully embrace military logic and ethos, not only to protect state sovereignty but also for civilian resistance. Security and development are presented as a quid pro quo arrangement where citizens participate in securing the state in return for development benefits. This is reinforced by public campaigns, such as Prime Minister Modi's request for people to write to soldiers on Diwali or applaud them at airports, bringing militarism into the social mainstream (Parashar, 2018).

In social media discourse, it is clear that the audience had an affective reaction to the films. Musarat (2024) says that wars are waged while surrounded and supported by *mediated discourse.* Generally, the language and communication enabling the discourse, and directing and influencing public perceptions about war, are tightly controlled. Here, emotions can legitimise war and gather support:

> *Emotions are a legitimate way of shaping individual and mob responses into an institutional form of identity. The political shaping of a mob and individual can be transformed linguistically to concentrate emotions for specific responses. (Musarat, 2024, p. 78)*

Military narratives weaponise emotion—celebrating bravery, patriotism, and battlefield sacrifice while showcasing families' raw grief. This pageant of noble suffering transforms war into virtue and brands peace as cowardice.

When the India-Pakistan ceasefire was announced after several days of *Operation Sindhoor*, Indian Defence Minister Vikram Misri was subjected to intense cyber-bullying and trolling, and blamed for ending the war. His daughter was doxed, and her phone number and other private details were leaked.[30] Minister Vikram Misri had to lock his X account after unceasing attacks on him and his family.

Hence, films with militaristic overtones contribute to a sense of national pride and technological achievement (Mishra & Desai, 2025). Public sentiment is amplified through videos of crowds celebrating military actions and emotional appeals from soldiers' families (Devakumar, 2025).

The military's superior authority and impunity in conflict zones is doubtlessly widely accepted. This is exemplified by acts like the Armed Forces Special Powers Act.

CONCLUSION

Women from various ethnic backgrounds, whether Hindu, Muslim, or Sikh, are often seen as properties of the nation-state and their communities, reinforcing male dominance (Daiya, 2008). Despite the Hinduisation of minority women in war films explored, they are simply relocated from one patriarchal regime to another nationalistic patriarchal regime.

For an outsider, India and Pakistan share a united heritage till the partition of 1947. The memory of that communal trauma is part of the common history that continues to define the present.

The constant geopolitical tug-of-war over Jammu and Kashmir keeps the enmity alive. Vehement communal rhetoric in Pakistan and India intensifies the othering of both sides. In the process of othering Pakistan, India also marginalises the Muslims in India, who are viewed as threats to nation-building.

These films reflect how national discourse, centred on military conflict, systematically villainises India's Muslim population—a pattern evident in vitriolic reactions to the recent Pahalgam tourist attacks. This aligns with the notion that "'individuals' hawkish stance on territorial disputes could be triggered by historical ownership frames" (Shelef & Koo, 2022, p. 424).

Nationalism is shaped and communicated through different cultural forms, influenced by state and media strategies to gain public support (Bharucha, 1999). This process involves creating a shared sense of identity and belonging among people.

The social media discourse around the war films explored here implies that the media is vital to spreading government-propagated ideological beliefs and values (Banga & Kapoor, 2020). A benevolent India dichotomously with an aggressive Pakistan (Verma & Roy, 2022) is positioned even in well-intentioned films like *Sita Ramam* (which has a far more positive depiction of Muslims and Hindu-Muslim brotherhood) and *Amaran,* which continue to reiterate the divisive binaries of Partition.

As we have seen, nations emerge through fictional worlds that feel intimately familiar to readers. This analysis explores national identity through exclusion—examining what 'our' leaves out rather than what 'our' contains.

Paradoxically, even strongly nationalist cinema can embed counter-narratives within its patriotic framework, challenging contemporary nationalist discourse. This duality appears clearly in the character contrasts: Geetha Varadarajan maintains a consistent counter-narrative voice, while Afreen's character ultimately aligns with hegemonic nationalism. Future research should explore audience perceptions of war and nationalism in cinema through deeper ethnographic methods.

REFERENCES

Althusser, L. (1971). Ideology and Ideological State Apparatuses (Notes towards an Investigation) (B. Brewster, Trans.). In *Lenin and Philosophy and Other Essays* (pp. 127–186). Monthly Review Press.

Anand, C. (Director). (1964). *Haqeeqat* [Film].

Anand, C. (Director). (1973). *Hindustan ki Kasam* [Film].

Anderson, B. R. O. (2006). *Imagined communities: Reflections on the origin and spread of nationalism* (Rev. ed). Verso.

Aravind, S. G. (2023). The gender of war: National masculinities and Hindutva in Bollywood war cinema. In *Historicizing Myths in Contemporary India* (pp. 81–101). Routledge India., https://www.taylorfrancis.com/chapters/edit/10.4324/9781003363149-5/gender-war-aravind DOI: 10.4324/9781003363149-5

Athique, A. (2012). Aggression and Transgression on the India-Pakistan Border. In *Filming the Line of Control* (pp. 21–39). Routledge India., https://api.taylorfrancis.com/content/chapters/edit/download?identifierName=doi&identifierValue=10.4324/9780203150351-3&type=chapterpdf

Banerjee, S. (2006). Armed masculinity, Hindu nationalism and female political... - Google Scholar. *International Feminist Journal of Politics*, *8*(1), 62–83. DOI: 10.1080/14616740500415482

Banga, B., & Kapoor, V. (2020). Representation of Political Ideology through Bollywood Movies: A Critical Study of the Movie "Uri-The Surgical Strike.". *Journal of Xi'an University of Architecture & Technology*, *XII*(III), 2899–2905.

Bharucha, R. (1999). *In the name of the secular: Contemporary cultural activism in India*. Oxford University Press., https://cir.nii.ac.jp/crid/1130282272976718720

Bhutia, U. (2017). *The Role of Media in the Emergence of Hindu Nationalism in Indian Politics* [Sikkim University]. https://dspace.cus.ac.in/jspui/bitstream/1/5985/1/Ugen%20Bhutia-History-PhD.pdf

Breuilly, J. (1985). Reflections on Nationalism. *Philosophy of the Social Sciences*, *15*(1), 65–75. DOI: 10.1177/004839318501500106

Carey, J. (1989). A Cultural Approach to Communication. In *Communication as Culture: Essays on Media and Society* (pp. 13–36). Unwin Hyman.

Chatterjee, P. (1989). Colonialism, nationalism, and colonialized women: The contest in India. *American Ethnologist*, *16*(4), 622–633. DOI: 10.1525/ae.1989.16.4.02a00020

Chatterjee, P. (1996). Whose Imagined Community? In Balakrishnan, G. (Ed.), *Mapping Nation* (pp. 214–225). Verso., https://journals.sagepub.com/doi/10.1177/030582989102000306O1

Chatterjee, P. (2012). Nationalism Today. *Rethinking Marxism*, *24*(1), 9–25. DOI: 10.1080/08935696.2012.635023

Chopra, Y. (Director). (2004). *Veer Zara* [Film].

Daiya, K. (2008). Re-Gendering the Nation: Masculinity, Romance, and Secular Citizenship. *Violent Belongings: Partition, Gender, and National Culture In.*

Dennis, E. (2020). *How Hinduism Rooted Itself in Film.* https://scholarlycommons.obu.edu/scholars_day_conference/2020/christian_studies/1/

Devakumar, Dr. T. (2025). Between the Headlines: A Comparative Study of Indian and International Television News Perspectives on the Pre-War India-Pakistan Tensions. *INTERNATIONAL JOURNAL OF SCIENTIFIC RESEARCH IN ENGINEERING AND MANAGEMENT, 09*(06), 1–9. DOI: 10.55041/IJSREM50221

Dhar, A. (Director). (2019). *Uri: The Surgical Strike* [Film].

Dutta, J. P. (Director). (1997). *Border* [Film].

Enloe, C. (2014). *Bananas, beaches and bases: Making feminist sense of international politics*. Univ of California Press. DOI: 10.1525/9780520957282

Fanon, F. (2000). The Wretched of the Earth. In Badmington, N. (Ed.), *Posthumanism* (pp. 23–26). Macmillan Education UK., DOI: 10.1007/978-1-137-05194-3_4

Gabriel, K. (1998). Manning the Border: Gender and War in 'Border'. *Economic and Political Weekly*, •••, 828–832.

Ganti, T. (2004). *Bollywood: A guidebook to popular Hindi cinema.* Routledge. https://www.taylorfrancis.com/books/mono/10.4324/9780203643945/bollywood-tejaswini-ganti

Gopinath, S., & Deshmukh, R. (2023). *Historicizing Myths in Contemporary India: Cinematic Representations and Nationalist Agendas in Hindi Cinema.* Taylor & Francis. DOI: 10.4324/9781003363149

Hall, S. (1993). Culture, community, nation. *Cultural Studies*, *7*(3), 349–363. DOI: 10.1080/09502389300490251

Iyer, J. B., & Das, S. (2025). Leveraging history to invoke nationalism: From the annals of history to social engineering of present and future in Hindi cinema. *National Identities, 27*(1–2), 155–174. DOI: 10.1080/14608944.2024.2365914

Kulasekaran, V. (Director). (2021). *Shershaah* [Film].

Lakho, M. B., Siddiqui, H. A., Bakhtawar, S., & Soomro, M. A. (2025). Media Narratives and Public Perception during Cross-Border Tensions: A Comparative Study of Pakistan and India. *Review Journal of Social Psychology & Social Works, 3*(2), 748–765. DOI: 10.71145/rjsp.v3i2.234

Livingstone, S. (2019). Reception Studies. In *The Blackwell Encyclopedia of Sociology* (pp. 1–3). John Wiley & Sons, Ltd., DOI: 10.1002/9781405165518.wbeosr031.pub3

Livingstone, S., & Das, R. (2013). *Interpretation/Reception* (pp. 9780199756841–0134) [Dataset]. DOI: 10.1093/obo/9780199756841-0134

Maniratnam, G. R. (Director). (1992). *Roja* [Film].

Metz, C. (1991). *Film language: A Semiotics of the Cinema* (Taylor, M., Trans.). Univ. of Chicago Press.

Mishra, S., & Desai, H. (2025). The Media Framing of India's Nuclear Weapons Discourse. *International Journal of Nuclear Security, 9*(1). Advance online publication. DOI: 10.7290/ijns09153822

Misra, A. (2011). The Missionary Position: Christianity and Politics of Religious Conversion in India. *Nationalism & Ethnic Politics, 17*(4), 361–381. DOI: 10.1080/13537113.2011.622635

Mulvey, L. (1989). Visual pleasure and narrative cinema. In *Visual and other pleasures* (pp. 14–26). Springer. DOI: 10.1007/978-1-349-19798-9_3

Musarat, Y. (2024). (De-)Legitimizing War: A Linguistic Analysis of Indian and Pakistani Civil and Military Conflict Discourses. *Critical Military Studies, 10*(1), 61–82. DOI: 10.1080/23337486.2023.2188997

Musolff, A. (2021). *National Conceptualisations of the Body Politic: Cultural Experience and Political Imagination.* Springer., DOI: 10.1007/978-981-15-8740-5

Nihlani, G. (Director). (1982). *Vijetha* [Film].

Ohri, R., & Sharma, R. (2014, September 18). Maulvi in Army allegedly censured for using slogan that sends 'message of extremism.' *The Economic Times*. https://economictimes.indiatimes.com/news/politics-and-nation/maulvi-in-army-allegedly-censured-for-using-slogan-that-sends-message-of-extremism/articleshow/42747816.cms

Pandit, S., & Chattopadhyay, S. (2017). Coverage of the surgical strike on television news in India: Nationalism, journalistic discourse and India–Pakistan conflict. *Journalism Studies*, *12*(2), 162–176.

Parashar, S. (2018). Discursive (in)securities and postcolonial anxiety: Enabling excessive militarism in India. *Security Dialogue*, *49*(1–2), 123–135. DOI: 10.1177/0967010617746527

Periasamy, R. (Director). (2024). *Amaran* [Film].

Prateek, G., Knopf, R., & Peterson, B. C. (2017). 779 Newspaper Articles Show How the Ruling and Opposition Parties Used the Narrative of the "Common Man" during Demonetisation. *Economic and Political Weekly*, *52*(36). https://www.epw.in/engage/article/analysis-779-newspaper-articles-shows-how-ruling-and-opposition-parties-used

Raghavan, S. (Director). (2012). *Agent Vinod* [Film].

Raghavapudi, H. (Director). (2022). *Sita Ramam* [Film].

Raj, S. J., & Suresh, A. K. (2022). Surgical Strikes on Screen: Narrations of Terrorism and Military Cross-Border Violence in Bollywood Cinema. In Choe, S. (Ed.), *The Palgrave Handbook of Violence in Film and Media* (pp. 169–186). Springer International Publishing., DOI: 10.1007/978-3-031-05390-0_9

Sagar, R. (Director). (1972). *Lalkar* [Film].

Saigal, R. (Director). (1950). *Samadhi* [Film].

Seth, A., De, S., Arya, A., Wilkinson, S., Singh, S., & Pal, J. (2022, April 8). Closed Ranks: The Discursive Value of Military Support for Indian Politicians on Social Media. *Proceedings of the 2022 International Conference on Information and Communication Technologies and Development*. Information and Communication Technologies and Development, Seattle. DOI: 10.1145/3572334.3572395

Sharma, H., & Pegu, U. K. (2023). Is the liberal media dead?: A critical analysis of Modi's Government and the performance of Indian Media. *India Review*, *22*(3), 219–248. DOI: 10.1080/14736489.2023.2201538

Sharma, S., & Pal, P. (2023). The rise of spy universe in contemporary Hindi cinema: Masculinity and nation-building in War (2019). *Intersections: Gender and Sexuality in Asia and the Pacific, 50*. https://www.academia.edu/download/113394978/The_Rise_of_Spy_Universe.pdf

Shelef, N. G., & Koo, A. Z.-X. (2022). Homelands and nationalism. *Nationalities Papers, 50*(3), 417–429. DOI: 10.1017/nps.2021.38

Sherfudeen, C., Mahmood, K., & Hossain, M. J. (2024). Cultural Nationalism or National Culture as Portrayed in Indian Popular Culture: The Case of Indian Visual Media. *International Peer Reviewed Journal: AWEJ for Translation & Literary Studies*, 115–130.

Shoesmith, B. (1988). Swadeshi cinema: Cinema, politics and culture: The writings of D.G. Phalke. *Continuum (Perth), 2*(1), 44–50. Advance online publication. DOI: 10.1080/10304318809359350

Staiger, J. (2005). *Media Reception Studies*. New York University Press., DOI: 10.18574/nyu/9780814786741.001.0001

Upadhyay, S. P., & Robinson, R. (2012). Revisiting communalism and fundamentalism in India. *Economic and Political Weekly*, •••, 35–57.

Verma, R., & Roy, A. G. (2022). Ideological positioning in the representation of borders: An analysis of recent Hindi films. *India Review, 21*(3), 307–328. DOI: 10.1080/14736489.2022.2086405

Viswanath, G. (2014). *The" nation" in War: A Study of Military Literature and Hindi War Cinema*. Cambridge Scholars Publishing.

Williams, A. L. (2002). *Film and nationalism*. Rutgers University Press.

KEY TERMS AND DEFINITIONS

Communalism: Allegiance to only one's own community (often based on religion in the South Asian context).

Discourse: The prevalent ideas, ideology and societal structures shaped by particular historical contexts.

Doxing: Revealing identifying details with malicious intentions.

Gender: Distinct from biological sex, but the sociocultural and psychological construction of being a man, woman, non-binary, etc.

Nation: A sovereign state with defined borders and strict territoriality.

Nationalism: An abstract idea of a land that is home, fostering loyalty and pride toward that nation.

Other: The process of differentiating a dominant group from a community, race, ethnicity, religion, gender, etc. to oppress and deprive.

ENDNOTES

1 In 1965, Pakistan launched Operation Gibraltar with orders for 7000 Pakistani Armed Personnel Guerrilla Freedom Fighters to infiltrate the ceasefire line to Indian Occupied Kashmir. They were to ignite a Muslim uprising against the Indian Government. Both countries consider this ill-planned. The operation failed and India retaliated, causing a full-fledged war. (Hassan, 2024, 2024; SI, 2011)

2 The women's question as laid out by Chatterjee will be discussed in the analysis section.

3 The enumeration revealed the domination by numbers.

4 The written records revealed differences in communities in India not just on religious lines but the social structures. The history of India as ancient, medieval and modern by the historian James Mill created formation of identities with connection to glorious pasts and past of oppression by invaders. The Indian historians tried to write about a past that glorifies Brahmanical Hindutva. (see Upadhyay & Robinson, 2012)

5 Hindus were the victims of Muslim aggression beginning from the resistance to brutal Zamindari system and the employment of slave peasants, and demand for land rights under Nizam rule blown into a communally-motivated massacre.

6 see Jaffrelot, C., & Therwath, I. (2012). The global Sangh Parivar: A study of contemporary international Hinduism. In *Religious Internationals in the Modern World: Globalization and Faith Communities since 1750* (pp. 343-364). London: Palgrave Macmillan UK. Sangh Parivar http://www.sanghparivar.org

7 Vedic Brahmanism is the Hindu religion derived from Vedas in the ancient period of India.

8 This was to explain the outer/material domain of Indian nationalism by Partha Chatterjee which stands relevant here.

9 See Staff, T. N. M. (2022, April 11). *Watch: Glimpse video from Dulquer, Rashmika and Mrunal Thakur's Sita Ramam is out*. The News Minute. https://

www.thenewsminute.com/flix/watch-glimpse-video-dulquer-rashmika-and-mrunal-thakurs-sita-ramam-out-162796

10 *THESE Countries Have Banned Hijab, Burqa—Are Egypt, France One Of Them?* (n.d.). Zee News. Retrieved August 3, 2025, from https://zeenews.india.com/photos/world/these-countries-have-banned-hijab-burqa-are-egypt-france-one-of-them-2926721

11 Ravana is the demon king of Sri Lanka who abducts Sita desiring her. Rama wages a war, killing him and rescuing Sita

12 Kurukresthra is a holy war between good and evil from the epic Mahabharata where good triumphs over evil.

13 Swayamvaram is a Hindu tradition mentioned numerous times in Epics like Mahabharata and Ramayana where the women choose the groom.

14 The dot on Hindu girls and women's foreheads.

15 See Mi Telugammayi |Voice overs on Instagram: "~ 1 year for seetharamam - Aug 5, 2022. Thank you Seetharamam team for making us feel the 'true essence of love' @hanurpudi @dqsalmaan @mrunalthakur Video script Written by: @rachana_ruchulu . . Intha Premaga matladthuna aina follow avvali anipinchadam ledha sumii .... #seetharamam #sitaramam #sitaramammovie #sitaramammovie #seetharamam #seetharamam #1yearforseetharamam #1yearforsitaramam #seetharamamwhatsappstatus #sitaramamreview #sitaramamsongs #sitaramam ." (2023, August 4). Instagram. https://www.instagram.com/telugintiammayi/reel/CvhuqD2t3nk/

16 See *I Cried | Pranav |#shorts#sitaramam#movie#review—YouTube*. (n.d.). Retrieved August 3, 2025, from https://www.youtube.com/shorts/VdPR6bI5Xn0

17 Watch The Wire (Director). (2025, May 6). *In Mourning, She Asked for Unity: How Himanshi Narwal is a Victim of Trolling | Arfa Khanum Sherwani* [Video recording]. https://www.youtube.com/watch?v=M2yAUVs9W3Q

18 *Watch | 39 Assaults, 19 Acts of Vandalism, 42 Incidents of Harassment Against Muslims Since Pahalgam: Report.* (n.d.). The Wire. Retrieved August 3, 2025, from https://thewire.in/communalism/apcr-report-muslims-targeted-after-pahalgam

19 See Pahalgam terror attack: Kashmiris face threats in other states, leaders seek government action. (2025, April 24). *The Times of India*. https://timesofindia

.indiatimes.com/india/pahalgam-terror-attack-kashmiris-face-threats-in-other-states-leaders-seek-government-action/articleshow/120572799.cms

[20] See India Today (Director). (2025, April 30). *15-year-old Muslim Boy Assaulted, Forced To Pee On Pakistan Flag In Aligarh | Pahalgam Terror Attack* [Video recording]. https://www.youtube.com/watch?v=iJY_Zyqz4fc

[21] See *ANI on X: "During War, Our Soldiers Fight sans Eating for 7-8 Days; so Can't We Do the Same for Our Nation?: Baba Ramdev #DeMonetisation Https://T.Co/7kDz7dipVW" / X*, n.d.

[22] A chant praising the Lord Hanuman.

[23] A chant in remembrance and praise of Lord Shiva.

[24] See the news article 'Maulvi in Army allegedly censured for using slogan that sends 'message of extremism'', The Economic Times, 18 September 2014. https://economictimes.indiatimes.com/news/politics-and-nation/maulvi-in-army-allegedly-censured-for-using-slogan-that-sends-message-of-extremism/articleshow/42747816.cms

[25] Vermillion worn by married Hindu women to signify their married status.

[26] It is the fabric of a saree that flows down freely from the shoulder of a woman draped in a saree.

[27] Doxing or Doxxing is the act of publishing private or indentify information about an individual, particularly with a malicious intention.

[28] tomatoSauceey. (2022, September 18). Buhkts watching sitaramam [Reddit Post]. R/Ni_Bondha. https://www.reddit.com/r/Ni_Bondha/comments/xhfjpa/buhkts_watching_sitaramam/

[29] A deliberate misspelling of *Bhakts* (devotees). Bhakts are described as subscribing to extremist right-wing ideology and belief in establishing India as Hindu Nation-State. Apart from this, they are also identified with extreme hatred towards Muslims. Bhakts is not used for all Hindus, only Hindu who holding extremist Hindutva and Islamophobic views. Meanwhile *Namazi* (One who does Muslim Prayers (Namaz)) is used to identify any Muslim and Muslim identity

[30] See Staff, T. N. M. (2025, May 11). *Foreign Sec Vikram Misri locks X account after facing abuse over India-Pak ceasefire*. The News Minute. https://www.thenewsminute.com/news/foreign-sec-vikram-misri-locks-x-account-after-facing-abuse-over-india-pak-ceasefire

Chapter 6
Echoes of War:
The Lingering Shadows of Conflict in Domestic Area

Aida Alinia
Istinye University, Turkey

Hasan Gürkan
https://orcid.org/0000-0002-3805-9951
Istinye University, Turkey

ABSTRACT

This study investigates how cinema portrays the intrusion of war into domestic and private spaces, transforming them into battlegrounds for psychological and emotional struggles. Through a combination of psychoanalytic film theory and spatial analysis, the research examines how the traumatic legacies of war shape personal identities, intimate relationships, and physical spaces, even when removed from direct combat. Focusing on three films—The Zone of Interest (2023), The Pianist (2002), The Reader (2008),—the study explores how these narratives construct symbolic and visual representations of private spaces as sites of unresolved conflict. By analyzing mise-en-scène, character psychology, and narrative structure, the research demonstrates how war's echoes reverberate within memory, family dynamics, and spatial design, blurring the boundaries between the personal and the political. The Pianist and The Zone of Interest focus on the Holocaust's legacy, while The Reader examines post-war guilt and reconciliation.

DOI: 10.4018/979-8-3373-7589-2.ch006

INTRODUCTION

In recent decades, the intersection of psychoanalytic theory, spatial analysis, and trauma studies has become an increasingly rich field for interpreting contemporary cinema. These interdisciplinary approaches offer tools for examining how films render the psychological aftereffects of historical violence, the architecture of repression, and the haunting return of traumatic memory (Traverso and Broderick, 2010; Wright, 2021). Particularly in films dealing with the legacy of war, private and domestic spaces become central to the representation of unspoken trauma and the interplay between interiority and historical conflict (Parziale, 2013).

This chapter focuses on the analysis of three films that engage directly with the psychological and spatial aftermath of the *Second World War: The Zone of Interest* (Jonathan Glazer, 2023), *The Pianist* (Roman Polanski, 2002), and *The Reader* (Stephen Daldry, 2008). While each of these films depicts different temporal and geographical contexts, they share an interest in how war inhabits intimate spaces—homes, bedrooms, cellars, bathtubs—and how these spaces, in turn, operate as containers of memory, repression, and unresolved guilt.

Unlike films that visualize trauma through graphic representations of violence or battlefield spectacle, the selected films articulate trauma spatially and emotionally, often through absence, silence, or spatial dissonance. In these films, trauma is not always externalized or spoken; rather, it lingers in the mise-en-scène, the framing of interiors, the choreography of movement, and the echoes of unoccupied rooms. in particular, is repeatedly positioned as a site where the personal and the political collide—where memory is inscribed and erased simultaneously

The theoretical framework of this study is constructed through a combination of three major paradigms: (i) Psychoanalytic film theory, particularly Freudian (1995) and Lacanian (1998) concepts such as repression, the unconscious, the gaze, and the return of the repressed; (ii) Henri Lefebvre's theory of space (1991), which conceptualizes space as socially produced and experienced through three interrelated forms—perceived, conceived, and lived space; (iii) Michel Foucault's heterotopia and spatial discourse (1986), which articulate how power structures and historical memory manifest in spatial arrangements, especially in institutional and domestic architecture.

Trauma studies, as an independent theoretical lens, offers important insights into how individual and collective suffering is culturally mediated and represented in film. According to Cathy Caruth (1995) and LaCapra (2001), trauma is not simply a psychological event, but a narrative and moral challenge. Trauma disrupts linear time, destabilizes memory, and demands new modes of testimony and witnessing. In this regard, it can be said that trauma is not just a personal issue but can become part of a broader cultural memory, where stories of loss and atrocities are transmitted and recreated across generations (Hirsch, 2012). By placing trauma studies alongside psychoanalysis and spatial theory, the aim of this chapter is to show that trauma is not only inscribed in the psyche or the built environment, but is also a discursive force that shapes how historical violence is remembered and recounted. As Kathy Caruth points out, trauma is not just an individual experience, but a phenomenon that recurs at different times through collective memory and cultural narratives (Caruth, 1996).

Trauma resists direct representation and often returns belatedly through disjointed memory, silence, or fragmented temporalities. Films such as *The Pianist* and *The Reader* enact this delay structurally, through their use of flashbacks, gaps in narration, and spatial metaphors of concealment and exposure. At the same time, space is not merely the backdrop against which psychological trauma unfolds. Drawing from Lefebvre, this study understands cinematic space as produced and productive—a site where power, ideology, and subjectivity intersect. In *The Zone of Interest*, the idyllic architecture of the Höss household in Auschwitz becomes a chilling symbol of spatial denial: war and genocide remain "outside the frame", yet are omnipresent in sound design and spatial tension. Similarly, in *The Pianist*, destroyed urban space becomes a living map of disintegration and survival, while in *The Reader*, the banality of post-war domestic interiors mirrors the inner conflict and ethical collapse of its characters.

Moreover, this chapter argues that spatiality in these films is not simply physical—it is also psychological. Private spaces, such as the home or the bedroom, function as what Freud (1995) might call "screen memories": surfaces onto which the unconscious projects censored memories. Following Lacan (1998), we can also see these spaces as symbolic structures that mediate desire, repression, and guilt. Foucault's concept of heterotopia further illuminates how spaces like prisons, archives, or even bathtubs operate as "other spaces," charged with paradox and displacement.

By combining these theoretical lenses, this study moves beyond conventional trauma analysis in film and attempts to offer a more embodied, spatialized reading of memory and repression. Each of the selected films presents trauma not as a resolved historical fact but as a lingering presence inscribed into the architectural and emotional landscapes of characters. This interdisciplinary approach seeks to uncover how spatial and psychological structures converge in the cinematic rendering of trauma, loss, and the enduring afterlife of war.

This study aims to explore how Holocaust and post-Holocaust cinema spatializes political violence and collective trauma within the most intimate of settings—domestic and architectural spaces. By moving beyond conventional depictions of war on battlefields, the selected films turn inward, portraying homes and interiors as fraught landscapes marked by repression, memory, and psychological suffering. Through close attention to mise-en-scène, set design, camera movement, framing, sound, and silence, this analysis investigates how spatial aesthetics and bodily presence become cinematic languages of historical burden. The characters' embodied experiences are read not only through visual and auditory cues, but also through the interplay of interiority and exterior space, suggesting a deeply inscribed relationship between trauma and environment. Integrating psychoanalytic theory, spatial analysis, and trauma studies, the chapter provides a multi-dimensional framework that reveals how film can serve as a site for processing and reimagining the affective legacies of historical violence.

LITERATURE REVIEW

If we take a brief look at war cinema, we will find that war cinema has always interacted with cultural values, social memory, and dominant politics. War films were often used as a tool of state propaganda during the First and Second World Wars, reinforcing nationalist and heroic ideas and thoughts (Paris, 2000). However, during the Vietnam War, this trend changed and filmmakers' perspective on this category changed. During this period, we were faced with a more critical view and the presentation of a realistic image that challenged the moral dimension of modern man and war (Dittmar & Michaud, 1990; Anderegg, 1991). According to Virilio (1989), the role of cinema in this was as a tool for representing the visual logic of war. Cinema tried to blur the boundary between spectacle and reality.

On the other hand, it is impossible to ignore how war narratives were transformed by censorship and cultural policies, and which of these narratives and in what form were incorporated into collective memory, and which were deleted or completely destroyed by the state system (Sorlin, 1994; Monaco, 2009). With this in mind, it is possible to look at contemporary films and address the question of how these films balance national mythmaking, traumatic memories, and critical representations of war.

Spatial Analysis, Psychoanalysis, and Trauma in Cinema

In this literature review, an attempt is made to analyze the multilayered connections between space, memory, body, and trauma in cinema by drawing on the theories of Lefebvre, Freud, Lacan, and Foucault. This approach is particularly significant in analyzing films set in wartime or post-war contexts, as such films often employ spaces such as homes, camps, prisons, or ruined cities not only as backdrops but also as stages of psychological trauma and mental pressure.

Henri Lefebvre, in his influential book *The Production of Space* (1991), views space not as a natural substrate but as a social product shaped within economic, political, and ideological relations. By introducing the "spatial triad" (perceived space, conceived space, and lived space), Lefebvre provides an important analytical tool for understanding spatial structure in art and everyday life (Lefebvre, 1991, p. 38). This triad has been particularly applied in film studies to examine space not just as a background but as an active element in constructing narrative and audience experience.

In cinema, this triad helps in understanding the viewer's experience of urban, domestic, or psychological spaces. For example, neorealist films such as *Rome, Open City* (Rossellini, 1945) represent the lived space of the post-war era by utilizing real locations (Brancaleone, 2021). Nick Jones, in the article *The Production of Cinematic Space*, argues that cinema is not merely a representation of space but a producer of it (Jones, 2015). Stephen Connolly, in the documentary *Space Machine* (2016), analyzes Detroit as a product of capitalist relations and highlights the connection between space, memory, and ideology (Connolly, 2018).

Studies such as Elden (2004) and Soja (1989) clearly demonstrate that space in cinema can act as a cultural and political mechanism—a space that both organizes collective memory and reproduces power structures. From this perspective, the space of the home, prison, shelter, or destroyed city in war-themed films gains additional significance.

Sigmund Freud introduces repression as a defense mechanism through which the mind attempts to set aside painful thoughts or memories from consciousness (Freud, 1915). These repressed materials may later return in altered forms as dreams, slips of the tongue, or neurotic symptoms (Freud, 1919). Freud's theory of "the return of the repressed" has found broad application in analyzing films with nonlinear structures and traumatic experiences.

Jacques Lacan, from a structuralist perspective, sees the unconscious as structured like a language and believes that trauma often creates a disruption in the symbolic order of language (Lacan, 1977). From his viewpoint, repression is governed by discursive structures, and cinema can reflect these structures. Lacan's theory has been widely used in film studies, especially in analyzing desire, the gaze, and the representation of the failure of meaning.

More recent studies, including works by Cathy Caruth and E. Ann Kaplan, show that trauma is not experienced at the moment of its occurrence but in its delayed return. Trauma is an event that is not processed at the time but returns later through scattered signs, dreams, or cinematic structures such as flashbacks, temporal jumps, or silence (Caruth, 1995; Kaplan, 2005). These structures enable the representation of the fragmented and disordered experience of the character. Janet Walker, in her study of trauma cinema, also emphasizes the importance of fragmented narratives and shattered memories (Walker, 2005).

The connection of Lefebvre's theory with psychoanalysis allows space to be analyzed as a carrier of the trauma, memory, and the cultural unconscious. From this perspective, domestic spaces, schools, prisons, or war-torn cities are not simply narrative contexts, but are themselves dynamic "stages" for the emergence of repressed memories and the reemergence of historical traumas (Bourazi, 2024). In Lefebvre's theory, space is not simply representational; it is constructed within the process of power production and the reproduction of social relations (Lefebvre, 1991). This idea aligns with Foucault's concept of heterotopia, where "deviant" or antagonistic spaces—such as unsafe homes or war shelters—offer an experience that is opposed to the dominant social order and exposes the gap between the symbolic order and lived reality.

Meanwhile, as Lefebvre points out, the body plays a very important role as a surface of the production of space, in that the body is present not only as a mass that occupies space but also as a mass that produces and experiences spatiality itself (Lefebvre, 1991, p. 200). The body also functions in psychoanalysis as a site of trauma—traumas that manifest in silence, stuttering, hallucinations, or involuntary movements and give meaning to the surrounding space. In this framework, the body and space in cinema form a two-way relationship, on the one hand, the body locates and signifies space with its movement and presence, and on the other hand, the arrangement and structure of the stage, sound, and camera movement involve the body in the structure of trauma and memory. Films like *Son of Saul*, *The Machinist*, or *Waltz with Bashir* are prime examples in which mental-physical spaces are not only displayed but also engage the audience in a sensory-spatial experience of neurosis.

Michel Foucault also considers space a tool for exercising power. In his theory of "heterotopia," spaces such as prisons, asylums, or museums are deviant spaces that both reflect and subvert the dominant order (Foucault, 1986). These spaces in cinema—especially in films dealing with political repression or historical traumas—function as archives of collective memories and traumas. Foucault (1986) also shows that memory, history, and place are not disconnected; rather, through institutions such as schools, prisons, and the military, a specific memory is constructed.

It is here that we can clearly see the connection between Foucault's theory of spatiality and Althusser's idea. Using the concept of ideological state apparatuses, Althusser shows how institutions such as the family, school, church or media shape and reproduce the memory and identity of the subject (Althusser, 1971).

From this connection, it can be understood that both heterotopian spaces and ideological institutions are structures that both produce meaning and, in cinema, become narrative scenes. It can therefore be said that the architecture of places such as the classroom, the prison cell or the backyard of the house—whether in historical narrative or in cinematic representation—plays a key role in shaping memory and trauma. Because these places are in fact places where the subject is located within the network of power, knowledge, and ideology (Foucault, 1986; Althusser, 1971).

In most film studies, either a psychoanalytic approach has been used or a spatial theory, but the combination of these two approaches with a focus on the representation of trauma in domestic spaces, prisons, and war-torn cities has received less attention. The present research, relying on the theories of Lefebvre, Freudian and Lacanian psychoanalysis, and Foucault's discourse of power, seeks to analyze the spatial-psychological structures of trauma experience in films such as *The Zone of Interest*, *The Pianist*, and *The Reader*. This chapter aims to show how space, as both material and symbolic substrate, shapes the emergence or suppression of memory and trauma, and how cinematic form can either reinforce or destabilize this process.

METHODOLOGY AND THE AIM OF THE STUDY

This study employs an approach that integrates psychoanalytic cinema theory with spatial analysis to investigate how war portrays its psychological impacts on people not actively engaged in combat. The study seeks to comprehend how the ramifications of war permeate public spaces and infiltrate domestic and private environments, transforming them into realms of psychological and emotional conflict.

While psychoanalytic film theory is used to reveal characters' subconscious traumas, repressed desires, and the effects of war on individual identity and family dynamics (Starks, 2002; Garland, 2018; Ullman and Brothers, 2013), spatial analysis aims to show how places such as houses, private rooms, prisons, and retreat areas in the film are represented as symbolic spaces bearing the traces of war (Machado 2016; Smith, 2014). This method is developed through character psychology, narrative structure, and mise-en-scène elements, revealing how the echoes of war echo in memories, family relationships, and spatial order.

In this context, the films *The Zone of Interest* (2023, dir.: Jonathan Glazer), *The Pianist* (2002, dir.: Roman Polanski) and *The Reader* (2008, dir.: Stephen Daldry) are examined, and how private spaces turn into conflict areas in these narratives, how they make the characters' traumas and the psychological legacy of war visible. In this way, it is revealed how private spaces are interrupted by political and historical contexts and how individual and collective memory shapes them.

These three films were selected for the way they address the psychological effects of war on individuals not directly on the front lines and how it infiltrates private spaces. Each of them has powerful narratives that show how the war bears not physical but psychological and spatial scars. These three films provide strong examples of both psychoanalytic and spatial analysis, allowing us to understand the repercussions of war outside the direct conflict area. The films discussed in this study, *The Zone of Interest, The Pianist, and The Reader,* while presenting different accounts of the trauma and psychological impact of World War II on human history, have also been widely received at the box office: *Zone of Interest* earned over $52 million, *The Pianist* earned over $120 million, and *The Reader* earned over $108 million worldwide. But what do these figures indicate? They suggest that themes such as repressed memories, domestic violence, and the return of the repressed are not just elitist concerns but have great potential to appeal to a general audience. As Kaplan (2005) points out, this widespread reception suggests that trauma and collective memory are becoming commodities in contemporary culture; In a way, the representation of historical violence, while being alarming, becomes part of the global entertainment industry. From a critical perspective, this situation reveals contradictory dimensions: on the one hand, such films circulate in the global market as cultural products; on the other hand, by watching them, the audience is involved in a process of rereading and confronting a past that has usually been marginalized (LaCapra, 2001). Therefore, the box office sales of these works should not be read only as financial success, but can also be seen as a sign of the commodification of war memory and at the same time, a sign of the cultural need to question the past and reconstruct historical traumas.

The benefits of these two methods to the study are as follows: (i) It allows us to consider the psychological and spatial representations of war holistically. (ii) It provides the opportunity for visual and narrative analysis. (iii) It explains the multi-layered penetration of war into private areas. Psychoanalytic film theory and spatial analysis offer an intersectional perspective that shows how war is intertwined with individual psychology and spatial arrangements. It reveals the connection between space and the unconscious, thus analyzing how traumatic events occur in the characters' inner world and how they are represented in film spaces. It shows how political and historical contexts shape private spaces and at the same time, how these spaces become scenes for the manifestation or repression of trauma.

For example, in *The Pianist*, the family apartment and the besieged streets of Warsaw become places that record the experience of survival and psychological wounds. In *The Reader*, the bedroom and courtroom spaces reflect the tension between individual desire and historical conscience and show how secrets and guilt are repressed or revealed in private spaces. Similarly, in *The Zone of Interest,* the Auschwitz commandant's house and the garden walls become symbols of the denial of violence and the embodiment of collective memory.

This research explains how war is "spatialized" in individual and collective memory, and how the body and architecture are involved in the narrative of historical wounds. Finally, it can be said that this study shows how personal and historical traumas shape the narrative structure of the film through spatial representation.

In this context, the films selected by the study offer a multi-layered analysis to understand the war's individual, spatial, and social repercussions by addressing its impact on individuals both psychologically and spatially, even though they are far from physical areas.

The research questions of the study are as follows:

RQ1: How do films depict the intrusion of war into intimate spaces?
RQ2: How do domestic settings become sites of trauma, memory, and psychological conflict?
RQ3: How do spatial design and mise-en-scène contribute to these representations?

Psychoanalytic film theory and spatial analysis based on the main themes of the selected films is summarized as follows:

1. The Home as a Site of Psychological Conflict
2. War Memory and Domestic Architecture
3. Spatial Disruption and the Breakdown of Private Life

FILM ANALYSES

To this end, this part offers a close study of three films – *The Zone of Interest (2023), The Pianist (2002), and The Reader (2008)* – in order to consider how war crosses from the public to the domestic sphere to remake domesticity as a site of psychological trauma and cultural memory. We argue that the film's mediations structure the representation of the 1994 genocide between personal trauma, domestic space, and collective past by examining psychoanalytic film theory (Freud, 1995; Lacan, 1998), spatial analysis (Lefebvre, 1991), and trauma studies (Caruth, 1995; Kaplan, 2005). Instead of treating trauma as the isolated phenomenology of battlefield-specific depictions of war, these films position traumatic experience as an internalized and spatialized phenomenon that travels beyond the domains of war, rendering indelible marks onto everydayness and intimate spaces. As one moves through each analysis, results and analysis flows directly with the research questions and method (refer to above), and with mise-en-scène, character psychology, sound design and spatio-temporal organization as moments for interpretation.

The Zone of Interest by Ganatan Glazer is about The commandant of Auschwitz, Rudolf Höss, and his family - they lived in a little villa on the edge of the camp in the grounds of which grew the lovely flowers, and they had the beautiful view: the dead bodies, the trainloads pouring in, the lashings. Their domestic life is the object of the narrative, and we feel the uncanny normality of their routines as the horrors of the Holocaust unfold outside their garden walls. The movie steers clear of explicit brutality; rather, it leans heavily on ambient sound design — the distant rumble of crematoria, muffled screams — offscreen horrors haunted by noise. This contrast of tranquil domestic picture and omnipresent hidden violence results in a chilling reflection on the banality of evil (Arendt, 1963). And in his decision to take the genocide out of the frame, Glazer only amplifies its omnipresence in the domestic sphere, to the point that everyday life becomes an ideological complicity in mass murder.

The Pianist is a film directed by Roman Polanski, which is based on the true life of Wladyslaw Szpilman the Polish-Jewish pianist who survived the ghetto of Warsaw. From the comfort and safety of his family's home, where they've lived a privileged life in the Jewish hub of Warsaw, Szpilman and his family are eventually uprooted, then shuffled to the ghetto, and finally, after they've lost so many of their family, to a life on the run, eating barely enough to survive. The movie vividly depicts urban wreckage, blown-out apartments, and hiding places, reflecting Szpilman's psychic disintegration with the physical disintegration of Warsaw. Music is both a badge of cultural identity and a psychological haven. By employing very long takes, little dialogue and chillingly sterile imagery, the film captures the post-traumatic fragmentedness of experience (Caruth, 1995) and demonstrates how war ruptures both the external terrain and the internal landscape.

The Reader, directed by Stephen Daldry, is a post war film which explores whether the new generation of Germans should be held accountable for the Holocaust of their fathers generation. It tells the story of Michael Berg, a young German boy who begins a passionate and secret affair with an older woman named Hanna Schmitz. Years later, Michael learns that Hanna had been an SS guard and is now on trial for war crimes. Memory, guilt, historical reckoning's moral ambiguities: the film deals in all the big stuff. Domestic spaces — Hanna's airless apartment, courtrooms, prison cells — are mute witnesses to buried atrocities and suppressed guilt. The film's temporal fluctuations and flashbacks create a story akin to the psychoanalytic notion of belated trauma (Kaplan, 2005) and reveal how individual memories are bound up with historical memory. It can also be concluded that silence and eyewitness testimony are two opposing dimensions of the process of remembering: silence, as LaCapra (2001) argues, sometimes acts as a defense mechanism against the re-invocation of trauma; while testimony is an attempt to make sense of suffering and builds a bridge between individual memory and collective narrative. In this film, Hannah's silence and the court testimonies simultaneously show how the unspeakable is inscribed in the architecture of space and how language becomes an archive of wounds.

In *The Zone of Interest*, Jonathan Glazer shows how the atrocities of war, while hidden in plain sight, manage to percolate through the domestic. To calculatively leave bloody scenes off-screen and also place the violence off-screen, the film fits a

Freudian (1919) definition of "the uncanny" (Unheimlich) which is that the familiar becomes unfamiliar by its latent horror.

The home of the Höss family is painted in all its remarkable bourgeois normality — well-kept gardens, orderly interiors, routine family life. But the aural landscape interrupts this visual peace. The constant background noise of the crackling fire, the screaming humans and warriors marching sends an acoustic reminder of genocide (Baron, 2006). Space, as Lefebvre (1991) contends, is not neutral, but is the product of social and political relations. Here, the sanctity of the home—a bastion of safety—becomes just as complicit in systemic violence.

Utilizing framespace compositions to visually restrain characters behind oppressive spatial barriers, Glazer's framing underscores containment and deprivation. This visible politics reflects "the conceived space" – in Lefebvre's language, a place which is organized, ideological, repressive of the traumatic "lived space" beyond (Lefebvre, 1991).

And the day-to-day doings of the Hösses — gardening, swimming, attending dinner — grotesquely resemble mundane life against the horror of genocide. This normalization resonates with Arendt's (1963) concept of the 'banality of evil', the fact that ordinary people carry out monstrous acts because of bureaucratic distance and ethical blindness.

One especially memorable moment shows Hedwig, Rudolf's wife, exultantly rising on the roof of her greenhouse as the sound of the crematories rumbles in the distance. The aural infiltration seems to operate as what Hirsch (2017) describes as "postmemory": that which is shared, though remotely, and physically across realms and generations. In their efforts to enclose themselves in a domestic utopia, the family fail to cut themselves off from the acoustic and moral pollution of the Holocaust.

In psychoanalytic terms, the structure of the film, like repression itself, embodies this process: The horror is expelled from the visual frame but always returns by way of sound, and by the dissonance of space, unsettling the characters' surface calm. The house is an "uncanny space" (Vidler 1992) in which comfort is haunted by the very violence it hopes to suppress. Violence is conveyed not through the direct image of bloody scenes but through media intermediaries such as sound, spatial architecture, visual dissonance, and even imperfect framing. In this way, it is cinema that "mediates" violence, allowing it to be embedded in visual and auditory layers without being directly present on stage, engaging the viewer with a sense of invisible threat (Kaplan, 2005; LaCapra, 2001). In this way, violence flows into the narrative at a symbolic and spatial level and is etched in the viewer's visual memory.

Glazer uses framing devices such as "frame-within-frame" compositions to visually situate his characters within oppressive spatial planes that underline containment and refusal. These visual approaches are expressive of what Lefebvre calls the "conceived space", a contained space that preserves ideological constructs and suppresses the traumatic "lived space" beyond (Lefebvre, 1991).

In the end, *The Zone of Interest* spatialises trauma, making a home that is a place of complicity as well as mute witness. With strategic visual lacks and ubiquitous auditory presences, Glazer not only makes the war absent but also everywhere, demonstrating the deep-rooted mental and spatial effects of historical violence.

The Pianist: Hiding Places, Urban Dereliction and Psychacity

Roman Polanski's *The Pianist* shows very well how hiding and urban ruins take on the psychic dimensions of trauma in time of war. The family's once-vibrant home is darkened by steadily advancing destruction as Szpilman's descent into bombed-out flats and finally into derelict ruins echoes the demise of self and society.

At first, the family of Szpilman — music itself — represent the vitality of culture. However, as the Nazi terror grows, it is through Szpilman's shrinking pathway to narrowing and darkening space that his life-world, in Lefebvre's (1991) concept of "lived space," is gradually enfolded into the discursive of political violence. His dark, empty rooms mirror the inner darkness of our unconscious feelings and thoughts (Freud, 1915).

Warsaw's ruined cityscape can be seen as memoryscapes in line with Bevan's (2007) position that shattered architecture represents collective trauma. Polanski's characteristic use of long, unmoving takes and muted, nickel grey colors foreground this hourglass city as a ghostly memory-scape. In trauma theory, this fragmentation and absence are construed as the deferred arrival of the traumatic memory (Caruth, 1995).

Spzilman's periods of concealment are punctuated with quiet, broken only by distant booms of gunfire or falling debris, constantly reminding him of his solitude and his peril. The crumbled buildings are prosthetic memories that allow contemporary audiences to experience historical trauma through a proxy (Landsberg, 2004).

Leni's MindShare The fact that Szpilman's physical state (his skeletal frame) reflects the ravaged world around him speaks to the reciprocity between body, space, and trauma. K's mental disintegration is spatially rather than verbally structured by his environment – in line with Lacan's (1998) understanding of the Real as the unsymbolizable trauma that shatters the symbolic order.

Ultimately, *The Pianist* recasts hiding places and city ruins as deeply symbolic sites of personal and social trauma, the war having entered and colonized not only the architecture but the psyche.

The Reader: How Memory, Guilt, and Post-War Domesticity Inform the Story

The Reader (Stephen Daldry, 2009) is about the ongoing psychological artifacts of guilt and memory in postwar, domestic spaces. In contrast to *The Zone of Interest* and *The Pianist*, where trauma is born of the ill that is done from outside, in The Reader trauma takes shape as an inward, late-glimpsed comprehension that fashions itself as an act of collective and personal moral failing.

Hanna Schmitz's sparsely furnished apartment, the place where her liaison with Michael takes shape, is a liminal space, mixing intimacy and secrecy. Indeed, as Hirsch (2017) argues, domestic interiors in Holocaust writing can exist as "zones of postmemory", where the trauma of the past makes itself felt in, and interacts dynamically with, the present. The austere design further highlights the unsaid aspects of Hanna's history, and the silence of the flat reflects the unspeakable horrors she harbours.

The narrative style of the film which deconstructs the linearity of time and often jumps back and forth–reminiscent of flashbacks–is a mirror to this psychoanalytic understanding of belated trauma (Kaplan, 2005) where the traumatic moment is realized only after its occurrence. Michael's subsequent returns to the flat recall Freud's (1919) notion of "the uncanny," that familiar ground is made bed fell through an alignment with repressed guilt.

Institutional places (courtrooms and the jail cell) outside of the person take the emotional weight to yet a greater distance. Indeed, though the courtroom stage is formal, it does not totally engage the emotional ambiguity of Hanna's guilt, and reveals what Fritzsche (2001) identifies as the postwar memory's 'banal archive'. It is stark and symmetrical, and in it we feel Vidler's (1992) "architectural uncanny," a space that punishes while memorializing.

Michael's battle to reconcile love and betrayal between Helen and a friend, encapsulates the tension between personal affection and historical obligation. The visits to Hanna in prison, and to her grave, demonstrate the merge of memory and spatial practice. As has been argued by Lefebvre (1991), space is a medium of (social) relationships and of memory and as a consequence Michael's personal quest cannot be separated from the places he goes to and the places where he returns to.

In the end, as *The Reader* demonstrates, home is not a refuge; it is a haunted terrain on which the penumbric writ of personal memory and collective guilt intersect. That the space is fractured — from Hanna's stark home to the prison's institutional rigidity — reflects the fragmented psyche of a generation flailing in the aftermath of moral catastrophe. The second generation here is engaged in the transmission of trauma and the discovery of truth that was buried in the silence of the previous generation and is now revealed in the architecture of place and physical memory

(Hirsch, 2012; Kaplan, 2005). Hirsch (2012) explains in the concept of "postmemory" that this generation did not directly experience the traumatic experience but reconstructed memory through broken narratives, objects, domestic spaces and incomplete documents. The traces of these stories are often represented in cinema by silence, narrative holes and empty spaces; that is, where the trauma remains invisible but tangible (LaCapra, 2001). In *The Reader*, Hannah's silence and delay in telling the truth lead the next generation to a situation where they must choose between moral judgment and understanding historical complexity. In this regard, the role of space as a psychological reservoir of this conflict is significant. For example, walls, rooms, and courts carry memories that are both individual and collective.

From the reading of *The Zone of Interest, The Pianist*, and *The Reader*, one can deduce that trauma, memory, and historical violence are not limited to the front, but are woven into the most private sphere of everyday life. Each film illustrates how domestic and urban environments become sites where personal identification and collective history intersect and collide.

Comparative Drama Structure, Space, and Memory in Three Films

Some overlap would always be expected considering that the material in each of these movies is so specific and that the films are so uniquely historical and narrative, though thematic meetings of *The Zone of Interest, The Pianist and The Reader* do bring them into this strange fold. None of the three films is interested in battlefield violence per se, but instead, traumatization as it seeps into domestic and urban spaces, shaping identity, memory, and emotional landscapes.

The Zone of Interest underscores the ubiquity of systemic violence in the most mundane domestic spaces and through ambient sound and spatial repression makes the Holocaust an unspoken, but omnipresent fact. In this film, war functions as an absent reference or "aporia" (Caruth, 1996); something that is pushed out of the visual frame but is always present through sound, architecture, and symbolic order. In *The Pianist*, this aporia is linked to physical destruction. The film's ruined city of Warsaw and makeshift shelters make Szpilman's physical and psychological collapse a visible and visible space, while the war and violence itself are largely absent but linger in the traces of ruins and the silence of the character (Kaplan, 2005). But in the film *The Reader*, trauma is institutionalized and internalized. In Hannah's silence lies war and guilt. The courtroom and domestic spaces become locations where the truth is never fully told and narrative gaps are always present. According to LaCapra (2001), aporia is inevitable in post-Holocaust memory.

All three films are methodologically situated in Lefebvre's (1991) socially produced space and Freud's (1919) uncanny. They show us how familiar settings

— homes, apartments, cities — metamorphose into haunted terrains, where the personal and the collective histories become enmeshed. Through sonic invasion, geographic disintegration, or chronological dispersion, each narrative frame points to the impossibility of containing historical trauma in spatial or psychic containers.

Thus, despite their differences, the films collectively articulate a cinematic geography where trauma persists beyond physical violence, embedding itself in the intimate architectures of everyday life.

The Zone of Interest: How War Is Conducted Beyond the Frame and Yet Dominates the Domestic Sphere

In *The Zone of Interest,* Glazer masterfully depicts how war's horror is nowhere on screen yet is somehow everywhere inside the home. Special-effects maker Roger Mayer agrees with Freud's (1919) concept of "the uncanny" ("Unheimlich"), as familiarity and ease are now overlaid with a sense of the uncanny, that is, familiarity becomes static on account of the restrained imagery and local of violence, and placing the violence offscreen, according to Freud, represents "the unheimlich".

Glazer use framing "framewih-inti a-frame" compositions openly to position his characters as visually trapped by oppressive spatial boundaries, visually to infer containment and denial. These visual tactics correspond to the Lefebvrian "conceived space", an artificial space upholding ideology (Lefebvre, 1991), excluding the traumatic "lived space" beyond.

The importance of ambient sound — of distant moaning or heavy rhythmic industrial clank or a low rumble of machinery — cannot be overstated. These aural cues function as much as part of the domestic mise-en-scène of the skin as visual mise-en-scène. The horror is off-screen, but not off- sense. Glazer deploys sound as disturbing as it drills through the comforting facade of the home, turning the auditory volume into a primary conduit of historical memory and of affective disquiet. Chion (1994) points out that sound in film can work autonomously of image and can, therefore, have narrative material all its own, especially when it suggests the unfelt and unseen.

In psychoanalytic terms the structure of the film replicates the process of repression: the horrors are pushed out beyond the visual frame, but return both through sound and through spatial dissonance, disturbing the apparent peace of the characters. The dwelling itself, in this way, turns into an "uncanny space" (Vidler, 1992), in which shelter encompasses violence from the which it is an iterance.

In the end, The Zone of Interest also creates space for trauma: it turns the home into a place of complicity – and mute observer. With strategic visual dominances and ubiquitous sound presences, Glazer makes the war both out of sight and ever present, to show how deeply psychological and spatially devastating historical violence can be.

The Pianist: Hiding spaces, urban ruins and psychological suffering

In *The Pianist*, Roman Polanski presents urban ruins and hiding spaces as the physical aftermath of a psychological response to wartime. Szpilman's gradual fall from his family's cheery apartment to bombed out flats and desolate ruins is the decline of the self and society. First, the Szpilman family, a bunch of musicians, is a place of cultural vivacity. However, as the Nazi occupation progresses, Szpilman's circulation into smaller and darker spaces represents what Lefebvre (1991) refers to as the "lived space" of political violence. The dark, sparsely-furnished concealed areas of his house offer a tangible representation of inner dread and inarticulacy (Freud, 1915).

These reminders take the form of mnemonic landscapes, in line with Bevan (2007) who asserts that destroyed architecture is the embodiment of collective trauma. Polanski's regular reliance on long, still takes and muted color palettes also foreground the city as a ghostly memory-scape. In trauma studies, such dismemberment and absence embody the belated nature of traumatic memory (Caruth, 1995).

Szpilman's concealed interludes are filled-in with silence only pierced by distant gunfire or reversing buildings as an echo of the solitude and lethal circumstances. The devastated buildings act as prosthetic memories (Landsberg, 2004), giving audiences the opportunity to feel metahistorical trauma. These spaces, through cinematic form, invite the viewer into a transhistorical experience of trauma. This experience is outside the viewer's direct memory but is activated through sensory participation and emotional engagement in the moment of viewing the film. In this way, the viewer engages with the silence, sound, and architecture of suffering and becomes a subject involved in the process of remembering (Kaplan, 2005).

Furthermore, that Szpilman's own dwindling body – his starving frame – mirrors the ruined environment, illustrates the co-extensiveness of body, space, and trauma. He is driven to madness not through articulation, but through spatialization of his environment, in Lacanian (1998) Real, the unsymbolizable traumatic that destroys symbolic order.

As Caruth (1996) explained, the traumatic is frequently resistant to being directly articulated and its expression belatedly emerges through what appear disconnected, sensations. The split chronology of The Pianist and Szpilman's dislocation in a moment is itself a way of registering this lateness with millimetric accuracy. The places he resides aren't fixed homes, but shifting mental conditions — spaces of delayed suffering, where memory and meaning come together only after the fact.

At its base, *The Pianist* turns hiding places and urban ruins in poignant piercings of personal and mass trauma, showing us how war can infiltrate both the architecture and the psyche. From an urban sociology perspective, the film also depicts the collapse of urban spatial order, where streets, buildings, and public spaces lose their primary function and become places of instability and isolation (Zukin, 1995). The destruction of Warsaw is represented in the film not only as a physical backdrop, but also as a living witness to social collapse, the elimination of residents, and the obliteration of urban memory. In this narrative, the city has become a dead city, devoid of human connections. The city, as Augé (1995) puts it, has become a kind of non-place, where identity, belonging, and time are disintegrated. In this way, the film shows how political violence affects not only the body and psyche, but also the urban fabric and social relations.

The Reader: Memory, Guilt and Post-War Domesticity in the Shaping of the Story

Stehphen Daldry's *The Reader* explores the ongoing psychological effects of guilt and memory in the post-war home. It's not like *The Zone of Interest* and *The Pianist*, where the trauma comes from forces worked upon the outside of us: The Reader speaks about trauma in terms of internalized, belated knowledge, direct from an actively constructed personal and collective ethics.

Feminist trauma theory emphasizes that the ways in which trauma is experienced and narrated vary according to gender (Kaplan, 2005; Radstone, 2010). In *The Reader*, Hanna's body and apartment function as "living archives" that carry the silent legacy of war: Silence, repressed guilt, and desire coexist simultaneously (Hirsch, 2012). Michael's return to these spaces years later evokes Freud's concept of the "uncanny" (1919), rendering the familiar unsettling; thus, individual passion collides with historical responsibility, and the domestic space transforms into a stage where trauma materializes.

Hanna Schmitz's sparsely furnished apartment, where she and Michael begin their relationship, is a liminal space, combining elements of privacy and intimacy. As Hirsch (2017) points out, domestic interiors in Holocaust literature often serve as "zones of postmemory," in which the trauma of the past vibrates in the present. This is how such environments become reserves of emotional inheritance to be transmitted, even though they did not in fact endure the original trauma. The transmission of identity is not always encountered in conscious, articulated terms, but rather, lingers in the silences of rooms, the rhythms of daily routines, and the symbolic of lack (Hirsch, 2012).

The bare-bones space underscores the unspeakability of Hanna's history, in and the quiet solo life she leads away from it in the apartment, reflects the unspeakable horrors she keeps within. So these spaces are both sanctuary and accusation — containing within them a history that is unspoken but palpable.

The narrative pattern of the film which utilizes timeline refrains from linearity along with the replays of memories is representative of the psychoanalytic concept of belated trauma (Kaplan, 2005) in the sense that the traumatic revelation is developed in retrospect. This revisiting of the apartment by Michael brings to mind Freud's (1919) notion of "the uncanny," where the familiar is rendered unsettling by its taint of repressed guilt.

Institutional spaces like courtrooms and cells only exacerbate the insistent exteriorization of the mental anguish. The devices of the courtroom with its protocol, meanwhile, does not recognize the emotional depth of Hanna's guilt, thereby exemplifying what Fritzsche (2001) has referred to as the "banal archive" of postwar memory. Hanna's prison cell, stark and perfectly symmetrical, is what Vidler (1992) would call an "architectural uncanny" in that it is a room that both punishes and commemorates.

Michael tries to reconcile his memories of love and betrayal in this particular ideological equation of personal affection versus historical accountability. We see this most visibly in his visits to Hanna in prison and to her grave, which epitomize the tendency to weld memory and spatial practice. As Lefebvre (1991) suggests, space mediates the social and the historical, and Michael's journey through feeling is also a journey through the spaces he occupies and returns to.

In the end, The Reader positions domesticity not as a sanctuary, but as a haunted terrain where personal memory and collective guilt intersect. The spatial disunity — from Hanna's stark quarters to the institutional cold of the prison — reflects the splintered psyche of a generation trying to come to grips with the aftermath of moral catastrophe.

Between them, these films ask us to reimagine what it is to experience war, not merely as a series of violent actions but as a condition that saturates the texture of everyday life. War doesn't end when the shooting stops; it reverberates in footsteps in empty hallways, in the shutting of doors, in quick glances across dinner tables. These are the echoes that cinema, with its sensory and spatial capacities, is uniquely positioned to capture.

The result of this study is not only a typology of war-affected spaces, but also a lexicon for comprehending trauma as architectural, domestic and intimate. That blurs the distinction between public and private suffering and implies that the deepest wounds from war need not be on bodies or landscapes, but instead be those found inside kitchens, bedrooms or bathtubs. Cinema, in this sense, serves not only as a mode of historical representation, but as a mode of historical processing, of allowing audiences to inhabit —if only briefly— the interior architectures of trauma.

A second important observation is how memory and spatial design interact. Both films deploy mise-en-scène and sound not merely decoratively, but also narratively. The streets in The Pianist (which paid host to a reduction in the volume of pedestrian traffic, in order to create the appearance of a war-torn city), the sterile symmetry of The Zone of Interest, the brutal minimalism of The Reader's prison sequences—all these function as narrative actors. They don't merely contain trauma; they produce it. This corresponds to Alison Landsberg's (2004) concept of the "prosthetic memory;" in the case of the invoked experiences, mediated experiences which render possible the emotional and ethical engagement with histories not directly lived.

And these works show how post-traumatic space refuses to cohere. Order is not re-gained by the end of any of these movies. Szpilman regains a relationship to music in The Pianist, but not to a home. In "The Reader," Michael comes to know, but he does not find peace. And that is what makes The Zone of Interest so compelling: There is no reckoning at all, just the chilling continuation of everyday life. Such endings are a reminder that trauma does not always conclude; it often gets stuck, trapped in the buildings around us.

DISCUSSION AND CONCLUSION

In exploring the research questions, the following themes are discussed:

(i) The Home as a Site of Psychological Conflict

In Holocaust cinema, the domestic space often emerges not as a safe haven but as a site where ideological violence is internalized and psychological conflict unfolds. As Kerner (2011) observes, the home becomes complicit in the ideological project of Nazism, reflecting how ordinary environments participate in systemic violence. In *The Zone of Interest* (2023), the Höss family's home, situated just outside the Auschwitz concentration camp, embodies this ideological complicity. Although the house is framed by Glazer's camera as a picturesque and orderly German dwelling, the ambient sound design—particularly the persistent noise of crematoria—disrupts any sense of pastoral calm. As Baron (2006) suggests, such audiovisual techniques transform the home into a kind of archival space, embedding it with the acoustic memory of genocide. In one scene, as Hedwig converses with her mother in a garden brimming with flowers, the muffled violence from the camp permeates the background, reconfiguring domesticity into a site of denial and normalization. This visual and sonic juxtaposition highlights what Hannah Arendt famously termed the "banality of evil," showing how horror is not external to the home but entwined with its very fabric.

In *The Pianist* (2002), the home initially serves as a site of cultural identity and familial intimacy. The Szpilman household, centered around the piano, is a hub of intellectual and emotional life. Yet as the Nazi regime encroaches, the house transitions into a zone of fragmentation and fear. The narrowing camera frames and creeping darkness—coupled with the sound of distant bombing—mirror the breakdown of family and memory. Sykes (1999) interprets the ruins of Warsaw's houses as physical manifestations of psychological trauma, emphasizing how spatial destruction mirrors the shattering of identity. The family's final breakfast together is not merely a farewell, but a ritualized surrender to the collapse of the home as a symbolic refuge. What remains are fragmented spaces of memory, where each room becomes a haunted chamber of loss.

In *The Reader* (2008), domestic interiors function as affective landscapes shaped by secrecy, guilt, and desire. The home where Michael and Hanna begin their sexual and literary relationship is infused with both intimacy and latent trauma. According to Hirsch (2017), domestic spaces in Holocaust narratives often act as zones of postmemory, where the trauma of one generation is transmitted to the next. In addition to oral narratives and family memories, these intergenerational traumas can be transmitted through objects, photographs, and living spaces that carry the psychological burden of the past (Hirsch, 2012). For example, the silence of surviving parents, incomplete archives, or even the architecture of old houses can keep the moral and emotional scars of war alive in the minds of the second generation. Films such as *The Reader* and even memoirs such as *Maus* (Spiegelman, 1986) or *Everything Is Illuminated* (Schreiber, 2005) show how subsequent generations carry the burden of unresolved questions, guilt, or an inability to understand the dimensions of violence (Hirsch, 2012; LaCapra, 2001).

Hanna's bedroom—sparse and austere—becomes a charged arena where silence speaks volumes. The absence of dialogue in several scenes intensifies the spatial weight of unspoken guilt. As Dwyer and Alderman (2008) argue in the context of memorial culture, memory is embedded not only in public monuments but also in everyday spaces, a notion clearly at play in how Michael later recalls the apartment. The home Hanna leaves behind lingers in Michael's memory as a palimpsest of erotic encounter and historical reckoning.

(ii) War Memory and Domestic Architecture

After the devastation of war, architecture emerges not merely as a physical construct but as a medium for encoding memory, denial, and political ideology. Fritzsche (2001) contends that domestic architecture in postwar Germany embodies a complex relationship between private life and collective memory. In *The Zone of Interest*, the symmetry of the Höss household—manicured gardens, sterile interiors, and geometric order—reflects Nazi aesthetic values. Yet this architectural precision is a visual metaphor for the regime's ideological violence. Glazer's frequent use of frame-within-frame compositions emphasizes both containment and repression, aligning with Kwint et al. (1999) assertion that materials and architectural surfaces serve as mnemonic devices. The house becomes a performative façade, shielding the family from the atrocities occurring just beyond the garden wall, and enacting a spatial form of denial. Vidler's (1992) theory of the "architectural uncanny" is particularly resonant here: what should be familiar—domestic space—is rendered eerie and morally hollow.

In *The Pianist*, architecture is the primary witness to historical trauma. The crumbling buildings of Warsaw, depicted in long static takes and low lighting, serve as monuments to a lost culture. As Bevan (2007) argues, architectural ruins can embody collective loss and institutional violence. Szpilman's solitary refuge in a bombed-out apartment is emblematic of such memory-infused spaces. The silence of the piano room—once vibrant with music—now signifies the death of cultural life. Landsberg's (2004) concept of "prosthetic memory" is applicable here, as viewers are invited to inhabit these damaged spaces through sensory engagement, generating empathetic memory that transcends direct experience. Gürkan (2015) reinforces this idea, noting how Holocaust cinema employs sensory disruption—sound, camera movement and spatial instability—to destabilize the viewer's perception and replicate the characters' trauma.

In *The Reader*, domestic and institutional architecture operates as both prison and archive. The university setting where Michael discovers Hanna's past is visually unremarkable yet symbolically dense, evoking what Fritzsche (2001) calls the "banal archive" of postwar memory—spaces that store trauma without acknowledging it. Hanna's prison cell, by contrast, is starkly minimal and symmetrical, yet charged with emotional weight. The walls, bare and mute, enact what Kwint et al. (1999) describe as material memory: the architecture becomes a silent container for guilt and historical consequence. Vidler's (1992) notion of the unhomely resurfaces here, as the cell's austerity and simplicity turn it into a psychological mirror of Hanna's internal reckoning. The visual austerity does not neutralize the past; instead, it paradoxically intensifies the gravity of what cannot be spoken.

(iii) Spatial Disruption and the Breakdown of Private Life

The spatial disintegration of the home in Holocaust cinema parallels the psychological breakdown of its inhabitants. Suk (2009) asserts that oppressive regimes politicize domesticity, transforming it into a battleground. In *The Zone of Interest*, Glazer's juxtaposition of aesthetic order with auditory horror destabilizes the domestic space, suggesting that comfort itself has become a form of violence. Gürkan

(2015) discusses how sound and camera angles alter spatial perception, and here the unseen horrors beyond the garden walls become felt through sonic pressure—turning the home into a sensory prison.

In *The Pianist*, the breakdown of private life is materialized through the progressive ruination of domestic interiors. Landsberg's (2004) theory of "prosthetic memory" helps explain how viewers empathetically inhabit Szpilman's disrupted world; the shattered windows and silent rooms mediate historical trauma into accessible emotional experiences. The home, stripped of its protective function, becomes a relic that both commemorates and conceals trauma.

In *The Reader*, spatial disruption occurs not through war but through the slow realization of past atrocities. As Michael revisits the spaces associated with Hanna—her apartment, the court, the prison—each site becomes a repository of unspoken guilt. Vidler's "unhomely" spaces offer a useful lens: these locations, though architecturally mundane, are imbued with emotional excess and spectral presence. The decline of intimacy in these environments signifies the difficulty of reverting to a state of innocent domesticity after historical intervention.

This is a study about how war, Holocaust most particularly, infiltrates the most intimate—supposedly shielded— fabric and turns our homes and private spaces into psychological battlefields. By using psychoanalytic film theory and spatial theories to analyze The Zone of Interest, The Pianist, and The Reader, this work demonstrates that trauma, guilt, memory, and repression do not rest solely on a military battlefield, but fathom in domestic spheres and at the level of the individual. The analysed films reveal that despite a general sense of 'remoteness', characters are not protected against the psychological/emotional effects of war. Instead, the implications of war are absorbed and intertwined with mundane models of space--homes as continuations of historical violence, 'family' dynamics as echoes of more general ideological fractures.

Through visual narrative, sound, and mise-en-scène, these films expose the lingering legacies of war embedded in private spaces. They challenge the boundary between battlefield and home by showing how domestic settings become sites where historical trauma endures. Integrating psychoanalysis and spatial theory clarifies how cinema maps the emotional and collective dimensions of memory. At the end, these films show how cinema not only represents historical violence but also makes visible its lasting psychological impact within everyday life.

The theory pursued throughout this chapter is that cinema offers a particularly affective language for expressing the enduring affect of war and trauma in the most protected of spaces. Through this sustained discussion of three separate but thematically interconnected films— *The Zone of Interest, The Pianist,* and *The Reader* — this chapter has provided an analysis of how filmic space operates not as just a theatrical backdrop, but as dynamic companion through a visual exploration of the representation of psychological and historical conflict.

Crucially, the chapter has followed the way in which war is not just ‘out there’, on the battlefield or at some faraway frontline. Instead, it has been exhibited to seep inside, to haunt the domestic sectors and upset the emotional and symbolic architectures of daily living. By way of psychoanalytic film theory and spatial analysis, we understand how trauma gets anatomized into walls, soundscapes, bits of furniture or even silence. Freud’s ideas of repression and the uncanny, Lacan’s sinthome and the symbolic order and the lack, and Lefebvre’s work on socially produced space have all played a role in explaining how film (not only) narrativizes what war does to people, but what it does to space—and how space in turn re-vamps and rebuilds the psyche.

The study also demonstrates how the combination of psychoanalytic theory and spatial analysis can open up new interpretive paths for reading trauma films, especially those that deal with war, genocide, and their lasting effects on private spaces. By approaching domestic and urban spaces not just as passive backdrops but also as active sites of psychological and cultural registration, this research demonstrates the value of interdisciplinary approaches to understanding the complex entanglements of memory, place, and the unconscious.

Future researchers could extend this framework by integrating other perspectives, such as urban sociology or cultural studies. For example, comparative studies might examine how different historical traumas, such as colonial violence or forced migration, are embedded in the architecture of everyday life. In this way, the multilayered approach used in this chapter offers a model for analyzing how film represents historical atrocities and unspoken emotions through the intimate geographies of the body, the city, and the home.

REFERENCES

Althusser, L. (1971). Ideology and ideological state apparatuses (Notes towards an investigation). In Althusser, L. (Ed.), *Lenin and Philosophy and Other Essays* (pp. 86–127). Monthly Review Press.

Anderegg, M. A. (1991). *Inventing Vietnam: The war in film and television*. Temple University Press.

Arendt, H. (1963). *Eichmann in Jerusalem: A report on the banality of evil*. Viking Press.

Augé, M. (1995). *Non-places: Introduction to an anthropology of supermodernity*. Verso.

Baron, J. (2006). The archive effect: Found footage and the audiovisual experience of history. *Film Quarterly*, *60*(2), 48–57. DOI: 10.1525/fq.2006.60.2.48

Bevan, R. (2007). *The destruction of memory: Architecture at war*. Reaktion Books.

Bourazi, A. H. N. (2024). *Trauma through the lens of attachment and resilience*.

Brancaleone, D. (2021). Framing the real. *AMPS Proceedings, 21*, 1–16.

Caruth, C. (Ed.). (1995). *Trauma: Explorations in memory*. Johns Hopkins University Press. DOI: 10.56021/9781421413525

Caruth, C. (1996). *Unclaimed experience: Trauma, narrative, and history*. Johns Hopkins University Press. DOI: 10.1353/book.20656

Chion, M. (1994). *Audio-vision: Sound on screen* (Gorbman, C., Trans.). Columbia University Press.

Connolly, S. (2018). *The spatial cinema* (Doctoral dissertation, University of Kent).

Dittmar, L., & Michaud, G. (1990). *From Hanoi to Hollywood: The Vietnam War in American film*. Rutgers University Press.

Dwyer, O. J., & Alderman, D. H. (2008). Memorial landscapes: Analytic questions and metaphors. *GeoJournal*, *73*(3), 165–178. DOI: 10.1007/s10708-008-9201-5

Elden, S. (2004). Understanding Henri Lefebvre. *Continuum*.

Foucault, M. (1986). Space, knowledge and power. In Rainbow, P. (Ed.), *The Foucault reader* (pp. 239–256). Pantheon Books.

Freud, S. (1915). Repression. In *The standard edition of the complete psychological works of Sigmund Freud* (Vol. 14, pp. 141–158). Hogarth Press.

Freud, S. (1919). The uncanny. In *The standard edition of the complete psychological works of Sigmund Freud* (Vol. 17, pp. 217–256). Hogarth Press.

Freud, S. (1995). The unconscious. In *The standard edition of the complete psychological works of Sigmund Freud* (Vol. XIV; Strachey, J., Trans.). Hogarth Press.

Fritzsche, P. (2001). The case of modern memory. *The Journal of Modern History, 73*(1), 87–117. DOI: 10.1086/319880

Garland, C. (2018). *Understanding trauma: A psychoanalytical approach.* Routledge. DOI: 10.4324/9780429484575

Gürkan, H. (2015). *Karşı Sinema.* Es Yayınları.

Hirsch, M. (2012). *The generation of postmemory: Writing and visual culture after the Holocaust.* Columbia University Press. DOI: 10.7312/hirs15652

Hirsch, M. (2017). Touching memory: Postmemory, art, and the Holocaust. *Esprit (Paris, France)*, (10), 42–61. DOI: 10.3917/espri.1710.0042

Jones, N. (2015). The production of cinematic space. Retrieved from https://www.academia.edu/11910555, Accessed date: 21.01.2025

Kaplan, E. A. (2005). *Trauma culture: The politics of terror and loss in media and literature.* Rutgers University Press.

Kerner, A. (2011). *Film and the Holocaust: New perspectives on dramas, documentaries, and experimental films.* A&C Black. DOI: 10.5040/9781628928716

Kwint, M., Breward, C., & Aynsley, J. (1999). *Material memories: Design and evocation.* Berg.

Lacan, J. (1977). *Écrits: A selection.* W. W. Norton & Company.

Lacan, J. (1998). *The four fundamental concepts of psychoanalysis* (Vol. 11). W. W. Norton & Company.

LaCapra, D. (2001). *Writing history, writing trauma.* Johns Hopkins University Press.

Landsberg, A. (2004). *Prosthetic memory: The transformation of American remembrance in the age of mass culture.* Columbia University Press.

Lefebvre, H. (1991). *The production of space* (Nicholson-Smith, D., Trans.). Blackwell. (Original work published 1974)

Machado, L. (2016). *A palimpsest-image: Place, space and film geographies* (Doctoral dissertation, University of Westminster).

Monaco, J. (2009). *How to read a film: Movies, media, and beyond*. Oxford University Press.

Paris, M. (2000). *From the Wright brothers to Top Gun: Aviation, nationalism, and popular cinema*. Manchester University Press.

Parziale, A. E. (2013). *Representations of trauma in contemporary American literature and film: Moving from erasure to creative transformation*.

Smith, I. S. (2014). *In lonely places: Film noir beyond the city*. McFarland.

Soja, E. (1989). *Postmodern geographies*. Verso.

Sorlin, P. (1994). *The film in history: Restaging the past*. Blackwell.

Spiegelman, A. (1986). *Maus: A survivor's tale*. Pantheon.

Starks, L. S. (2002). "Remember me": Psychoanalysis, cinema, and the crisis of modernity. *Shakespeare Quarterly*, *53*(2), 181–200. DOI: 10.1353/shq.2002.0025

Suk, J. (2009). *At home in the law: How the domestic violence revolution is transforming privacy*. Yale University Press.

Sykes, C. J. (1999). *The end of privacy: The attack on personal rights at home, at work, on-line, and in court*. St. Martin's Press.

Traverso, A., & Broderick, M. (2010). Interrogating trauma: Towards a critical trauma studies. *Continuum (Perth)*, *24*(1), 3–15. DOI: 10.1080/10304310903461270

Ulman, R. B., & Brothers, D. (2013). *The shattered self: A psychoanalytic study of trauma*. Routledge. DOI: 10.4324/9780203766408

Vidler, A. (1992). *The architectural uncanny: Essays in the modern unhomely*. MIT Press.

Virilio, P. (1989). *War and cinema: The logistics of perception*. Verso.

Walker, J. (2005). *Trauma cinema: Documenting incest and the Holocaust*. University of California Press. DOI: 10.1525/9780520937932

Wright, E. (2021). *PTSD and narration: Traumatic stories' impact on realism in film and literature* (Doctoral dissertation, University of Maryland, Baltimore County).

Zukin, S. (1995). *The cultures of cities*. Blackwell.

Chapter 7
Maritime Myths, Monsters, and Mysteries:
Crafting ‘Us’ vs. 'Them' in ‘Captain Phillips’ and ‘A Hijacking’

Jessica Kate Simonds
https://orcid.org/0000-0003-0019-221X
American University in the Emirates, UAE

ABSTRACT

This chapter examines how cinema constructs capitalist spatial orders at sea through Captain Phillips and A Hijacking. Using spatial theory, it introduces "cinematic territorialization" to show how film renders the ocean governable, securitized, and economically productive. These films offer contrasting views: Captain Phillips celebrates American power against racialized maritime threats, while A Hijacking exposes detached corporate logics of risk and negotiation. Linking this to a longer history of maritime myth-making, the chapter shows how cinema reproduces binaries of order/disorder and North/South. While acknowledging Somali perspectives, it focuses on how dominant film narratives naturalize capitalist power. Ultimately, it argues film actively shapes how we imagine ocean space, urging renewed critique as climate change and resource extraction open new oceanic frontiers.

INTRODUCTION

The sea has long occupied a powerful place in our cultural and political imagination. It has been portrayed as a space of both wonder and danger, where myths are born, monsters lurk, and modern-day threats continue to emerge. Far from being an empty or neutral expanse, the sea is a deeply politicized and actively territorialized

DOI: 10.4018/979-8-3373-7589-2.ch007

space, one that has been shaped by centuries of storytelling, legal regimes, and capitalist interests. This chapter examines how maritime spaces are produced and represented through dominant narratives that construct binaries of 'us' and 'them', and how these constructions work to legitimize particular spatial orders while excluding others. The focus here is not on how piracy has been understood across time, but on how its representation in literature, film, and historical record has served to reinforce broader geopolitical and economic structures. Drawing on a wide range of examples, from ancient sea monsters and religious texts to cinematic portrayals in *Captain Phillips* and *A Hijacking*, the chapter traces a clear thread: those who control the narrative of the sea often control the sea itself. Whether it is the myth of the siren, the figure of the pirate, or the glorified heroism of Western corporate negotiators, each narrative plays a role in shaping who is seen as legitimate at sea, and who is excluded, criminalized, or silenced.

By approaching the sea as a socially and politically constructed space, the chapter unpacks how media and cultural production intersect with spatial philosophy and global power dynamics. Concepts like spatial consciousness, territorialization, and the spatial economy are used to understand how representations of the sea have consistently aligned with the interests of dominant actors, from colonial empires to contemporary corporate powers. The analysis moves beyond simplistic good-versus-evil binaries to ask who gets to speak, who gets to be seen, and who is rendered invisible in the process of narrating maritime security. Rather than neutral depictions, representations of maritime threats are deeply political. Whether drawing on myth or modern media, they serve to uphold dominant understandings of who belongs at sea, and who does not, is shaped by histories of capitalism, exclusion, and power.

PRODUCING POLITICIZED SPACES

Interrogating the production of politicized spaces before they are presented to us as media representations is a process where math and science meets politics and the arts. The process of spatial planning reflects how power and plural knowledge(s) are produced as a practice of representation and extension of ownership and dominance over the earth's geography. Vuillemin argues that "history enters into the considerations of theoretical philosophy through mathematics" (Vuillemin as cited in Grozholz, 2007, p. 228), meaning that the mathematical conceptualization of space is in itself a social process that seeks to display the world relative to the measurements that have been constructed by the mathematician and their political perception of the space in which they wish to represent. As philosophers who were considerably influential to Ancient Greek and Renaissance French political thinking, both Euclid and Descartes respectively drew on their knowledge and expertise of the political in their development of mathematical spatial practices. These practices were then mobilized (and continue to this day) in making previously unmarked spaces and places connected and visible.

Prior to the dissemination of folklore and myths, and the release of blockbuster movies, the representation of the spaces they represent must be thoroughly crafted to portray a narrative to the audience. This involves a carefully considered process of spatial production, which is what happens before a tale is told. Social and mental spaces are defined by storytelling, ideas, experience, and political narratives. They are important in spreading a message through redefining a person's consciousness relative to emerging phenomena. A spatial consciousness is first produced in the mind of an individual and connects a topological path that connects an individual's mental interpretation of space with material or social phenomena to create a personalized perspective of *place*. Lefebvre, as a key spatial philosopher refers to this as the "locus of knowing" (Lefebvre, 1974). When producing a spatial consciousness, a person draws on familiarity, such as sensations, emotions, and memories to reconstruct a space in their mind as formative to their consciousness. This takes place when a person hears about a new place, experiences something for the first time, or reads about it in a book. Central to this chapter, is understanding how the sea has been produced in the spatial consciousness of those who have experienced media representations of risk.

Galland and Gronning (2019) go further to define a spatial consciousness as an "individual or collective awareness about real-world spatial phenomena or processes" (2019, p.1). The collective aspect of their definition demonstrates the power of representation in reconstructing or *producing* a particular dominant perspective of space that can be experienced by consuming media narratives. Other key philosophers who conceptualize the sociologies of space by drawing on distributions of power often leave out the importance of individualized experiences of reality and thus individualized contributions of knowledge in orienting how a space is constructed. Lefebvre argues that "Foucault never explains what space is, nor does he bridge the gap between the epistemological realm and the practical one (the mental and the social)" (Lefebvre, 1974: 4). Massey argues that in his later life, Foucault commented on traditional understandings of space as something "we thought about… as inert, flat, and without dynamism" (Massey, 2013) but did not expand upon this observation or offer any contribution beyond. Similarly, Chomsky is critiqued for his inability to bridge the gap between mental space and social space (Lefebvre, 1972: 5). He instead focuses on the relationship between language and practice as a mobilizing factor to identify spatial discourse (Chomsky, 1957). More recently, Stuart Elden has prompted us to think about space as 'volumetric' rather than a flat plane in which to observe from above (2021). Consequently, each spatial philosopher has used 'space' metaphorically, rather than providing a clear conceptual distinction between the mental construction of space based on experience interaction with the media and does not go any step further to demonstrate how it links to the physical *performance* of space in everyday life.

Once an individual has produced a personal, mental interpretation of *place,* the narrative is reproduced and codified through the connections formed between individuals. This can come in many forms relative to popular cultures, such as the shared love a movie, the viral status of a short video or meme, and the shared connotations that surround them. These 'social spaces' are dependent on the activation of the aforementioned mental spaces through communication and human interaction. A common assumption in normative political studies is that social spaces are exclusively the places of bureaucratic decision making and are confined to the formal physical places where transactions of power are perceived to occur, such as the Palace of Westminster debating chamber, or the UN Security Council. Alternatively, and as proposed throughout this chapter, social spaces can be reimagined as a product of small moments between people who are not traditionally powerful actors (such as world leaders, or movie producers!) but are instead those with experience and the power to spread a message that can be realistically received by an audience.

In understanding how media representations shape commonly received perceptions of space, we can draw upon the concept of a *spatial economy*. A spatial economy can be understood as the assumed (or sometimes overtly defined) rules of participation within space. When tracing back who is portrayed as a 'hero' and who is portrayed as a 'villain' it is often due to the former upholding an agreed perception of how spaced should be *legitimately* used, whilst the latter serves to *disrupt* and challenge the spatial economy. Lefebvre defines the spatial economy as a set of rules that:

Valorizes certain relationships between people in particular places and thus gives rise to connotative discourses surrounding these places; these in turn generate 'consensus' or conventions according to which consequently, that place is seen as trouble free (Lefebvre, 1974, p. 56).

The construction of a spatial economy is a form of behavioral regulation that is evident when a spatial consciousness is projected into the material world as a practice of territorialization. These practices are both social and physical as they serve to reinforce one another through the design of specific physical architectures that perpetuate or repress the performance of certain behaviors. Some rules are unspoken, yet still apply; for example, it is socially conditioned that you are quiet in spaces such as libraries and churches, and you do not need to read a guide to adhere to the spatial economy that these places embody. Yet, some spatial economies require guidance to ensure conformity to a specifically designed spatial order. Such as key performance indicators (KPIs) in the workplace, prescribed safety measures in a warehouse, or carefully recited school rules in the classroom. Whether spoken or unspoken, the spatial economy seeks to regulate a dominantly established order in which space is governed, and media serves as a tool to perpetuate these narratives.

CINEMATIC TERRITORIALIZATION OF THE SEAS

Modern scholarship in International Relations (IR) speaks mostly of territorialisation as a terrestrial, landed concept, falling into what Agnew and Corbridge (1995) term the *territorial trap*, the "set of assumptions about the territorial nature and boundaries of political and economic life that permit state-centrism in theoretical approaches" (p. 98). This trap limits territorial production to the state, reinforcing an exclusive understanding of political space as flat, static, and solid. Elden (2013) adds that territory is frequently assumed to be "self-evident in meaning," grounded in the traditional IR fixation on bounded national borders tied to the material solidity of land. In response, contemporary scholars have sought to rethink territorialisation in ways that foreground the sea's dynamic and volumetric qualities. Benton's (2010) analysis of early modern maritime corridors of control, Ryan's (2015) work on fluid spatial spheres, and Steinberg and Peters' (2015) notion of "wet ontologies" each disrupt the flat ontology of IR by repositioning the sea as a politically charged, stratified, and multilayered space. These works conceive of the ocean not merely as a surface of transit or exception, but as a domain that is governed, securitised, and increasingly territorialised, albeit through logics distinct from those applied to land.

A key site through which such maritime territorialisation is enacted is visual culture, particularly cinema. The concept of cinematic territorialisation refers to the ways in which films and visual media represent, construct, and normalize geopolitical imaginaries of space (Lee, 2025). In the context of the sea, cinematic territorialisation makes the ocean visible as a securitised, governable, and bordered arena. In other maritime spaces, scholars such as Vélez-Serna & Stauff have sought to contextualise how places such as ports have come to embody representations of capitalism as a cultural symbol (2023).These representations work to territorialise the sea by overlaying its fluidity with stories of threat, sovereignty, and heroic intervention. However, the process of cinematic territorialization is not merely symbolic. As argued by Tawil-Souri (2014) Palestinian cinema functions as a "structure of feeling" enabling films to transgress and reimagine geopolitical and psychological boundaries imposed by the Israeli occupation—what she calls the "territorial trap". In this sense, cinematic territorialisation becomes a powerful mechanism not only for representing space, but for materially shaping how it is governed, contested, and emotionally experienced. Cinema also performs a social act of inclusion and exclusion through its casting and characterisation. It installs into audiences a sense of who belongs at sea, and who does not, through racialised and nationalised representations of actors. In films such as *Captain Phillips* (2013), Western naval officers and corporate seafarers are portrayed as rational, legitimate agents of order, while Somali pirates are framed as lawless, faceless threats. Such casting is not accidental. As Shapiro (2009) notes, cinematic narratives often encode security onto white, Anglophone bodies while attaching risk, deviance, or illegality to racialised others, particularly African, Arab, and Southeast Asian actors. These embodied performances reinforce a visual hierarchy of sovereignty, producing a moral cartography in which security and state legitimacy are visibly aligned with Western norms and military capacity. Drawing on Foucault's concept of governmentality, cinematic representations such as *Captain Phillips*, *Battleship* (2012), and even animated features like *Moana* (2016) serve not merely as entertainment but as pedagogical tools. They visually instruct audiences in who belongs at sea, who commands its legitimacy, and who threatens its order. As Steinberg and Peters (2015) suggest, vision and volume become central to the sea's political reimagination: the ocean is not just seen, but layered, mapped from above, patrolled on its surface, and surveyed beneath.

Territory, as Raffestin (as cited in Elden, 2013) notes, emerges through the territorialising actions of social actors who assign symbolic and strategic meaning to natural or constructed spaces. Media representations, especially cinematic ones, thus become performative tools in the practice of territorialisation. They translate the fluid geographies of the sea into fixed narratives of inclusion and exclusion, sovereignty and piracy, legitimacy and threat. These symbolic mappings transform the sea into a lived and known space, actively situated within political imaginaries. This process constructs a dichotomised seascape of safe and dangerous spaces, authentic heroes and alien others. Historically, elites, be they political, military, or commercial, have used stories, poems, maps, and now cinema, to produce the sea as a place of sovereign control or contested incursion. These cultural texts draw on and reinforce existing geopolitical hierarchies, reinforcing who has the right to territorialise the sea and who must be excluded. Thus, cinematic territorialisation not only visualises but legitimates ocean governance. It becomes a central mechanism through which the sea is framed not as a global common, but as a space of stratified access, constant surveillance, and expanding state and corporate sovereignty.

MYTHICAL MARITIME HEROES AND VILLAINS

Following the discussion of cinematic territorialisation, which reveals how films and visual media inscribe geopolitical meaning onto the sea, it is important to recognize that this practice of constructing spatial consciousness through narrative is not new. Long before cinema, political and cultural elites used stories, poems, maps, and religious texts as epistemological tools to territorialise the sea in the popular imagination. These narratives transformed the ocean from an abstract expanse into a site of symbolic danger, mystery, and exclusion, often mobilised to provoke compliance, set boundaries, and legitimise authority on land and sea.

As established, spatial consciousness, one's mental mapping of risk, access, and belonging, relies on experiential or mediated engagement with space. Yet direct experience of the sea has historically been limited to a select few: seafarers, fishermen, and those in religious or political power. The sea, inhospitable and mysterious to most, was thus largely imagined through mediated stories. This allowed elites to concentrate symbolic power by constructing fearsome representations of the sea and its inhabitants. These constructions have shifted over time, adapting to cultural anxieties and political needs, and repurposing old monsters into new moral and territorial messages. One of the most enduring tools in this myth-making process has been the production of monsters, embodied fears that transform the sea into a domain of peril. In the sixth century BCE, Jewish religious authorities framed the sea as a threatening frontier to discourage emigration and maintain religious community cohesion. Snyder (1999) notes that in this context, the sea became a metaphor for imperial threats, exemplified in the Book of Daniel, where the "real enemy of the Jewish people was a monster coming from the sea," representing the empires of Babylon, Persia, Greece, and Rome (p. 11). The sea monster, in this formulation, served both theological and territorial ends, warning against dispersal and reinforcing collective identity through shared spatial fear.

Similarly, the Kraken emerged in Norse storytelling as a monstrous deterrent. Initially described in *Konungs Skuggsjá* (c. 1250) as an immense beast whose body was surrounded by fish, the Kraken became a way to territorialise the North Sea by fear. Seafarers and coastal communities used the monster as a narrative device to warn outsiders and justify exclusive claims to resource-rich areas. Eighteenth-century accounts described the Kraken as being over a mile in circumference, bursting from the depths to drag entire vessels beneath the waves (Heuvelmans, 2003). Wallenberg's (1717) warnings for fishermen to avoid the Kraken's resting place reflect how monster myths became functional political tools, producing a spatial consciousness of risk that reinforced territorial control.

In addition to monstrous villains, the sea has been heavily gendered in classical mythologies, often embodying dangerous femininity. The myth of the Siren, as famously portrayed in Homer's *Odyssey*, constructs the sea as a site of seductive peril, threatening to undo male seafarers through temptation. Homer writes:

"[Sirens] spellbind any man alive… no sailing home for him… round them heaps of corpses rotting away, rags of skin shrivelling on their bones" (*Odyssey*, Book XII, p. 48–57).

These sirens symbolize a fusion of sex and danger that served to enforce the heroic masculinity of the mariner who resists them. Rubarth (2015) observes that male-authored literary accounts from antiquity consistently valorised masculine navigation as a triumph over feminised chaos. The Siren, then, functions not just as a monster but as a moral geography, mapping the sea as a feminised danger zone requiring masculine dominance to traverse. This binary between the masculine hero and feminised villain was central to Ancient Greek gender ideology. As Hanson and O'Rawe (2010) argue, the Siren embodies a "femme fatale" archetype: "otherness to the temporarily wandering man," shaped by heterosexual patriarchy and gendered exclusion (p. 10). The Siren's fatal allure contrasts sharply with the violent masculinity of Poseidon, the god of the sea, who, though often wrathful and destructive, is portrayed as a figure to be revered, not feared. As Bernard (2012) notes, Poseidon's aggression, greed, and entitlement to women are naturalised through his divine masculinity, legitimising his control over the feminised dangers of the sea.

This mythic coding had deep social consequences. In Ancient Athens, success at sea was closely tied to masculine virtue, while femininity was equated with weakness, seduction, and instability. The Siren myth thus does more than entertain, it territorialises the sea through gendered boundaries, casting male sailors as rightful navigators and feminine figures as destabilising threats. Even in later epics such as the *Argonautica* and the *Iliad*, maritime danger is consistently externalised onto monstrous or feminised figures, whose control can only be mastered by a masculine hero supported by divine (often male) guidance (Snyder, 1999, p. 13). These narratives, whether ancient myth or modern cinema, do more than simply describe the sea; they create it. By ascribing moral, gendered, and political meaning to maritime space, they territorialise it in the minds of audiences and publics. As with cinematic territorialisation, these myths shape perceptions of who belongs at sea, who controls it, and who threatens it. In both visual and literary traditions, the sea becomes not a neutral space, but a canvas for projecting cultural anxieties and reproducing hierarchies of power.

PLUNDER, PIRATES, AND DISRUPTION

Each of the villains covered in the previous sections have ceased to be believable maritime threats as knowledge of the sea and unfamiliar places have been more broadly documented to account for the geography and wildlife of global planet earth. Yet, the human 'pirate' has been produced and reproduced to perpetuate the binary of 'insider' and 'outsider' at sea, in the same way that mythical monsters were appropriated to define risky and hostile maritime territories, and ringfence resource rich areas for the use of one group over another. This section will focus on the 'pirate' and argue that its status as the "perpetual enemy of all mankind *(hostis humani generis)*" (Thomson, 1994, p. 186) draws on a momentary reproduction of *who* is labelled as the pirate and their disruptor in a dominant spatialization of the sea. By covering how piracy has been reconstructed to address challenges to dominant spatial constructions at sea in different moments of history that have been highlighted to reveal that the act of plunder has remained a constant thread that ties each reimagination of the threat together. The application of the piratical status reveals how dominant actors in different moments of history have sought to further establish their claim to maritime space based on constructing a binary between their participation at sea, and the establishment of a spatial economy designed to facilitate the exclusion of the 'pirate'. By exploring a genealogy of the *pirate,* the modern representation of the Somali pirate can be placed within a longer history of plunder at sea.

The *illegal migrant,* the *pirate,* and the *terrorist* are all examples of terms that have been appropriated to construct a pejorative and exclusionary status to those who threaten the rhythmic production of politicized spaces. They are presented as outsiders who violate space by virtue of their existence, and measures to mitigate their mobility are entwined in the spatial economy of most political regimes. The illegal migrant is constructed and reproduced as the perpetual outsider to sovereign territory both socially and physically. A stark example of this is the US immigration regime (amongst others), which through practices such as border checks and deportation are facilitated by the architectural design of detention centers and walls to propagate the idea that those held there are unwelcome, undesirable and in many instances, awaiting their removal (Johnson, 2014, p. 133). By virtue of the combination of social practice and the architecture of detention centers, illegal migrants are physically removed and excluded from the space of the state based on the spatial economy of difference and exclusion that these physical designs perpetuate. These binaries are further propagated through media appeals. Most recently (2025), the United States under the leadership of President Donald Trump has launched a series of media appeals with the goal of dissuading potential migrants to enter the country illegally. A video released on YouTube by the US Department of Homeland Security (2025) depicts a series of videos and photographs of migrants who have been arrested or in the process thereof who are mostly black or Latin American (2025). As an example of systematic racism, the depiction of these images only exacerbates arguments made by scholars such as Menjívar (2021), Figueroa-Caballero (2019), and Armenta (2017) that the criminalization of immigration status only serves to perpetuate racist attachments of crime and migration. Therefore, actors such as the *illegal migrant*, the *pirate* and the *terrorist* will always be reproduced to legitimize practices of exclusion – with representations of each, drawing on knowledges and interpretations of a threat to distinguish who belongs, and who does not.

Exploring ancient experiences of piracy reveals how acts of plunder were historically recorded to define who was deemed authentic or illegitimate at sea. Plunder, understood as the profit-driven theft of goods that disrupts the spatial economy, was particularly threatening when the sea served as a vital trade route (Blackmore, 2009). Ancient societies did not use the term 'piracy' themselves; instead, modern scholars have retrospectively labelled certain acts to construct binaries of righteous versus disruptive maritime actors. The earliest recorded evidence of maritime plunder dates back to 1350 BCE, when a ship attacked coastal communities off North Africa under Pharaoh Akhenaten. This clay tablet inscription portrays how early Egyptian rulers framed maritime raiders as outsiders who threatened their spatial and political order (Vallar, 2009). A later example is the 'Sea Peoples,' recorded during Ramses III's reign in 1179 BCE. They are often described as one of the first groups to develop a distinct sea-based identity, treating the Mediterranean and their ships as home (Johnson, 2020). Inscriptions from Ramses III's tomb describe how these raiders dismantled coastal settlements, marking them as the archetypal maritime 'other' whose existence disrupted established political life on land and sea (Aruz et al., 2014). The Sea Peoples' legacy demonstrates how maritime communities could be cast as existential threats to spatial order and credited with the decline of Bronze Age societies (Johnson, 2020).

In Ancient Greece, piracy further illustrates this shifting binary. The historian Polybius was the first to use the term 'pirate' around 140 BCE to describe maritime theft in the contested Mediterranean (Hassan and Hassan, 2017). Yet, early Greek perceptions were complex. Until 331 BCE, coastal raids and ship plundering were often viewed as legitimate extensions of warfare. Homer's epics, Thucydides, and Aristotle all reflect a time when piracy carried no stigma but was considered akin to hunting or fishing (Nisbet, 2009; Spiegal, 1991). This changed with Alexander the Great's efforts to unify Greek city-states, when piracy became framed as an illegitimate threat to political cohesion (Vallar, 2009). Despite this shift, the Greek 'pirate' remained connected to the act of leistes, meaning plunderer, whether operating by sea or raiding coastal villages (Heebøllholm, 2013). These ancient accounts demonstrate that the meaning and legitimacy of piracy have always depended on who tells the story and who controls the space. Retrospective labels obscure how plunder historically blurred the boundaries between land and sea, order and disorder.

CONTEMPORARY SOMALI PIRACY

The modern construction of Somali piracy shows how the pirate continues to occupy the role of a perpetual maritime outsider through acts of plunder and refusal to comply with the Western capitalist territorialisation of the sea. Contemporary Somali piracy typically involves hijacking vessels in the Horn of Africa, the wider Indian Ocean, and the Gulf of Aden, detaining crews and cargo until ransom payments are secured. This pattern, which escalated from around 2000 onwards, mirrors the same spatial practice of plunder and disruption discussed in earlier sections (Burlando et al., 2015, p. 525). In the contemporary era, the sea has become the ultimate metaphor for globalisation, with people, capital, goods, and ideas constantly circulating along transnational currents (Benton, 2013, p. 104). Such flows rely on the idea of an ordered sea where shipping routes remain open and secure. Modern maritime law, including the United Nations Convention on the Law of the Sea (UNCLOS), carves up the world's oceans into zoned areas that regulate ownership, access, and transit rights. These zones exist to enable the smooth operation of the global merchant shipping industry (United Nations, 1982). Characters who disrupt this order, like the Somali pirate, are constructed as threats to both maritime space and the capitalist systems it sustains.

However, the story of Somali piracy cannot be reduced to criminality alone. Its roots lie in the legacies of foreign exploitation and chronic poverty that followed the collapse of Somalia's central government in 1991. With no navy or coast guard, Somalia's 3,300-kilometre coastline became vulnerable to illegal industrial fishing by foreign fleets, mainly from Europe and Asia. These fleets are estimated to have stolen up to $300 million worth of fish annually from Somali waters, undermining local livelihoods (Hansen, 2009; Murphy, 2011). Alongside illegal fishing, reports have documented how foreign companies dumped toxic and radioactive waste along the Somali coast, worsening environmental degradation and pushing already struggling coastal communities further into poverty (Lehr and Lehmann, 2007). Against this backdrop, piracy emerged as an act of survival and resistance. Many early groups described themselves as 'coast guards' defending their waters from foreign trawlers. For communities with limited alternatives, hijacking commercial vessels became a way to profit from the same maritime space that had been exploited by powerful external actors. This context is critical when recognising that Somalia remains one of the poorest countries in the world, with over 69% of the population living on less than $1.90 per day (World Bank, 2022).

Despite this, dominant legal and political frameworks depict Somali pirates purely as violent criminals, ignoring the broader structures of inequality and environmental harm that helped shape modern piracy in the region. The figure of the Somali pirate is therefore produced through media and legal narratives as a disruptive force that must be excluded to protect the free flow of global trade. This representation supports the logic of naval securitisation and justifies increased surveillance and military presence across the Indian Ocean, all to maintain the Western capitalist spatial order at sea. By understanding Somali piracy in this way, it becomes clear that the pirate remains a deeply political figure. They expose the contradictions of a maritime space that is legally ordered for the benefit of global trade yet historically and materially shaped by unequal access, extraction, and exclusion.

CAPTAIN PHILLIPS AND AMERICAN MARITIME EXCEPTIONALISM

The mobilization of imagery and interpretations of space is not exclusive to the legal sphere. In understanding modern threats and challenges, literature and film play a crucial role in shaping how audiences understand places that are physically, politically, or experientially inaccessible, such as outer space, the deep sea, or restricted countries. Through narrative, imagery, and emotional resonance, these mediums offer imaginative access to environments that lie beyond direct human experience. They do not merely depict such places; they interpret them, framing how we are meant to feel and think about these spaces. For example, outer space is often portrayed not just as a physical frontier but as a symbol of isolation, discovery, or existential awe, as seen in films like *Interstellar* (2014) or *The Martian* (2015). Similarly, the ocean becomes a metaphor for depth, mystery, or danger in works like *Moby Dick* (1851) or *The Lighthouse* (2019). In the case of politically restricted regions, literature and film can provide rare insight into daily life, resistance, or oppression, as seen in memoirs from North Korean defectors or censored Iranian cinema. These portrayals often reflect the cultural and ideological perspectives of their creators, meaning they can humanize, critique, romanticize, or reinforce stereotypes depending on their framing. Technological advances such as CGI, IMAX, and virtual reality further enhance the immersive potential of film, making inaccessible worlds feel tangibly real. As such, literature and film function not only as artistic expressions but also as cultural artefacts that influence how we imagine and engage with the unknown, the forbidden, and the remote.

Paul Greengrass's *Captain Phillips* (2013) serves as a compelling case study in the cinematic reinforcement of American exceptionalism and the portrayal of non-Western 'others' as threats to global order. Based on the 2009 hijacking of the MV Maersk Alabama by Somali pirates, the film dramatizes a real-world event through a narrative that emphasizes American heroism and moral superiority, while casting the Somali pirates as desperate and dangerous outsiders. The film's protagonist, Captain Richard Phillips, portrayed by Tom Hanks, embodies the archetype of the American everyman hero, stoic, rational, and self-sacrificing. His calm demeanor and ethical fortitude under duress align with traditional narratives of American exceptionalism, wherein American individuals are depicted as natural leaders and moral compasses in chaotic global contexts (Markus, 2021).In contrast, the Somali pirates are depicted through a lens that emphasizes their volatility and otherness. While the film offers brief glimpses into the socio-economic hardships that may drive individuals toward piracy, these moments are overshadowed by portrayals that focus on their aggression and unpredictability. This dichotomy reinforces a simplistic binary between the civilized West and the chaotic non-West, a common trope in

Western media that serves to legitimize Western interventionism (Nilmaola, 2017). The film further amplifies this narrative through its depiction of the U.S. military response. The arrival of the Navy SEALs and their efficient, technologically advanced operation to rescue *Captain Phillips* underscores the theme of American military superiority and moral righteousness. This portrayal aligns with the notion of the United States as a global enforcer of order and justice, a central tenet of American exceptionalist ideology (Sy & Smith, 2013). By representing maritime space as a place of inclusion and exclusion through film, it posits a sense of legitimacy to the audience on *who* is considered a 'hero' in the context of maritime security, which in the case of '*Captain Phillips*' is the American crew.

As a story that brought the reality of piratical capture home to everyday Americans, public outcry saw a political response that fused the threat with that of terrorism (Simonds, 2025, p. 5). Consequently, the film's visual and narrative choices contribute to the construction of the Somali pirates as the 'other.' The use of contrasting settings, the orderly, modern environment of the American ship versus the impoverished, chaotic depiction of the Somali village, serves to highlight cultural and socio-economic disparities. This contrast not only reinforces stereotypes but also simplifies complex geopolitical realities into a digestible narrative of good versus evil (Markus, 2021). Critics have noted that while *Captain Phillips* attempts to humanize the Somali pirates, it ultimately falls short by not fully exploring the systemic issues that contribute to piracy, such as illegal fishing and toxic waste dumping by foreign entities in Somali waters. By omitting these factors, the film perpetuates a narrative that isolates piracy from its broader context, thereby absolving global actors of responsibility and maintaining the focus on American heroism (Sy & Smith, 2013).

Beyond its role as entertainment, *Captain Phillips* has had a significant impact on how the American public understands piracy and, more broadly, how people from the Global South are perceived in relation to global security and order. As with many Hollywood films based on real events, *Captain Phillips* does more than dramatize *A Hijacking*, it also functions as a cultural artefact that shapes audience perceptions, particularly by reinforcing reductive binaries between the Global North and South, between order and disorder, and between moral civility and violent desperation. The film's portrayal of Somali pirates as impoverished, thin, and erratic men driven by desperation serves to frame piracy as an individual criminal pathology rather than a symptom of larger structural and geopolitical realities. While it includes moments of humanization example, Muse's line "I am the captain now" has become iconic and is often used mockingly in media and online discourse, it still overwhelmingly associates Somali identities with violence, instability, and threat. In doing so, the film contributes to what Edward Said (1978) termed *Orientalism*: the process by which Western cultures define and represent non-Western peoples as backward, irrational, and inherently threatening. For American audiences unfamiliar with the complexities of Somali history, maritime law, or the legacy of Western exploitation of African maritime spaces, the film offers a singular narrative: the Global South, represented by the pirates, is a lawless zone that necessitates intervention from the disciplined and moral forces of the United States. The pirates are not explored as products of failed statehood, global inequality, or environmental exploitation, issues well-documented in piracy scholarship (Hansen, 2009; Murphy, 2011), but as alien figures who must be subdued.

This simplification not only decontextualizes the roots of Somali piracy but also reinforces harmful stereotypes about African and Global South populations as primitive, unstable, and in need of Western control. Moreover, *Captain Phillips* was released at a time when piracy off the Horn of Africa was already a prominent concern in international media and political discourse. The film amplified public anxiety about maritime threats and fed into broader fears about borderlessness, terrorism, and failed states. Its mass appeal, bolstered by Tom Hanks' star power and the film's critical success meant that it reached a wide audience, many of whom may not have engaged with piracy as a complex geopolitical issue before. As a result, the film helped to entrench a popular understanding of pirates not as economically marginalized actors in a global system, but as violent criminals who threaten global trade and American lives. This framing aligns with a long-standing tradition in U.S. cinema of externalizing threats and reinforcing national identity through conflict with a racialized or cultural "other" as discussed in relation to the current U.S immigration regime (Shaheen, 2001).

The visual contrast between the American crew and the Somali pirates highlights differences in body type, language, hygiene, and even tone of voice, also serves to construct the Global South as a space of danger and disorder. In this way, the film reinforces a specialized form of security discourse in which places like the Indian Ocean are presented as voids of law and civilization, awaiting pacification by the technologically and morally superior West. The American military's final intervention, depicted as clean, justified, and efficient, reinforces a sense of U.S. exceptionalism as not just a national identity, but as a global responsibility. Ultimately, *Captain Phillips* did more than tell a gripping story; it contributed to a cultural and political narrative in which the Global South is defined primarily through risk, instability, and moral inferiority. Such representations not only influence public attitudes but can also shape policy discourse, particularly in areas of maritime security, international development, and humanitarian engagement. *Captain Phillips* exemplifies how mainstream cinema can reinforce American exceptionalism by constructing narratives that emphasize American moral and military superiority while depicting non-Western individuals as threats to global stability. This portrayal not only influences public perception but also contributes to the broader discourse on international relations and cultural representation.

SOMALI CRITIQUES OF *CAPTAIN PHILLIPS*

The film Captain Phillips offers a striking example of how mainstream cinema perpetuates simplistic, binary representations that erase the deeper historical and socio-economic contexts of formerly colonized societies. A key post-colonial critique, voiced by Somali academic Abdi Latif Ega, is that the film reduces Somali pirates to chaotic, violent criminals while ignoring the structural legacies of foreign over-fishing, toxic dumping, and economic desperation that drove coastal communities to piracy in the first place (Latif Ega, 2019). On diaspora forums like SomaliNet, Somalis have criticized how the film depicts pirates as "literally shouting and yelling the whole time" with "no plan, no semblance of hierarchy," reinforcing stereotypes of irrational Black violence (Somalinet, 2013). Additionally, Somali-Americans in Minnesota have expressed frustration that the story silences the real history of how illegal foreign trawlers depleted Somali waters, pushing local fishermen to take up arms in defense of their livelihoods. While there is pride in Barkhad Abdi's performance, many Somalis feel the film ultimately locks their story within a Western hostage drama that upholds US heroism while rendering Somali voices marginal, if not disposable. This narrative is further explained by Omar Jamal, then chargé d'affaires of Somalia's UN mission:

"The movie really only depicts these kids as criminals, going after money… So the movie completely ignores that part of the story. It's the story told through the eyes of Captain Phillips…" (Jamal as quoted in Walsh, 2013)

This imbalance highlights how post-colonial media analysis must challenge such narratives and advocate for stories that center local perspectives and historical truth.

A HIJACKING: REPRESENTING A CAPITALIST SOLUTION TO A CAPITALIST PROBELM

While *Captain Phillips* offers a dramatized, Hollywood portrayal of Somali piracy centred on action and heroism, the Danish film *A Hijacking* (*Kapringen*), which premiered a year earlier at the 69th Venice International Film Festival, presents a more subdued and procedural depiction of the negotiation process. Avoiding spectacle, it focuses instead on the psychological strain and bureaucratic impasses that unfold in the wake of a ship's seizure. Directed by Tobias Lindholm, *A Hijacking* centres on the cargo vessel MV Rozen, hijacked by Somali pirates while transiting the Indian

Ocean. The film's narrative unfolds primarily through the experiences of the crew held hostage and the executives back in Denmark navigating ransom negotiations.

The filmmakers' commitment to realism is notable. Lindholm incorporated individuals from the maritime industry into the production: a real shipping company CEO plays the company negotiator, and seafarers portray the crew. Much of the dialogue is improvised, and real crisis consultants, including former British military hostage negotiator Skjoldmose-Porter, were brought in to guide the portrayal (Cinema, 2013). The result is a highly authentic depiction of the psychological toll and institutional dynamics involved in hostage negotiations. Unlike *Captain Phillips*, *A Hijacking* avoids glorifying any one figure or depicting overt heroism; instead, it depicts the slow bureaucratic grind of decision-making, pressure from shareholders, and the emotional devastation experienced by both hostages and executives. However, the film's critical focus on corporate realism comes at the expense of representing the Somali pirates with equal depth. While the pirates are visibly present throughout the film, they are largely rendered voiceless, and the absence of subtitles for Somali dialogue reinforces their role as an unknowable threat. Their motivations, socio-political contexts, and perspectives are left unexplored, presenting them more as obstacles to Western business continuity than as individuals shaped by structural conditions such as illegal fishing, toxic dumping, or economic exclusion. This narrative choice has been critiqued for reinforcing binaries of West vs. non-West, order vs. chaos, and for marginalising the Somali perspective in a film that otherwise strives for verisimilitude (Al-Mahadin, 2018).

What distinguishes A Hijacking (2012) is not its action but its institutional focus. The film lays bare the capitalist machinery underpinning maritime security: shareholders' interests, PR risks, and the monetisation of life in ransom negotiations. It captures the emotional fragmentation of those involved in corporate crisis response, showing how decisions are driven less by moral imperative and more by financial calculation. In this way, A Hijacking offers a subtle critique of capitalist responses to piracy, revealing not only who gets to speak and act in these crises, but also who suffers in silence. Yet, by leaving the Somali pirates' stories underdeveloped, the film risks sustaining the very binaries it might otherwise disrupt. The pirates become a background force—threatening, unpredictable, and functionally interchangeable. They are not given voice, history, or context, and this absence of narrative complexity reproduces the same spatial and racial exclusions seen in more overtly ideological films. Holm and de Bruijn (2016) argue that the filmmakers use an "affective-discursive approach" to construct Somali pirates as monstrous and irrational, denying them any meaningful complexity or historical grounding. One striking choice, as they note, is that all Somali dialogue is left untranslated, effectively marginalising the pirates' voices and rendering their motivations inaccessible to non-Somali audiences (Holm & de Bruijn, 2016, p. 113).

This narrative silence is not neutral: it erases the deeper realities of overfishing, toxic dumping, and economic desperation that have pushed some coastal communities into piracy (Holm & de Bruijn, 2016). Instead, the pirates become a visceral spectacle—"little better than savages"—while the trauma and moral dilemmas of Western characters are foregrounded. Thus, even as A Hijacking avoids the overt exceptionalism of Captain Phillips and offers a more nuanced portrayal of corporate maritime governance, it ultimately falls into a representational trap: centring Western trauma and institutional interests while sidelining the very actors whose lived experience of marginalisation fuels piracy in the first place. Like many maritime hostage dramas, A Hijacking shows how popular cinema continues to silence Somali perspectives through post-colonial narratives. The Somali pirate remains a figure of disruption rather than a voice of critique—reflecting the broader structures that determine whose pain is recognised, whose lives are valued, and whose subjectivities remain invisible in the geopolitical narratives of the sea (Holm & de Bruijn, 2016).

CAPTAIN PHILLIPS AND *A HIJACKING* IN CRITICAL COMPARISON

When studied together, Captain Phillips and A Hijacking demonstrate how mainstream cinema constructs the sea not simply as a physical frontier but as an ideological space where narratives of order and disorder, inclusion and exclusion, are actively reinforced to sustain Western capitalist logics. Both films engage the Indian Ocean as a zone in which the moral and economic order of the Global North collides with the perceived threat of the Global South. Although each film deploys distinct cinematic strategies, they converge in reproducing binary worldviews and a commodified approach to maritime insecurity that privileges Western interests and perspectives.

In Captain Phillips, this ideological project is explicit. The film mobilises well-worn tropes of American exceptionalism, positioning the stoic, rational figure of Captain Phillips as a moral anchor set against the backdrop of desperate, chaotic Somali pirates. Through the stark visual and narrative contrasts between the sleek modernity of the Maersk Alabama and the impoverished Somali coastline, the film renders the sea a site of threat that must be contained by American heroism and technological superiority. The intervention of the U.S. Navy is depicted as a clean, justified restoration of order, with little room for exploring the deeper structural drivers of Somali piracy such as foreign overfishing, toxic dumping, or the legacies of failed statehood. In doing so, the film naturalises the sea as a domain in which the Global South is imagined primarily through criminality and risk, requiring constant Western surveillance and control. A Hijacking, by contrast, avoids overt militaristic spectacle in favour of procedural realism, focusing instead on the psychological toll and corporate calculations involved in ransom negotiations. This framing exposes the capitalist machinery that underpins contemporary maritime security: the influence of shareholders, insurance premiums, and brand image in determining the value of human life at sea. The sea is thus rendered as a commercial artery whose flow must be safeguarded against disruption. However, despite its critical lens on corporate power, A Hijacking reproduces familiar binaries by leaving the Somali pirates largely voiceless and inscrutable. The absence of subtitles for Somali dialogue and the lack of meaningful context for the pirates' motivations effectively cast them as an amorphous, unknowable threat. Their silence reinforces the same representational exclusions seen in Captain Phillips, where the Somali pirate remains a figure defined through disruption rather than through any deeper exploration of the structural violence that fuels maritime crime.

Both films ultimately participate in what Edward Said identifies as the orientalist project: the Global South is depicted as unstable and inherently threatening, while the Global North is positioned as the locus of rationality, moral authority, and order. Crucially, both narratives treat the ocean as an economic space to be secured and exploited, aligning with long-standing patterns of capitalist spatialisation that render the sea a conduit for trade but also a site for the projection of Western security regimes. In both films, the Somali pirate is stripped of history and voice, framed primarily as a problem to be solved through military force or corporate negotiation rather than as a subject with legitimate grievances and agency. This comparative reading illustrates how cinema does not merely represent maritime spaces but actively shapes how audiences imagine them. Whether through the spectacle of American military rescue or the subdued realism of boardroom negotiation, both Captain Phillips and A Hijacking reinforce binaries of West versus non-West, order versus chaos, and moral civility versus violent desperation. They remind us that the sea, far from being a neutral expanse, is continually constructed in cultural texts as a zone where capitalist interests must be defended and where non-Western actors are rendered disposable. In doing so, these films contribute to broader discourses that legitimise unequal power relations and entrench stereotypes that continue to shape how maritime security is understood and governed.

CONCLUSION

This chapter has demonstrated that the sea is not a neutral or passive space but a contested and symbolically charged arena through which cinema articulates and reinforces the spatial logic of global capitalism. By analysing *Captain Phillips* and *A Hijacking*, the chapter explored two distinct but interrelated cinematic approaches to representing capitalist power at sea. *Captain Phillips* constructs a narrative of American heroism and technological superiority, portraying the sea as a site where global trade must be defended from external, racialised threats. In contrast, *A Hijacking* foregrounds the economic and psychological cost of maritime labour under capitalism, revealing how corporate decision-making transforms human lives into assets and liabilities. Together, the films present a fuller picture of how capitalism operates across maritime space: through both spectacular militarism and detached economic calculation.

The chapter's key conceptual contribution is the idea of cinematic territorialisation. This concept captures how cinema works to fix the fluid, borderless nature of the sea into emotionally and politically meaningful spaces. Through framing, narrative, and aesthetics, these films delineate zones of risk, order, and sovereignty, shaping public perceptions of who belongs at sea and under what conditions. While traditional understandings of territorialisation have focused on land-based sovereignty, the films discussed here highlight how cinematic narratives extend that logic across oceanic spaces, making the maritime legible and governable in the interests of capitalist expansion. This analysis also reveals what is left out. Somali voices are largely absent or instrumentalised, and the deeper structural causes of piracy—including environmental degradation, unequal trade relations, and postcolonial neglect—are glossed over. These omissions underscore the ideological function of cinematic territorialisation, which not only maps the sea but also defines whose stories matter and whose are erased.

By bringing together media studies, critical geopolitics, and maritime history, this chapter offers two contributions. First, it provides a framework for understanding how film participates in the symbolic and strategic ordering of the sea. Second, it uses piracy narratives to interrogate the broader assemblage of shipping companies, insurers, security firms, and media producers that shape maritime capitalism today.

REFERENCES

Agnew, J., & Corbridge, S. (1995). *Mastering space: Hegemony, territory and international political economy*. Routledge.

Al-Mahadin, S. (2018). The politics of the untranslated. *Journal of African Cultural Studies, 30*(3), 307–322. DOI: 10.1080/13696815.2018.1427557

Armenta, A. (2017). *Protect, serve, and deport: The rise of policing as immigration enforcement*. University of California Press. DOI: 10.1525/luminos.33

Aroesti, R. (2014, February 28). *Barkhad Abdi: Captain Phillips star's rise from limo driver to Oscar's contender*. The Guardian. https://www.theguardian.com/film/2014/feb/28/barkhad-abdi-captain-phillips-oscar-contender

Aruz, J., Graff, S. B., & Rakic, Y. (2014). *Assyria to Iberia at the dawn of the Classical Age*. Metropolitan Museum of Art.

Benton, L. (2010). *A search for sovereignty: Law and geography in European empires, 1400–1900*. Cambridge University Press.

Benton, L. (2014). *Law and colonial cultures: Legal regimes in world history, 1400–1900*. Cambridge University Press.

Bernard, A. (2012). *Gods and mortals: Ancient myths and modern life*. Profile Books.

Blackmore, H. (2009). *Pirates and emperors, old and new: International terrorism in the real world*. University of California Press.

Buchanan, J., & Buchanan, E. (2013). *Impossible odds: The kidnapping of Jessica Buchanan and her dramatic rescue by SEAL Team Six*. Atria Books.

Burlando, A., Cristea, A., & Lee, S. (2015). The trade consequences of maritime insecurity: Evidence from Somali piracy. *Review of International Economics, 23*(3), 525–557. DOI: 10.1111/roie.12183

Chomsky, N. (1957). *Syntactic structures*. Mouton. DOI: 10.1515/9783112316009

Cinema. (2013, May 9). *The Herald Scotland*. https://www.heraldscotland.com/life_style/arts_ents/13103812.film-consultant-real-life-experience-piracy-high-seas/

Craze, J. (2016). Mercenarism and state formation: The rise and fall of a private military market in the Horn of Africa. *African Affairs, 115*(461), 645–667.

de Souza, P. (1999). *Piracy in the Graeco-Roman world*. Cambridge University Press.

Latif Ega, Abdi . "The Great Somali Apologia." *Africasacountry.com*, 26 Mar. 2019, africasacountry.com/2019/03/the-great-somali-apologia. Accessed 11 July 2025.

Elden, S. (2013). *The birth of territory*. University of Chicago Press. DOI: 10.7208/chicago/9780226041285.001.0001

Elden, S. (2021). Terrain, politics, history: Thinking the volume. *Political Geography, 87*, 102431. Advance online publication. DOI: 10.1177/2043820620951353#

Figueroa-Caballero, A. (2019). The criminalization of immigration in the United States: Examining the effects of news coverage linking undocumented immigrants with criminality. *Communication Monographs, 86*(1), 112–132. DOI: 10.1080/03637751.2018.1505049

Gabbert, J. (1986). Piracy in the early Hellenistic period: A career open to talents. *Greece & Rome*, 33(2), 156–163.

Galland, D., & Gronning, M. (2019). The social production of spatial knowledge. *Environment & Planning A, 51*(2), 287–304.

Gounaris, B. C. (2017). The privatization of war: Privateers in the Mediterranean and beyond. *Mediterranean Historical Review, 32*(2), 135–152.

Greengrass, P. (Director). (2013). *Captain Phillips* [Film]. Columbia Pictures.

Hallwood, P., & Miceli, T. J. (2015). *Maritime piracy and its control: An economic analysis*. Palgrave Macmillan. DOI: 10.1057/9781137461506

Hansen, S. J. (2009). *Piracy in the greater Gulf of Aden: Myths, misconceptions and remedies. Norwegian Institute for Urban and Regional Research*. NIBR. DOI: 10.7577/nibr/rapport/2009/29

Hanson, H., & O'Rawe, C. (2010). *The femme fatale: Images, histories, contexts*. Palgrave Macmillan. DOI: 10.1057/9780230282018

Harding, A. (2016). *The mayor of Mogadishu: A story of chaos and redemption in the ruins of Somalia*. St. Martin's Press.

Heebøll-Holm, T. (2013). *Ports, piracy, and maritime war: Piracy in the English Channel and the Atlantic, c.1280–c.1330*. Brill. DOI: 10.1163/9789004248168

Heuvelmans, B. (2003). *In the wake of sea serpents*. Kegan Paul International.

Holm, E., & de Bruijn, M. (2016). Pirates and the discursive construction of the Somali coast. In Biersteker, T. J. (Eds.), *International Political Sociology*. Routledge.

Homer. (n.d.). *The Odyssey* (Book XII). (R. Fagles, Trans.). Penguin Classics.

Johnson, J. C. (2014). *Spaces of detention: Spatiality and the organization of immigration detention*. Routledge.

Johnson, J. C. (2020). *Seafarers, statecraft, and the rise of maritime frontiers in the ancient Mediterranean*. Oxford University Press.

Kingston, R. (2014). *The pirate's nest: The rise and fall of Blackbeard's pirate haven*. University of North Carolina Press.

Larson, E. M. (1717). *Wallenberg's account of the Kraken*. [Primary Source].

Lee, S. (2025). *Cinematic Landscapes and Geopolitical Imaginaries*. Numberanalytics.com. https://www.numberanalytics.com/blog/cinematic-landscapes-geopolitical-imaginaries

Lefebvre, H. (1972). *The production of space*. Blackwell.

Lefebvre, H. (1974). La production de l'espace. *Anthropos: Studie z Oboru Anthropologie, Paleoethnologie, Paleontologie a Kvartérní Geologie*.

Markus, H. (2021). Maritime masculinity and the myth of *Captain Phillips. Journal of Cultural Studies*, *15*(1), 78–93.

Massey, D. (2013). *Space, place and gender*. University of Minnesota Press.

Mills, R. (2014). Letters of marque and reprisal. *Journal of Maritime Law and Commerce*, *45*(2), 201–215.

Murphy, M. (2011). *Small boats, weak states, dirty money: Piracy and maritime terrorism in the modern world.* Hurst.

Nilmaola, S. (2017). Deconstructing cinematic piracy: Representing Somali pirates in *Captain Phillips. Critical Studies in Media Communication*, *34*(4), 347–362.

Pevere, G. (2013, August 16). *A Hijacking: Hostage thriller turns focus to the cost of saving lives*. The Globe and Mail. https://www.theglobeandmail.com/arts/film/film-reviews/hostage-thriller-turns-focus-to-the-cost-of-saving-lives/article13802725/

Rubarth, S. (2015). Masculinity and the sea in Greek myth. *Classical World*, *108*(3), 323–343.

Said, E. W. (1978). *Orientalism*. Pantheon Books.

Shaheen, J. G. (2001). *Reel bad Arabs: How Hollywood vilifies a people*. Interlink Publishing Group.

Sherry, F. (1986). *Raiders and rebels: A history of the golden age of piracy*. Harper & Row.

Simonds, J. (2025). *Maritime risk, insurance, and seafarer safety: Navigating piracy in the Indian Ocean*. Routledge. DOI: 10.4324/9781032664736

Somalinet. (2013, October 12). "Captain Phillips! - SomaliNet Forums." somalinet.com/forums/viewtopic.php?t=336546#gsc.tab=0.

Spiegal, S. (1991). *The pirate myth: Genealogies of an imperial figure*. University of California Press.

Steinberg, P. (2001). *The social construction of the ocean*. Cambridge University Press.

Steinberg, P., & Peters, K. (2015). Wet ontologies, fluid spaces: Giving depth to volume through oceanic thinking. *Environment and Planning. D, Society & Space*, *33*(2), 247–264. DOI: 10.1068/d14148p

Steitkasimova, E. (2019). Dangerous femininity: Sirens and the patriarchal Athenian society. *Classical Mythology Review*, *12*(3), 44–57.

Sy, T., & Smith, J. (2013). The exceptional nation: Framing of the U.S. in *Captain Phillips*. *Journal of American Popular Culture*, *12*(1), 58–72.

Tawil-Souri, H. (2014). Cinema as the Space to Transgress Palestine's Territorial Trap. *Middle East Journal of Culture and Communication*, *7*(2), 169–189. DOI: 10.1163/18739865-00702005

Thomson, J. E. (1994). *Mercenaries, pirates, and sovereigns: State-building and extraterritorial violence in early modern Europe*. Princeton University Press.

Vallar, C. (2009). *Pirates and plunder: A history*. Maritime Archives Press.

Vélez-Serna, M., & Stauff, M. (2023, June 7). Ports: On the material and symbolic mediation of global capitalism - NECSUS. NECSUS. https://necsus-ejms.org/ports-on-the-material-and-symbolic-mediation-of-global-capitalism/

Vuillemin, J. (as cited in Grozholz, P.). (2007). History enters into the considerations of theoretical philosophy through mathematics. *Philosophy of Mathematics Review*, 28(3), 228–232.

Walsh, J. "Does Feel-Good "Captain Phillips" Film Unfairly Depict Somali Plight?" *MinnPost*, 16 Oct. 2013, www.minnpost.com/politics-policy/2013/10/does-feel-good-captain-phillips-film-unfairly-depict-somali-plight. Accessed 11 July 2025.

ADDITIONAL READING

Ansong, J. O., Isaksson, I., Suárez de Vivero, J. L., & Varjopuro, R. (2021). *MSPglobal international guide on marine/maritime spatial planning* (IOC Manuals and Guides No. 89). UNESCO-IOC. https://www.researchgate.net/publication/355117157

Benton, L., & Bailey, C. (2020). *A search for sovereignty: Law and geography in European empires, 1400–1900* (2nd ed.). Cambridge University Press.

Bueger, C., & Edmunds, T. (2020). Blue crime: Conceptualising transnational organised crime at sea. *Marine Policy, 119*, 104067. DOI: 10.1016/j.marpol.2020.104067

Bueger, C., & Edmunds, T. (2021). Maritime security: The politics of transnational interventions. In Bueger, C., & Edmunds, T. (Eds.), *Maritime security: The politics of interventions and the transformation of international order* (pp. 1–25). Routledge.

Menjívar, C. (2021). The racialization of "illegality". *Daedalus, 150*(2), 91–105. DOI: 10.1162/daed_a_01848 PMID: 37168289

Menzel, A. (2022). Fragmentation or effective governance? The regime complex of counter-piracy in Asia. *Politics and Governance, 10*(2), 145–154. DOI: 10.17645/pag.v10i3.5380

Nwalozie, C. J. (2020). Exploring contemporary sea piracy in Nigeria, the Niger Delta and the Gulf of Guinea. *Journal of Transportation Security, 13*(1), 31–55. DOI: 10.1007/s12198-020-00218-y

Prins, B. (2021). How history predicts COVID-19's impact on maritime piracy, and what America can do to help. *Howard H. Baker Jr. Center for Public Policy Brief.* https://baker.utk.edu/publication-nsfa-maritime-security/how-history-predicts-covid-19s-impact-on-maritime-piracy-and-what-america-can-do-to-help/

Rafael, T., Cabral, H., Mourato, J., & Ferrão, J. (2024). Marine spatial planning: A systematic literature review on its concepts, approaches, and tools (2004–2020). *Marine Policy, 158*, 106864. DOI: 10.1016/j.marpol.2023.106864

Shapiro, M. J. (2009). *Cinematic geopolitics*. Routledge.

Storey, I. (2022). Piracy and the pandemic: Maritime crime in Southeast Asia, 2020–2022. *ISEAS-Yusof Ishak Institute Perspective*, 2022(61), 1–12. https://www.iseas.edu.sg/articles-commentaries/iseas-perspective/2022-61-piracy-and-the-pandemic-maritime-crime-in-southeast-asia-2020-2022-by-ian-storey/

United Nations. (1982). *United Nations Convention on the Law of the Sea*. Retrieved from https://www.un.org/depts/los/convention_agreements/texts/unclos/unclos_e.pdf

Velázquez, M. C. (2023). *Cultural representations of piracy in England, Spain, and the Caribbean: Travelers, traders, and traitors, 1570 to 1604*. Routledge. https://doi.org/DOI: 10.4324/9781003151715

KEY TERMS AND DEFINITIONS

Buccaneer: Originally referring to hunters on the Caribbean islands, "buccaneer" came to describe semi-lawless sea raiders who attacked Spanish ships and settlements in the 17th century. Though sometimes possessing vague or local commissions, buccaneers operated largely outside formal naval structures and were known for their brutal tactics. They were especially active in the Caribbean and along the Spanish Main.

Cinematic Territorialization: The use of film to produce, legitimize, and emotionally charge specific narratives of space. Films such as *Captain Phillips* and *A Hijacking* are discussed as tools that shape public understanding of maritime order, highlighting whose experiences are centered and who's excluded.

Filibuster: In a maritime and 19th-century political context, a filibuster was an adventurer who launched unauthorized military expeditions into foreign territories, often with the intent of inciting rebellion or annexing land. Unlike privateers, filibusters acted without any formal government approval, targeting regions like Latin America. Famous filibusters include William Walker, who briefly seized control of Nicaragua.

Hostis Humani Generis: A legal term meaning "enemy of all mankind." Historically used to describe pirates, this designation marks certain actors as beyond the protection of law and order, legitimizing their exclusion from maritime and legal space.

Mythical Monsters: Symbolic creatures such as the Siren or Kraken used historically to mark the sea as dangerous or unknowable. In this chapter, these figures are analyzed as ideological tools that instill fear, enforce boundaries, and reproduce gendered and political hierarchies in maritime space.

Privateer: A privateer was a privately owned vessel granted legal authority by a government (via a letter of marque) to attack and capture enemy ships during wartime. Unlike pirates, privateers operated under state sanction and were expected to target only enemy nations. Their activities blurred the line between lawful combat and piracy, especially when state backing was weak or abused.

Spatial Consciousness: The mental and emotional awareness of space, shaped by personal experience, narrative, and media. In this chapter, spatial consciousness is explored as a socially constructed phenomenon that allows individuals to understand and relate to maritime spaces, even those they have never physically encountered, through stories, myths, and visual culture.

Spatial Economy: The informal and formal rules that govern how space is used, who is allowed to occupy it, and who is excluded. This concept explains how capitalist systems reward certain spatial behaviors (like trade and corporate navigation) while criminalizing others (like piracy), reinforcing dominant power structures at sea.

Territorialization: The process through which space becomes politically claimed, controlled, and ordered. Traditionally applied to land, the concept is reworked in this chapter to include the sea as a contested site of symbolic, legal, and narrative control through mechanisms such as film, policy, and myth-making.

'Us' vs. 'Them' Binary: A representational divide used to distinguish legitimate insiders from deviant outsiders. This binary is embedded in maritime narratives and media portrayals, often aligning the West with order and morality, and non-Western actors with chaos and illegitimacy.

Chapter 8
Contemporary WWII Films and Television Series:
Constructing and Deconstructing Heroes and Villains

Zitianxiao Xu
https://orcid.org/0009-0007-6883-8642
University College London, UK

Muhammad Yaqoub
https://orcid.org/0000-0002-4188-0273
Tongling University, China

ABSTRACT

This study analyzes how World War II–themed films and television productions from the 1990s to the present represent and deconstruct the archetypal hero–villain dynamic within the framework of traditional heroism. Centering on Saving Private Ryan, Band of Brothers, and The Thin Red Line, the discussion explores how these works both adhere to and challenge conventional heroic narratives. While Saving Private Ryan and Band of Brothers reinforce traditional heroic and villainous tropes through established heroism—thereby bolstering nationalist and patriotic discourses that legitimize American military endeavors and address past national traumas—The Thin Red Line offers a philosophical deconstruction of traditional heroism. Director Terrence Malick's distinct style transcends Hollywood's commercial mainstream, providing audiences with a contemplative lens on the nature of war.

DOI: 10.4018/979-8-3373-7589-2.ch008

INTRODUCTION

The genre of the war film is not confined merely to representations of actual warfare; rather, it encompasses historical events, futuristic imaginaries, and mythical reenactments, reflecting the genre's expansive temporal and spatial scope as well as its cultural inclusivity (Eberwein, 2010, p.12). From its inception, the war film has functioned as a collective cultural practice responding to public demands and constructing identity, thereby fulfilling significant roles in social communication and ideological dissemination (Eberwein, 2010, p.18). Bronfen's (2012) study on Hollywood war cinema suggests Hollywood war films, on the one hand, constitute America's national cultural possessions, on the other hand, audiences, confronted with Hollywood's emotionally engaging cinematic reconceptions of war, are unable to remain impartial and 'are called upon to respond' (2012, p. 20). Eberwein (2010, p.18), drawing upon Thomas Schatz's research on the origins of genre in general, argues that genre films—such as war films—function similarly to myths within society. When confronted with collective contradictions and anxieties, genre films provide structured narratives that interpret conflicts and regulate collective emotions. Consequently, genre cinema frequently relies on such structured narrative frameworks to address contemporary issues, offering audiences both emotional reassurance and cognitive resolution. Specifically, heroic figures in war films often exhibit this structural inclination; however, an over-reliance on traditional narrative structures risks entrapment within conventional forms of heroism, reinforcing simplistic binary oppositions. This narrative pattern fails to genuinely alleviate societal anxieties and instead may foster illusions of national exceptionalism and encourage outward projections of hostility. Similarly, Schönfeld discusses German war films about World War II, arguing that German war cinema, like its American counterpart, functions as a tool for constructing historical memory. It also serves as a vehicle for re-education, being shaped by Germany's reflection of Nazi past and by audiences' desire to avoid repeatedly and directly confronting wartime guilt (2022, p.200, p.207). For instance, post-war German 'rubble films' employed visual strategies that veiled direct responces to war, such as increasing camera-subject distance and utilizing flashback to mitigate emotional resonance with audiences, which has long set the tone for German filmic engagement with the memory of World War II (Schönfeld, 2022). Therefore, although German World War II films, like their Hollywood counterparts, the strategies used in German cinema for constructing heroic figures significantly deviate from traditional heroism.

World War II (WWII) has long served as a vital source for war narratives, occupying a central position in films and television series. The World War II combat film has a clear, specific marking and recognizable feature, making it an ideal case for analyzing the emergence, transformation, and continuity of a film genre (Basinger,

2003, p. 27). Over time, influenced by evolving contemporary ideological discourses, screen representations of WWII have developed diverse narrative styles. Even today, WWII remains a key reference point for war-related visual narratives, with exemplary works in this genre continuing to reflect the operative logic of national ideology, implicitly exerting normative and disciplinary effects upon the public. There are three general phases in their historical development based on Misek's (2008, p. 117) chronology of WWII combat films. The first phase, from the 1940s to the 1950s, was dominated by films extolling the justice of war and traditional heroism, such as *Twelve O'Clock High* (1949) and *The Dam Busters* (1954). These works typically focused on aerial or naval warfare, circumventing ground combat's graphic violence and confusion. The second phase, spanning from the mid-1950s to the 1970s, witnessed a shift toward revisionist tendencies. Directors began to depict the harsher realities of ground warfare and its psychological toll on soldiers, moving away from purely heroic tropes. Though Robert Aldrich's *Attack* (1956), Nicholas Ray's *Bitter Victory* (1957), and Don Siegel's *Hell is for Heroes* (1962) maintained aspects of valorizing the heroic figure, they increasingly revealed the grim nature of warfare. Nonetheless, while these films exposed the brutality of war, they did not necessarily adopt an entirely anti-war stance. Entering the 1970s through the 1990s, the Vietnam War and its accompanying anti-war sentiment in the United States made traditional WWII heroism appear increasingly out of step with contemporary attitudes. In 1968, John Wayne's *The Green Berets* (1968) was widely criticized for importing WWII's heroic narrative into the Vietnam context, leading to what critics have termed 'the metaphorical death of John Wayne in Vietnam.' The resulting cultural climate fostered a wave of films that questioned the legitimacy of warfare and highlighted psychological trauma. While many Vietnam War films critiqued the political dimensions of war, they still retained visually intense combat sequences for cinematic effect. Against this backdrop, some WWII productions also absorbed influences from the Vietnam era. Films like *Patton* (1970) and *Cross of Iron* (1977) questioned the political justification of war while still delivering visually spectacular battle sequences.

From the 1990s onward, WWII-themed works have regained heightened attention, largely because its morally resonant 'good war' narrative (Paget & Lipkin, 2009, p. 93) resonated with post–Gulf War American triumphalism and the national project of restoring patriotic pride. Despite the surge of WWII works in this period, only a handful delved deeply into core wartime questions. Among these, distinctions arose between critical and conservative camps: some directors used historical events to critique American militarism, while others sought to reinforce heroic soldier imagery (Hodgkins, 2002, p. 74-76).

The conservative camp aims to revitalize traditional heroism, thereby restoring the American soldier to his rightful status as a heroic icon. Steven Spielberg is a prototypical example of this conservative perspective (Hodgkins, 2002, p. 76). His conservatism emerges in his narrative 'regression': a resurgence of John Wayne–style heroism, albeit with modernized narrative techniques. Spielberg also employs moral absolutism, simplifying war as a 'good vs. evil' confrontation. His works emphasize heroism and American honour, framing history through a lens that elevates heroic sacrifice and national identity (Dibattista, 2000, p. 223). In *Saving Private Ryan*, Spielberg employs a secularized Jeremiad narrative to address the crisis of American national identity after Vietnam (Owen, 2002, p. 249). The film acknowledges this existential crisis of American national identity and strategically resolves it through a heroic war narrative, offering viewers a symbolic 'homecoming' to a 'mythic America.' *Band of Brothers* had a notable influence on American war culture post-9/11. While primarily a historical docudrama, it resonated with contemporary national sentiments, reinforcing themes of camaraderie, sacrifice, and resilience in warfare. In a post-9/11 context, it was interpreted as a reaffirmation of military ideals and national unity (Paget & Lipkin, 2009, p. 93). Compared to other World War II-themed films, Spielberg's work displays distinctiveness in several significant aspects. Compared *Pearl Harbor* (2001) closely replicates the traditional heroic archetypes familiar to 1940s wartime audiences (Dick, 2022, p. xxiii), and *Hacksaw Ridge* (2016), despite its apparent subversion, remains rooted in character-building logic inherited from 1940s war films (Dick, 2022, p. xxvii), *Saving Private Ryan* (1998) introduces a structural reversal by positioning its central narrative before the climactic battle (Dick, 2022 p. xxviii). The structural inversion created by the juxtaposition of the cemetery scene at the beginning and the mission scene at the end seems a departure from traditional narrative patterns. However, the film's formal deviation is confined to the overall narrative structure and does not alter the specific stages of the hero's development, thereby underscoring the resilience of the traditional heroic narrative, which persists beneath new narrative frameworks rather than being entirely subverted.

As a cinematic form, war films can either reinforce conventional war narratives or challenge militaristic ideology (Hodgkins, 2002, p. 74-76). The critical camp focuses on interrogating militarism and technological alienation. These works call attention to the concept of 'sanctioned killing,' illuminating how military mechanisms systematically destroy human life (Hodgkins, 2002, p. 74). Terrence Malick's *The Thin Red Line* (1998), for instance, adopts an existential lens to reinterpret WWII memory, emphasizing personal spiritual struggles and the philosophical weight of warfare (Dibattista, 2000, p. 226-227). Its narrative structure and character portrayal stand in clear contrast to *Saving Private Ryan*, sparking ongoing scholarly debate and comparison. Cohen (1999, p. 241) argues that *The Thin Red Line* falls short both as a war film and as an adaptation. From his viewpoint, it suffers from 'sentimentality and sheer lack of comprehension,' departing from the essence of

James Jones's original novel. Consequently, the author's raw depiction of warfare becomes overshadowed by 'mock heroics and stylized histories.' As suggested by this analysis, *The Thin Red Line* constitutes a subversion of the traditional heroic narrative paradigm.

By focusing on Hollywood World War II films and television series, this paper analyzes the representational strategies of heroes and antagonists to reveal how genre-based narrative structures sustain or challenge traditional models of heroism in response to shifting ideological and cultural demands. This article mainly adopts a narrative perspective to examine how *Saving Private Ryan* and *Band of Brothers* represent traditional heroes and antagonists in cinematic and television contexts, exploring how these portrayals reaffirm or subvert the 'Hero's Journey' paradigm. It then investigates how *The Thin Red Line* deploys natural elements to deconstruct traditional hero narratives and the hero-villain dualism. Furthermore, the article will delve into the structural factors underpinning the narrative styles of these three works and analyze the distinct audience experiences they generate. By tracing these differences, the study aims to uncover the underlying logic of contemporary Hollywood WWII-themed cinematic storytelling and its broader cultural implications.

1. The Righteous War in *Saving Private Ryan* and *Band of Brothers*

1.1 The Hero's Journey and the Binary Model

Drawing on the theory of the Hero's Journey, this section examines how WWII films and TV dramas construct and deconstruct hero and villain images. According to Campbell's (2008) canonical model, the Hero's Journey consists of three core stages: Separation, Initiation, and Return. In the Separation stage, the hero is often summoned by an external or internal call but may briefly reject it out of fear or self-doubt (2004, p. 32). Subsequently, driven by supernatural or external forces, the hero crosses the Threshold, saying goodbye to his or her established identity and way of life. In the traditional Hero's Journey, the 'separation' phase is typically triggered by a call of fate, a supernatural event, or a chance discovery (Campbell, 2004, p. 34). However, in war films set in World War II, this call often comes directly from the orders of superiors or the will of the state. Due to the constraints of the military hierarchy and military status, the protagonist's 'refusal' mostly stays at the psychological or verbal level, and it is more difficult to resist the order through actions. Taking *Saving Private Ryan* as an example, Captain John H. Miller of the 2nd Ranger Battalion questions the mission to rescue Ryan, not out of open disobedience, but rather from a sense of helplessness, but he still chooses to fulfil his duty as a soldier in the end, showing his loyalty and discipline as a soldier. In addition, heroic characters are often excellent at physical fitness and leadership, able to carry out strategic missions as well as maintain morale and order in their ranks, thus demonstrating the core characteristics of the traditional hero.

During the initiation stage, the hero encounters various trials and challenges and may receive external aid that enables him to transcend his limitations and develop further (Campbell, 2004, p. 35). For example, in *Band of Brothers*, Winters demonstrated excellent physical fitness when he led the training of soldiers as a platoon leader in his unit during the recruit training period, and since then, he has even demonstrated strong leadership and military qualities in various battles. For example, on D-Day, after parachuting in, Winters was able to calmly gather his team and use the compass to quickly find the direction of the march and led everyone to victory when he encountered a German squad. This also demonstrates that the hero possesses strong leadership skills. According to Campbell's theory, the hero also encounters archetypal female figures, often representing either spiritual guidance (the 'Goddess') or temptation and fear (the 'Temptress') (2004, p. 72, 77). Conversely, some characters might symbolize temptation or fear; their negative energy must be overcome or integrated for the hero to face patriarchal imagery—symbolizing rationality, order, and authority—and achieve spiritual integration (Apotheosis) (Campbell, 2004, p. 91). In the initiation phase of war narratives, traditional symbols such as 'Goddess' or 'Temptress' are often replaced or transformed due to the absence of female characters on the frontline battlefields. For example, the Guiding Woman (maternal wisdom) often appears as a wife or lover in memories, injecting the hero with the power of humanity and warmth; the Seductive and Fearful Woman is generally no longer a specific female character, but rather the hero's inner fear, vulnerability, and fear of losing his masculinity. If the hero can overcome these symbolic weaknesses, he can reaffirm his warrior identity and symbol of fatherhood. *Saving Private Ryan*, on the other hand, indirectly reinforces the evil image of the villain through the presentation of the hero's fears. Private Mellish, the only Jewish soldier in the squad, not only carries the historical trauma of the Holocaust, but also presents the role of Jews in the totalitarian Nazi regime through symbols such as the Hitler Youth knife, the silence of his teammates, and the memorial jar. He not only bears the historical trauma of the Nazi Holocaust but also embodies this suffering through key symbols, such as the Hitler Youth knife and the silence of his teammates. These elements collectively illustrate the Jews' plight under Nazi totalitarianism (Ehrenhaus, 2001, p. 325). Thus, in the narrative, the memory of the Holocaust is rekindled through the portrayal of the Jewish heroic character. The internal suffering endured by this character serves as a conduit, indirectly reinforcing the depiction of Germans as entirely malevolent antagonists. The villain is further characterized as a ruthless war machine in the scene where Miller is fatally shot by a German prisoner of war—one he had previously spared, despite the latter's pleas for mercy. The German soldier chooses to continue fighting after returning to his unit and shoots Miller in the final battle to hold the bridge, resulting in his death. This portrayal deepens the villain's image as a ruthless enforcer of war.

In the return phase, the hero returns to society with the 'treasure' (Boon) acquired during the initiation phase. However, in the context of war narratives, although the hero may be assisted by allies or external forces (reinforcements) during the return phase, the hero often pays with his or her life because of the gruelling nature of war (Campbell, 2004, p. 35). For example, Miller is shot during the duel and before he dies, he gives Ryan the spiritual handover of being a good man. Ryan returns to Miller's grave in his old age to pay homage. This demonstrates that in war narratives, even when heroes meet their demise, their legacy persists through the 'return' of their spirit, influencing society and future generations. This also further shapes the image of the hero, reflecting that the hero always sticks to justice, preferring to sacrifice himself rather than violate his beliefs and values. This linear narrative mode often maintains the stability of the story structure with the help of a single centralized point of view, thus amplifying individual heroism and relatively weakening the growth and subjectivity of the group of characters. Sometimes the hero's stability and reliability are even emphasized by the weakness of other characters. This also means that the hero rarely shows vulnerability or weakness, symbolizing absolute strength and victory. The lack of in-depth portrayal of their inner struggles and moral conflicts also results in a less-than-three-dimensional portrayal of the hero.

The relative absence of women's roles also creates another phenomenon: the child often replaces the woman as part of the 'Goddess' function. Through the vulnerability and innocence of the child, the hero displays fatherhood and protectiveness in addition to his otherwise cold warrior status, thus accomplishing some sort of emotional or moral redemption. Taking *Band of Brothers* as an example, the fourth episode of the series depicts the liberation of Eindhoven, the Netherlands, during Operation Market Garden in 1944 and the subsequent fighting: in the scenes of public punishment for women who had sex with the enemy during the occupation, the soldiers are mostly bystanders, otherizing these women; when the soldiers meet with the mother with her suffering child, they take the initiative to provide supplies and show paternal concern, shaping a more complex moral and psychological transformation. For the stigmatized women, once they are connected to motherhood (nurturing children), it is easier for them to regain the sympathy and help of the soldiers, indirectly reflecting the female roles waiting for redemption in war narratives.

The World War II war narrative in the integration of the 'hero's journey' mode, both retained the classic separation-initiation-return three-part structure, by the influence of the traditional heroes of the narrative of the two works of the heroes of the role of the sense of justice, obedience, willingness to sacrifice, powerful physicality and leadership. The villains, on the other hand, are othered, marginalized, or even invisible, usually appearing collectively and existing only to advance the plot or highlight the hero. This echoes Baudrillard's critique of the Gulf War, during which the U.S. media constantly emphasized the heterogeneity of Iraqis such as their religious beliefs rather than trying to make the public understand these cultural backgrounds. This otherness resulted in the war becoming more acceptable as the 'enemy' was portrayed as a completely different and incomprehensible object (Baudrillard, 2006). As a result, *Saving Private Ryan* and *Band of Brothers* are characterized by the dichotomy of heroes and villains through the reproduction of a linear narrative of traditional heroism.

It can also be seen that contemporary Hollywood films and TV dramas in the TV network era share a similar logic of heroism in politicized hero narratives, but there are still significant differences in narrative structure and content arrangement. In traditional theatrical film *Saving Private Ryan*, the hero's growth line usually revolves around a single protagonist or a few central characters who have distinct personalities and are differentiated from each other to enhance audience recognition. Cinematic narratives reinforce the protagonist's centrality, ensuring their dominance in key moments, thereby heightening their symbolic stature and dramatic impact. In contrast, the narrative mode of TV dramas pays more attention to group portrayal, where heroism is no longer confined to the individual, but is more of a group image of teamwork and collective sacrifice. For example, war documentary dramas such as *Band of Brothers* adopt a group narrative that breaks away from the single heroism model and constructs a comprehensive depiction of war through the experiences of diverse characters (Paget & Lipkin, 2009, p. 96). The continuity of the episodes provides a longer cycle of character growth, allowing the audience to gain insight into the psychological changes and war experiences of multiple characters.

In general, however, neither war films nor war-themed television dramas have been able to completely escape the framework of heroism. Whether it is individual heroism in films or group heroism in TV dramas, individual or group heroes are to a certain extent endowed with righteous qualities, forming a stark opposition to the villains, to strengthen the moral legitimacy and political meaning of war.

1.2 National Honour

Ferro (1988) argues that war films often enter 'psycho-socio-historical zones,' a conceptual space where ideology, history, and reality intersect to shape audience perceptions of war. Thus, the visual and thematic elements of war films often reflect the prevailing political climate, and in the 1970s, the United States underwent a period of profound national identity crisis. On the one hand, the 1971 Pentagon Papers revealed the government's deceitfulness over the Vietnam War, shaking the public's trust in the official narrative; on the other hand, the Watergate scandal further revealed the Nixon administration's corruption and abuse of power, deepening the public's disappointment and anger. The United States' defeat in the Vietnam War (1975) dealt a significant blow to its global hegemony, exacerbating a national identity crisis that persists today (Bates, 1998).

For many Americans, the 1991 Gulf War stands as one of the rare 'perfect wars' post-WWII: lasting only 42 days, incurring minimal casualties, and receiving widespread multinational support. This 'clean' victory contrasted sharply with the quagmire of the Vietnam War. It became an opportunity for the United States to rebuild its military self-confidence and escape from the shadow of the Vietnam War. In the aftermath of the Gulf War, the U.S. government was given a new opportunity to restore military glory and advance patriotic narratives. World War II cinema was an important vehicle for expressing this (Hodgkins, 2002, p. 79). According to Zinn (1998, p. 39), governments and the media induce the public to voluntarily support wars by portraying them as just. If war is 'justified', it can gain social acceptance even if it is brutal. In the late 1990s, war films shifted their focus from political legitimacy to the moral dimension, tending to present them as 'moral wars', emphasising the soldiers' convictions rather than their purely political stance (Gates, 2005, p. 298).

Saving Private Ryan emerged within this context—by centering on a heroic narrative, it seeks to reaffirm a sense of moral clarity and duty, steering the audience away from the skepticism and moral confusion surrounding Vietnam and back toward the 'American myth' of World War II, where the purpose of war was presented as just and unambiguous. The defeat in Vietnam led many Americans to doubt the legitimacy of the war, and the victory in the Gulf War provided an opportunity to rebuild the military's reputation. On an emotional level, the just-heroic narrative of *Saving Private Ryan* resonates with the rhetoric of justice surrounding the Gulf War, thereby attempting to legitimize governmental actions. Apart from that, *Saving Private Ryan* used high-precision special effects and documentary aesthetics to create an immersive and hyper-realistic depiction of the gruesomeness of D-Day. However, in doing so, it also transformed battlefield carnage into an intense cinematic spectacle, much like the Gulf War's media coverage turned destruction into a 'magical spectacle' ((Taylor, as cited in Hodgkins, 2002, p. 76; Virilio, as cited in Hodgkins, 2002, p. 76). Both employ technological spectacle to obscure the political complexities of war, thereby reinforcing the myth of 'American technological invincibility'. *Saving Private Ryan* is also a film that uses cinematic techniques and affective narratives to reimagine 'the good name of war'.

Entering the 21st century, the events of 9/11 once again changed the narrative strategy of the United States towards war. Based on the need to combat terrorism, the Bush administration rapidly launched a worldwide military campaign, which was in dire need of social consensus. In this context, the U.S. government and the media have joined forces to reshape the narrative of the legitimacy of war through various means of propaganda. WWII films and television series were once again imbued with political and cultural significance, with *Band of Brothers* standing as one of the most emblematic series of this period (Paget & Lipkin, 2009, p. 93). In *Band of Brothers*, the dichotomy of heroism and villainy is not only a convention of war narratives but also an invisible propaganda tool to justify war. By emphasising collective sacrifice, comradeship and moral condemnation of the enemy, the show reinforces the idea of a 'just war' within the framework of the World War II narrative, reinforcing the justification for warfare in the context of the US war on terror.

Under certain historical events and political contexts, war films are not only a form of art, but also a narrative tool with ideological influence, subconsciously shaping the audience's perception of war. When political leaders and popular culture collaborate to mythologize war, the public inevitably embraces the binary logic of 'our side is just,' thereby overlooking the underlying causes of war such as resource competition, hegemonic interests. Such narratives not only simplify history but also undermine critical thinking about war through emotional mobilisation such as heroism. The presentation of national heroism is crucial in this narrative structure. Through the positive portrayal of World War II heroes and the stereotypical degradation of the enemy group, works of World War II can simultaneously reinforce the legitimacy of war and its sublime significance. Consequently, war is no longer framed merely as an act of state violence but is instead elevated into a sacred symbol of national memory, serving to reshape national identity and legitimise governmental authority in the post-war period.

1.3 Violence and Nostalgia

This clear dichotomy reinforces the moral legitimacy of war but also provides a rich mechanism for the pleasure of violence. Maslin (1998) argues that the Normandy landing scenes in *Saving Private Ryan* depict war violence with horrific realism. In contrast, traditional war films often rely on visual spectacle, rapid editing, multichannel sound effects, and other technical elements to heighten audience engagement and intensify the thrill of combat. Panofsky (2012, p. 39–54) argues that audiences' attraction to depictions of violence stems from an innate psychological drive, which he attributes to a 'primordial instinct for bloodshed and cruelty'. This primal attraction to combat imagery explains why war films frequently exploit graphic violence as a key source of engagement, reinforcing the pleasure mechanism in traditional war cinema. The pleasure derived from violent spectacle is often contingent on an 'us vs. them' dichotomy, wherein violence against 'the other' is legitimized as a form of gratification (Misek, 2008, p. 116). The clear definition of this binary makes violence against the other justified and necessary, which further enables the audience to indulge in the sensual pleasure of violence without the moral dilemma or emotional struggle of sympathising with or identifying with the villain, thus enhancing the drama and spectacle of war narratives. However, when war narratives are simplified into a binary struggle between good and evil, war films risk prioritizing violent spectacle over critical engagement, thereby becoming mere entertainment products rather than thoughtful reflections on the nature of war. Violence operates not only through the hero-villain dichotomy but also through the distancing effect of technology. The mediated representation of war—through cinematic techniques creates a sanitized experience that allows audiences to consume violence without confronting its true horror. War is transformed into a form of entertainment, reinforcing the inevitability of American victory (Hodgkins, 2002, p. 76). At this point, the war narrative not only loses its critical nature and becomes an entertainment product but can also become a tool for ideological shaping and gaming.

It is worth noting that *Saving Private Ryan* captured the public's nostalgia for the memory of the Second World War before its release and cleverly integrated it into the film's cultural communication strategy (Hasian, 2001, p. 342). Doherty (1998) points out that the filmmakers created the effect of historical reenactment and noble motives through the esthetic of realism, thus weakening the audience's questioning of its entertainment nature and further shaping it as a commercial film with a historical commitment. Lowenthal (as cited in Hasian, 2001, p. 340) argues that people's quest for a simple and stable past is a psychological refuge mechanism that allows them to seek security 'from the turbulent and chaotic present'. Through its use of technical realism—such as desaturated color palettes, handheld cinematography, and immersive battle sequences—alongside an extensive marketing campaign that emphasized historical authenticity, Saving Private Ryan successfully cultivated a sense of both historical and emotional identity in the public mind. This, in turn, mitigated skepticism about the film's political and commercial motivations.

However, Caldwell (1998) points out that *Saving Private Ryan* focuses too much on individual moral sentiments and fails to give the war a sufficiently national dimension, which prevents the film from adequately presenting wider values such as national freedom and may even weaken the expression of the justness of war. Similarly, Rothkopf (1999, as cited in Hasian, 2001, p. 347) points out that it avoids an in-depth exploration of the legitimacy of war, resulting in a film that feels 'vacancy' and does not adequately present the gravity of the fascist threat and the beliefs of the soldiers. From this perspective, works such as *Saving Private Ryan* and *Band of Brothers*, as works of constructing memories of the Second World War, can to a certain extent satisfy the demand for nostalgia.

James Agee points out that war films are a projection of collective dream, habit, or desire (as cited in Agee, 2000, pp. 23-24). Contemporary World War II film and television works such as *Saving Private Ryan* and *Band of Brothers* continue to exhibit the characteristics identified by James Agee in Hollywood wartime entertainment films of the 1940s, functioning as an epitome of the era. In other words, they respond respectively to nostalgic fantasies of war memory, the structured cultural archetype of traditional heroism, and a profound societal longing for the pleasure derived from violence.

Through the aesthetics of violence or heroic epics, war not only satisfies the audience's fascination with dangerous experiences but also shapes societal attitudes toward war through preconceived cultural narratives, ultimately reinforcing a shared understanding of war within a given historical and ideological context. Works such as *Saving Private Ryan* and *Band of Brothers*, as fusions of national identity shaping, heroic narratives, and visual spectacle, have often been subjected to discussions in the academy of the politicisation of narrative and the entertainment of violence. However, these critical discussions aside, such films undoubtedly provide audiences with a path of emotional nostalgia. Through the visual reproduction of historical events, the films provide a certain degree of psychological comfort to contemporary audiences, allowing them to project their imaginations and identities of the past era in the video narratives.

2. Philosophical Reflection in *The Thin Red Line*

Since the early 1990s, some war films have reflected on and deconstruct the traditional heroic narrative. The 1992 film *A Midnight Clear* serves as an example, offering a 'counterhistory or a counteranalysis' that critiques the conventional war film narrative (Hodgkins, 2002, p. 78). Moreover, Terrence Malick's *The Thin Red Line* (1998) subverts the traditional narrative model on a deeper level. Unlike Saving Private Ryan, which presents World War II from an official perspective, and *A Midnight Clear*, which critiques the morality of war, *The Thin Red Line* transcends the binary debate of 'justice/injustice' and instead explores the fundamental conflicts between humanity and nature, as well as among individuals. It jumps out of the simple binary discussion of the 'justice/injustice' of war and focuses on the relationship between human beings and nature, between human beings and human beings. Instead, it breaks away from the simple binary discussion of 'justice/injustice' in war and deconstructs the traditional hero-villain dichotomy by focusing on the nature of conflict between man and nature, and between man and man.

2.1 Animals, Plants and the Deconstruction of Time

Malick blurs the boundaries between time and prompts audiences to question whether war is confined to a specific historical moment or represents a perpetual condition of human existence (Hodgkins, 2002, p. 81). Here he uses the time of plants and animals to deconstruct the linear flow of time, breaking down the traditional heroic narrative and further breaking the binary model of hero and villain.

In World War II films, non-linear, non-causal visual recording methods focused on non-substantive temporality are frequently employed as narrative strategies that transcend the simplistic opposition between heroes and villains typical in traditional war cinema. This reflects the concept of 'haecceity,' a mode of individuality independent of subjects or physical entities, characterized fundamentally by the fluid states of time, motion, and sensation (Deleuze & Guattari, 1987, p. 283). In *The Thin Red Line*, Terrence Malick employs long takes capturing intersecting shadows among trees, shifting patterns of natural light on animals, and subtle movements of the wind, along with non-narrative montage editing. Through these techniques, he constructs an existential order independent of human or non-human subjects. Elements such as light, wind, and fluid natural forces are presented visually as complete and individualized entities that neither attach themselves to objects nor depend on subjects; rather, they autonomously exist as non-material forms coexisting alongside physical entities. This cinematographic strategy does not center upon the materiality of entities; instead, it emphasizes temporal flux and perceptual intensity inherent to immateriality, thus creating a decentralized visual structure. Within this structure, the human entity and its heroic functions are de-emphasized, allowing the war film to move beyond its traditional reliance on heroic narratives and to enter a perceptual domain characterized by greater fluidity, multiplicity, and openness. From this perspective, Burch (2022, p. 242) suggests that *The Thin Red Line* can be seen as an embodiment of what Robert Sinnerbrink has called 'ecological cinematic poesis'. As a form of filmmaking, ecological cinematic poesis refuses to 'wound the world' on both formal and thematic levels (Luhmann, 1990), and by replacing traditional narrative modes with sound and image (Sinnerbrink, 2006). According to Sinnerbrink's (2006) definition of cinematic ecological wounding, when non-human activity is confined to a stereotypical narrative framework, its complexity and autonomy cannot be fully realised, and the natural landscape is reduced to a background for human motivational schemas. *The Thin Red Line* conveys the idea of anti-war by expressing the impact of war on human beings as well as the impact on nature, while at the same time carrying the intention of protecting the ecology. Therefore, the director uses a lot of natural white noise as background music and many meaningless, de-subjective close-up shots of parrots, crocodiles, and leaves instead of the traditional linear film narrative. The traditional hero narrative, as part of the traditional narrative, focuses on the game of good and evil between the hero and the villain, and the jungle is used as a place for them to ambush and fight, rather than an ecosystem with complex meanings. Therefore, the traditional hero-villain dichotomy can be seen here as a kind of cinematic ecological wounding, which is not employed in the film to better convey a thoughtful and critical stance on ecological conservation.

In cinematic narration, establishing "lines of segmentarity" followed by the deterritorialization process indicated by the "line of flight" precisely reveals the dynamic mechanism of decentralizing and deconstructing traditional heroic images and binary structures. A rhizomatic structure inherently comprises lines of segmentarity—forces of categorization, codification, and structural organization—and simultaneously contains lines of flight—pathways that escape such coding structures. When lines of segmentarity explode within the rhizome, lines of flight are consequently generated, fostering internal confrontation, transformation, and decentralization of structures (Deleuze & Guattari, 1987, p. 30). In *The Thin Red Line*, character backgrounds and military affiliations naturally embody segmentarity lines, manifesting clearly structured systems of identity, organizational hierarchy, and assigned roles. Yet, as the absurdity of war increasingly emerges, these structures gradually fracture, producing lines of flight. Heroes cease to quickly manifest traditional heroic temperaments and are no longer consistently unwavering in their pursuit of justice; villains, likewise, are no longer mere embodiments of evil, but instead present more profound emotional and psychological complexities. Such deterritorialization liberates characters from predetermined binary frameworks, revealing more fluid and non-hierarchical forms of existence. Hodgkins (2002, p. 82) points to a scene in *The Thin Red Line* that highlights a reversal of this paradigm: after shooting a Japanese soldier, a young American soldier is first excited by the tacit encouragement of killing by the institution of war, and then enters into a deep moral struggle, going so far as to state bitterly that 'The worst thing you can do. Worse than rape.' This suggests that in this narrative context, killing the enemy is no longer synonymous with bravery in the traditional sense; the hero's emotional turmoil and moral thorniness are also far removed from the archetypal hero of the past. The opening sequence of the film also maps the characteristics of 'deterritorialization': Private Witt of the U.S. Army voluntarily escapes from the military to live with Melanesian aborigines in the South Pacific. This choice is virtually absent in mainstream heroic narratives, as it fundamentally subverts the core heroic values of loyalty, obedience, and self-sacrifice. Although Witt is subsequently found and demoted by First Sergeant Welsh, Welsh continues to recognise his physical and operational abilities. There is a clear contradiction here: despite his traditional hero potential, Witt voluntarily withdraws from the battlefield and is no longer bound by military discipline. This behaviour disrupts the established framework of the traditional hero and signals a breakdown in the 'perfect' hero's image - even with superior fighting ability, an individual can refuse to be part of the 'war machine'. Although Witt ultimately dies protecting his comrades, the motivation for this action is very different from traditional heroism: he is not following the oppressive orders of the state and the military, nor is he motivated by institutionalised responsibility, but rather by an inner idealism and goodwill to redeem those around him. This choice downplays the institutionalised model of the hero and returns to a reflection on the nature of individual humanity. The death of Witte not only fails to enhance the legitimacy of war but also highlights the absurdity of war and the impermanence of individual destiny, further deconstructing the singular portrayal of heroes in traditional war narratives.

2.2 Bare Life

The image symbols of plants and animals are not only the presentation of naturalistic elements but also a deeper mapping of power structures, violence mechanisms and biopolitical issues. This situation resonates with Giorgio Agamben's (1998) concept of 'bare life' —as a form of life that is excluded yet simultaneously governed, it constitutes the core unit of modern political operation (1998, p.8)—particularly in light of Pick's (2011) distinction between the Agambenian notion of absolute bare life and the shared vulnerability she attributes to both humans and animals. Pick emphasizes that all life is exposed in the sense of being 'susceptible to the interventions of power' (as cited in Pick, 2011, p. 15). This suggests that under certain extreme conditions, human beings themselves can also be reduced to a state of 'bare life', with war being a prominent example.

The existential situation of these non-human subjects often becomes a projection of human destiny, visualising the power relations in the political game. Many traditional war films tend to shape heroic narratives, reinforcing the bravery of individual heroes and the despicable evil of villains; however, this model tends to inadvertently contribute to the imagination of the rationality of war. When films consistently centre on human heroes and villains, they tend to overlook the structural issues behind war, such as how the ruling class sees war as a tool for consolidating power or promotion, and how soldiers are exploited, consumed and even sacrificed. When the film uses the situation of non-human subjects such as plants and animals as a metaphor for power structures, the true face of war is further revealed. This not only helps to break the single narrative of heroism in war cinema and prevent the audience from being domesticated in the visual narrative but also uses animals and plants to reveal how military organisations objectify soldiers as expendable living beings, how they systematically manage and dominate life through war, and how they even shape individual compliance and controllability on a psychological level.

The Thin Red Line not only reveals the absurdity of war by deconstructing the images of heroes and villains but also further demonstrates the mechanism of the power structure behind war. A key scene in the film conveys this intuitively: a group of soldiers tie up a crocodile, which can only passively accept the soldiers' teasing. This scene transcends mere natural imagery, serving as an explicit metaphor for the plight of ordinary soldiers—subjugated by military hierarchy, stripped of autonomy, and reduced to pawns in a larger strategic apparatus. This metaphor is further reinforced in the film's war scenes. For example, in the film, Colonel Call orders his soldiers to continue charging and eliminating enemy machine gun positions even though they have suffered heavy casualties during the battle. He further emphasises that taking the island will allow the bombers to land and will have a direct impact on the Guadalcanal Campaign. This logic transforms the lives of ordinary soldiers into purely strategic resources, whose sacrifices are not based on personal will, but rather on the fulfilment of a higher-level power configuration. This role insinuates the personal ambitions of many real generals in war, and in many cases, war decisions are based not only on military necessity but on personal glory and political gain (Hodgkins, 2002, p. 82). Within this political structure, ordinary soldiers are instrumentalized as passive entities, subjected to external forces much like the plants and animals that populate the battlefield. While not entirely outside the legal order, their autonomy is severely curtailed, and they are subordinated to the demands of the state, existing in a liminal zone between political agency and expendability. Modern democracies, despite their formal commitment to human rights, perpetuate the production of 'bare life' through mechanisms such as states of emergency, militarization, carceral systems, and refugee policies, wherein sovereignty operates by suspending legal protections for specific groups. Within this system, soldiers' lives are no longer regarded as individual tragedies but as statistical variables within military strategy and national interest. Their existence is institutionalized and instrumentalized, reducing them to expendable units within the calculus of sovereign power.

In this way, *The Thin Red Line* transcends the heroism of traditional war films and instead presents the hidden, brutal political logic behind war. Like *A Midnight Clear*, the film also reveals how the ruling class uses emotional symbols to obscure the violent nature of war through the deconstruction of the ideological myth of the 'family'. Hodgkins (2002, p. 83) points out that in Malik's narrative, the protagonist's memories of a dying grandmother and a wife who cheated on him at home suggest that the so-called 'family unity' is not a universal truth of war, but more of an idealised construct. When officers construct the military 'family ethic along the lines of the 'father/son rhetoric, they are replacing the military hierarchy with a metaphor of blood, transforming state violence into a moral obligation to 'protect the family' and making soldiers accept the 'family' through an emotional connection. Through emotional ties, soldiers are made to accept sacrifices, and the image of heroic soldiers who are loyal, obedient, and courageous to sacrifice is domesticated, thus obscuring the macro geopolitical game and the power ambitions of individual officers. By the film's conclusion, the soldiers come to the stark realization that 'family' is nothing more than an ideological construct. As this illusion crumbles, so too does their faith in the traditional war ethics of heroism, loyalty, and obedience. This chain reaction from the dismantling of the family myth to the questioning of the meaning of war ultimately dissolves the characteristics of the traditional hero in the heroism narrative and exposes its essence as a tool of domination.

The Thin Red Line breaks the traditional war narrative that portrays the villain as a monolithic and instrumental model. First, the villain is no longer just an antagonist to the hero's growth but is given an independent narrative space. *The Thin Red Line* delicately captures the state of the Japanese army in many scenes, and these details are not simply to show the heroism of the American army, but to allow the audience to enter the perspective of the villains to empathise with their emotions and situations. Secondly, the image of the villain is no longer treated collectively and anonymously but rather shows the differences between individuals. After the U.S. troops capture the bunker, different Japanese soldiers show very different reactions to their defeat - some accept their deaths calmly, some sob in sadness, some pray in silence with their hands clasped together, some are stubborn and silent, and some pound the ground angrily. Through these details, the director deliberately breaks down the stereotypical image of the enemy in war narratives, no longer allowing the villains to serve merely as a vague, symbolic whole, but giving them individualised lives and psychological states. While the Japanese soldiers are hardly ever directly visible in the first half of *The Thin Red Line*, as the battle begins, the visibility of the enemy forces gradually increases, and in the POW scene after the battle the enemy forces gain a great deal of physicality in their portrayal, resulting in a process of moving from the abstract, to the invisible, to the anonymous, and gradually becoming embodied ((Misek, 2008, p. 119). This contrasts with the traditional anonymisation of enemy troops from start to finish in war films, a convention that *The Thin Red Line* breaks with, ultimately giving the enemy troops individuality. In addition, the villain is also given a more complex emotional level, breaking through the cold, heartless and sinister stereotypical image. After the U.S. army successfully captures the hill, the medium shot focuses on a Japanese soldier who hugs his fallen comrade and cries bitterly. At this moment, he is no longer an enemy on the battlefield, but an ordinary man who has lost a comrade. In the second half of the film, when Witt is surrounded by Japanese soldiers on a mission, a Japanese soldier points his gun at him and after a brief hesitation conveys his intention 'I don't want to kill you' in

Japanese. This scene reverses the simple dichotomy of the war genre, transforming the villain from a mere symbol of hostility into a complex human subject, making the audience realise that everyone in war is in a moral and emotional dilemma, not just good or evil.

Finally, the film goes one step further by completely deconstructing the identity of the traditional villain through a state of no identity. Just as the opening shot of *A Midnight Clear*, in which a young soldier runs screaming and naked, maps out a tribute to *The Deer Hunter*(1978)and the psychological trauma of the Vietnam War (Hodgkins, 2002, p. 79), *The Thin Red Line* features several similar sequences. During the final phase of the capture of the hill, the American and Japanese soldiers fall into mental numbness and helplessness after a long period of artillery bombardment. At this point, there are some disturbing and profound images on the battlefield: some Japanese soldiers who have lost their weapons are naked and laughing, while others cover their ears and hiss 'no', their state is neither life nor death, but on the verge of a mental breakdown. They are no longer soldiers because they have lost the will to be soldiers, but they cannot return to civilian life because they are still bound by the power mechanisms of the state structure. Their bodies are still clad in military uniforms, symbolising their former affiliation with a country and an army. But their spirits have collapsed, and they have become unsupported in the frenzy and killing of war. They have entered a 'liminal state' that cannot be defined in terms of country, army or position, and are outside the boundaries of identity, highlighting a situation of 'interminable survival'. McMahon (2019, p. 110) uses 'interminable survival' to portray the extreme exhaustion of human and animal labour in a capitalist environment. Just as workers and animals within the capitalist system are deprived of control over their own labor and are exhausted to the point of depletion, ordinary soldiers in warfare are similarly stripped of their agency over their own lives. Ordinary soldiers are killed repeatedly on the battlefield until they become numb, physically and mentally, and become consumables of the war machine. Though they are not yet dead, they are already incapable of continuing to live a full life, and their exhaustion is not only physical but also rooted in a total psychological collapse. This situation reveals the absurdity of war: it not only destroys life but also deprives subjectivity, leaving the individual in an existential dilemma of not belonging. By presenting both sides of the war in a similar state, the villain is freed from his inherent ideological labels and is transformed into a victim who is also consumed by the war machine. Through this reconstruction of the villain's image, *The Thin Red Line* breaks through the traditional narrative of good triumphing over evil and points to a deeper truth: war is the abyss of all people's identity and self-loss.

2.3 Genius, Heidegger and Nature

Instead of adopting the heroic narrative of mainstream war films, *The Thin Red Line* shows soldiers' fears, doubts and psychological breakdowns during the war. These narrative challenges the dominant ideology of war cinema by shifting the focus from triumphant heroism to the psychological and existential struggles of soldiers. It is mentioned that both *Saving Private Ryan* and *Band of Brothers* adopt the traditional mechanism of pleasure, which emphasises the violence itself as well as the opposition between the enemy and the enemy camp. *The Thin Red Line*, on the other hand, tends to subvert the traditional pleasure mechanism by giving the enemy a more specific identity through re-individuation, so that the viewer cannot simply see him or her as 'the other' (Misek, 2008, p. 116). This avoids the glorification of violence that reduces war films to a mere form of entertainment. Instead, it immerses the viewer in a more complex emotional experience, prompting deeper reflection on human existence, violence, and ethics.

Part of Malick's ability to distance himself from traditional heroic narratives and the mechanics of violence in his work stems from his relative detachment from the mainstream production system. Victory in the Gulf War largely enhanced the prestige of the U.S. military and caused Hollywood to take a more cautious approach to making war films, avoiding anything that might question the legitimacy of the U.S. military or challenge U.S. military operations. This made it difficult for films with anti-military or anti-war tendencies to gain mainstream support (Hodgkins, 2002, p. 78). Hollywood at the time generally shied away from making any films that could be labelled as 'anti-military' or 'anti-war' so as not to affect the post-war pride of the US military and the public (Ansen, 1992). Malick, who is widely recognised as a 'genius' director, disappeared for nearly two decades after the release of *Days of Heaven* in 1978, staying away from Hollywood and not directing any more films (Furstenau & MacAvoy, 2007, p. 179). This deepened his 'hermit genius' mystique and somewhat distanced him from the ideology of the mainstream studios. Rather than adhering to conventional commercial storytelling, Malick prioritizes visual aesthetics and philosophical inquiry, disrupting linear narrative structures. Nonetheless, Hollywood has chosen to collaborate with Malick to capitalise on the reputation of certain artists and to gain a certain cultural and artistic aura for itself, which is a form of commercial strategy (Furstenau & MacAvoy, 2007, p. 180). *The Thin Red Line* remains in the style of Malick's original work, embodying a traditional narrative non-linear structure that differs from the ideology of mainstream Hollywood war films and is a manifestation of a particular philosophical tradition.

Malick studied philosophy at Harvard and Oxford, taught philosophy at MIT, and translated Heidegger's *The Essence of Reasons*. Since the beginning of his film career in 1969, Heidegger's philosophy has had a profound influence on his style and narrative logic, and *The Thin Red Line* has been regarded as one of the masterpieces of Heideggerian cinema (Furstenau & MacAvoy, 2007, p. 181), which explores the fundamental propositions of survival, nature and death through the field of war. In Heidegger's view, poetry is not only an aesthetic expression but also has a profound mission to reveal the nature of the world. In 'destitute times' when the world is dominated by rationality, technology, and instrumentalised thinking, it loses its mystery and the poet takes on the responsibility of a philosopher (Furstenau & MacAvoy, 2007, p. 182). In his theoretical framework, the modern world is facing a crisis of obscuration by representation, whereby when people understand the world in terms of images or language, they tend to overlook the constructed-ness and limitations of these representations themselves. Malick's films differ from purely narrative films in that they tend to break with the function and utility of traditional narratives. For example, *The Thin Red Line* does not arrange the storyline according to the traditional war narrative but dismantles and transforms the representation of war that serves certain purposes into a discussion and presentation of war itself, thus deconstructing traditional heroism and dualism outside of the political context. This creative practice is not merely an artistic expression but a philosophical critique of modernity. By exposing the essence of war, it rejects the notion of war as an embodiment of 'instrumental rationality' in modern society—that is, the way in which warfare is organized and executed through rational calculation, while disregarding humanitarian and ethical values. Malick employs a non-linear, de-anthropocentric cinematic language to recontextualize human-nature relations, the individual's place within the state, and the existential link between violence and survival.

Heidegger argues that to understand Dasein (human existence), one must grasp it as "ahead-of-itself-Being-already-in-(the-world) as Being-alongside (entities encountered within-the-world)." In other words, Dasein implies an awareness that human beings and non- human creatures are always already situated within the world, oriented toward an unrealized future, and simultaneously coexist alongside other entities encountered in the world. Heidegger calls this structural mode of human existence "Sorge" (care), which does not refer to the everyday meanings of worry [Besorgnis] or carefreeness [Sorglosigkeit], but rather signifies a fundamental mode of existence (Heidegger, 1978,p. 237). 'Thrownness,' meanwhile, refers to the continuous existential condition of an entity being cast into "Being-in-the-world." *The Thin Red Line* offers a cinematic embodiment of Heideggerian Dasein, wherein the visual language that blurs boundaries between humans and animals constructs a shared uncertain future faced by both humans and animals under wartime conditions, collectively articulating a philosophy of coexistence with all beings. In this sense, war serves as a catalyst of human "thrownness," creating a context for contemplating human and other entities' existence, rather than merely serving the anthropocentric frameworks of narrative operation.

Through its de-anthropocentric perspective and non-linear narrative structure, *The Thin Red* Line fundamentally challenges the hero-villain dichotomy that underpins traditional war cinema. Unlike conventional war films, which often frame conflict through clear moral binaries, Malick's film eschews simplistic oppositions, presenting war as a deeply existential and philosophical phenomenon rather than a mere spectacle of heroism or villainy. One of the film's most striking features is its delicate and poetic portrayal of nature—plants, animals, and landscapes are not merely passive backdrops to human suffering but autonomous entities with their presence. By granting non-human subjects an almost sacred significance, Malick deconstructs the grand narrative of war, which typically relies on stark contrasts between good and evil, friend and enemy, civilization and wilderness. This reframing suggests that war is not only a human affair but an ecological catastrophe, reinforcing the interwoven fates of mankind and the natural world. At the same time, the film offers a meditative reflection on the fate of the individual soldier, interrogating the tension between the state and the individual. It incorporates philosophical concepts from Heidegger, foregrounding existential questions about being, violence, and sovereignty. The soldiers' introspective monologues, fragmented perspectives, and the film's dreamlike pacing urge the audience to confront the fundamental conflict between existence and destruction—a conflict heightened by Malick's juxtaposition of shocking wartime brutality with hauntingly beautiful, poetic imagery.

3. CONCLUSION

Following the Gulf War victory, the 1990s saw a resurgence of World War II-themed films as a means for the United States to reaffirm national identity and reignite patriotic sentiment. This revival also served to distance the American war narrative from traditional identity and evoke patriotic sentiment. The resurgence of traditional heroism in these narratives further facilitated the government's objective of utilizing World War II films to reinforce these ideals. *Saving Private Ryan* and *Band of Brothers* adhere to the classical *Hero's Journey* structure, emphasizing the binary opposition between good and evil and reinforcing the core characteristics of traditional war films. By portraying war as a moral confrontation between noble heroes and irredeemable villains, these films reinforce a binary worldview that legitimizes armed conflict. The extensive use of hyper-realistic depictions of violence and advanced visual effects generate a so-called 'pleasure mechanism,' intensifying the audience's sensory experience. This binary framework not only enhances the efficacy of the pleasure mechanism but also, to a certain extent, obscures the underlying political agenda, ultimately serving the broader political narrative. However, *Saving Private Ryan* and *Band of Brothers*, while grounded in traditional heroism, also function as vehicles of historical nostalgia, allowing audiences to romanticize World War II as a simpler and more righteous conflict.

By contrast, Terrence Malick's *The Thin Red Line* subverts conventional war narratives by dismantling the traditional hero-villain dichotomy through fragmented perspectives, poetic imagery, and existential contemplation. The film employs shifting perspectives and alternating internal monologues while incorporating extensive imagery of nature—plants and animals alike—to destabilize human centrality within the war narrative. This approach offers a profound critique of the conventional heroism found in traditional war films. *The Thin Red Line* not only depicts soldiers' moral struggles and emotional conflicts regarding acts of violence but also imbues the portrayal of the enemy with greater emotional depth. The antagonist is no longer a one-dimensional embodiment of 'evil otherness'; instead, individual enemy soldiers are shown experiencing sorrow, fear, and helplessness. Meanwhile, American soldiers, too, are depicted as vulnerable to moral crises, experiencing emotional breakdowns rather than unwavering heroism. By situating individual lives within the broader context of war, the film compels audiences to reflect on the political structures and existential implications underlying war, thereby deconstructing the rigid hero-villain dichotomy that traditionally dominates the genre.

Saving Private Ryan and *Band of Brothers* represent more traditional hero narratives, which use intense violence and clear-cut antagonism to construct the legitimacy of war, thus serving the mainstream national ideology. *The Thin Red Line*, on the other hand, makes a more subversive attempt at aesthetics and narrative structure, advocating a more complex philosophical reflection on war and human nature through de-anthropocentricity and the deconstruction of the hero archetype. An analysis of these three works reveals a dual trajectory in contemporary World War II cinema: on one hand, the persistence of traditional pleasure mechanisms and moral binaries; on the other, an evolving exploration of non-linear storytelling and plural subjectivities.

REFERENCES

Agamben, G. (1998). *Homo Sacer: Sovereign Power and Bare Life*. Stanford University Press., https://search.alexanderstreet.com/view/work/bibliographic_entity|bibliographic_details|4714514

Agee, J. (2000). Agee on film: criticism and comment on the movies. In *Agee on Film: Criticism and Comments on the Movies*. Modern Library., https://openlibrary.org/books/OL45445M/Agee_on_film

Ansen, D. (1992). Wars Domestic and Military: Three striking independent films—Irish, American and Australian—enliven the scene. *Newsweek, 119*(19), 77. https://www.proquest.com/magazines/wars-domestic-military/docview/1879164987/se-2?accountid=14511

Basinger, J. (2003). *The World War II Combat Film: Anatomy of a Genre*. Wesleyan University Press., https://www.weslpress.org/9780819566232/the-world-war-ii-combat-film/

Bates, M. J. (1998). *The Wars We Took to Vietnam: Cultural Conflict and Storytelling*. University of California Press., DOI: 10.1525/9780520917521

Baudrillard, J. (2006). The Gulf War Did Not Take Place. In *Hollywood and War, The Film Reader* (1st ed.). Routledge., DOI: 10.4324/9781003416661-30

Bronfen, E. (2012). *Specters of war : Hollywood's engagement with military conflict*. Rutgers University Press., https://www.jstor.org/stable/j.ctt5hj3rx

Burch, P. W. (2022). Receptivity, Simultaneity: The Thin Red Line as Ecological Cinematic Poesis. *Film-Philosophy*, *26*(2), 242–266. DOI: 10.3366/film.2022.0197

Caldwell, C. (1998, October). Spielberg at War. *Commentary (New York, N.Y.)*, *106*(4), 48–51. https://link-gale-com.libproxy.ucl.ac.uk/apps/doc/A21191193/AONE?u=ucl_ttda&sid=bookmark-AONE&xid=db2fdd6b

Campbell, J. (2008). *The Hero with a Thousand Faces* (3rd ed.). New World Library., https://newworldlibrary.com/product/the-hero-with-a-thousand-faces

Cohen, E. A. (1999). The Thin Red Line: Let it Slip Away. *SAIS Review (Paul H. Nitze School of Advanced International Studies)*, *19*(2), 237–241. DOI: 10.1353/sais.1999.0031

Deleuze, G., & Guattari, F. (1987). *A Thousand Plateaus: Capitalism and Schizophrenia* (1st ed.). University of Minnesota Press., https://www.upress.umn.edu/9780816614028/a-thousand-plateaus/

Dibattista, M. (2000). Steven Spielberg, Robert Rodat and Frank Darabont (1998), Saving Private Ryan; Terence Malick (1998), The Thin Red Line. *Rethinking History*, *4*(2), 223–228. DOI: 10.1080/13642520050074885

Dick, B. F. (2022). The Star-Spangled Screen: The American World War II Film. In *The Star-Spangled Screen*. Lexington, Kentucky: The University Press of Kentucky. DOI: 10.2307/j.ctv2ks7284

Doherty, T. (1998). Saving Private Ryan. *Cinéaste (New York, N.Y.)*, •••, 68–71. https://link.gale.com/apps/doc/A53709342/AONE?u=ucl_ttda&sid=bookmark-AONE&xid=0aab3062

Eberwein, R. (2010). *The Hollywood War Film*. The Hollywood War Film., DOI: 10.1002/9781444315103

Ehrenhaus, P. (2001). Why We Fought: Holocaust Memory in Spielberg's Saving Private Ryan. *Critical Studies in Media Communication*, *18*(3), 321–337. DOI: 10.1080/07393180128089

Ferro, M. (1988). *Cinema and History*. Wayne State University Press., https://wsupress.wayne.edu/9780814319055/

Furstenau, M., & MacAvoy, L. (2007). Terrence Malick's Heideggerian Cinema: War and the Question of Being in The Thin Red Line. In *The Cinema of Terrence Malick : Poetic Visions of America* (2nd ed., pp. 179–191). London : WallFlower Press. https://search-ebscohost-com.libproxy.ucl.ac.uk/login.aspx?direct=true&AuthType=ip,shib&db=nlebk&AN=781073&site=ehost-live&scope=site&custid=s8454451&ebv=EB&ppid=pp_179

Gates, P. (2005). "Fighting the Good Fight:" The Real and the Moral in the Contemporary Hollywood Combat Film. *Quarterly Review of Film and Video*, *22*(4), 297–310. DOI: 10.1080/10509200590475788

Hasian, M. (2001). Nostalgic Longings, Memories of the "Good War," and Cinematic representations in Saving Private Ryan. *Critical Studies in Media Communication*, *18*(3), 338–358. DOI: 10.1080/07393180128083

Heidegger, M. (1978). *Being and time*. Wiley-Blackwell., https://www.wiley.com/en-cn/Being+and+Time-p-9780631197706

Hodgkins, J. (2002). In the Wake of Desert Storm: A Consideration of Modern World War II Films. *The Journal of Popular Film and Television*, *30*(2), 74–84. DOI: 10.1080/01956050209602842

Lowenthal, D. (1989). Nostalgia Tells It Like It Wasn't. In Chase, M., & Shaw, C. (Eds.), *The Imagined Past: History and Nostalgia* (1st ed., pp. 18–32). Manchester University Press., https://www.amazon.co.uk/Imagined-Past-History-Nostalgia/dp/0719028752

Luhmann, N. (1990). The Cognitive Programs of Constructivism and a Reality That Remains Unknown. In Krohn, W., Küppers, G., & Nowotny, H. (Eds.), *Selforganization: Portrait of Scientific Revolution* (pp. 64–85). Springer., https://doi.org/https://doi.org/10.1007/978-94-017-2975-8_5 DOI: 10.1007/978-94-017-2975-8_5

Maslin, J. (1998). Saving Private Ryan. *The New York Times*, E1–E28. https://archive.nytimes.com/www.nytimes.com/library/film/072498ryan-film-review.html

McMahon, L. (2019). *Animal Worlds: Film, Philosophy and Time* (1st ed.). Edinburgh University Press., DOI: 10.1515/9781474446402

Misek, R. (2008). Exploding Binaries: Point-of-View and Combat in The Thin Red Line. *Quarterly Review of Film and Video*, *21*(1), 116–123. DOI: 10.1080/10509200601074710

Owen, A. S. (2002). Memory, war and American identity: *Saving Private Ryan* as cinematic jeremiad. *Critical Studies in Media Communication*, *19*(3), 249–282. DOI: 10.1080/07393180216565

Paget, D., & Lipkin, S. N. (2009). 'Movie-of-the-Week' docudrama, 'historical-event' television, and the Steven Spielberg series *Band of Brothers. New Review of Film and Television Studies*, *7*(1), 93–107. DOI: 10.1080/17400300802603047

Panofsky, E. (2012). Style and Medium in the Motion Pictures. In Cardullo, B. (Ed.), *Stage and Screen: Adaptation Theory from 1916 to 2000* (pp. 39–54). Continuum., DOI: 10.5040/9781628928983.ch-004

Pick, A. (2011). *Creaturely poetics : animality and vulnerability in literature and film*. Columbia University Press., https://cup.columbia.edu/book/creaturely-poetics/9780231147866/

Rothkopf, J. (1999, April 11). Shellshock. *In These Times*, 28–29.

Schönfeld, C. (2022). Memories of World War II in German Film after 1945. In Bragança, M., & Tame, P. (Eds.), *German Film after 1945". The Long Aftermath: Cultural Legacies of Europe at War, 1936-2016* (pp. 200–218). Berghahn Books., DOI: 10.1515/9781782381549-015

Sinnerbrink, R. (2006). A Heideggerian Cinema? On Terrence Malick's The Thin Red Line. *Film-Philosophy*, *10*(3), 26–37. DOI: 10.3366/film.2006.0027

Taylor, P. M. (1998). *War and the Media: Propaganda and Persuasion in the Gulf War* (Revised). Manchester: Manchester University Press. https://manchesteruniversitypress.co.uk/9780719055508/

Virilio, P. (1989). *War and Cinema: The Logistics of Perception*. Verso., https://www.versobooks.com/en-gb/products/1264-war-and-cinema?_pos=1&_sid=ab635b484&_ss=r

Zinn, H. (1998, October 10). Private Ryan Saves War. *The Progressive*, *62*, 38. https://link-gale-com.libproxy.ucl.ac.uk/apps/doc/A21200696/AONE?u=ucl_ttda&sid=bookmark-AONE&xid=15cedc36

Chapter 9

Revealing Shadows of War Through Unmasking Hero and Villain Archetypes

Harshini Nalam
https://orcid.org/0009-0006-8594-4802
PES University, India

Shruti Soudi
https://orcid.org/0000-0001-7463-0724
PES University, India

ABSTRACT

The chapter critically examines how Hero and Villain archetypes materialize in war stories through historical, cultural and psychological lenses. Drawing from Jungian psychology, moral philosophy, media studies and social psychology, exploring how symbolic narratives transform complex geopolitical conflicts into stories of moral dichotomy between good and evil. It considers cultural and religious roots shape of monomyth along with concepts such as moral injury and virtuous violence, to investigate how propaganda, national identity, and collective memory uphold strict dualistic classifications, reinforcing ideological agendas. Through comparative examples, it disputes the oversimplified frameworks dominating wartime narratives. It seeks to demonstrate how simplistic stereotypes legitimize violence, shape historical memory and impact public understanding where moral framing shape consent. Media representations recycle these binaries, influencing opinions through familiar narrative tropes. The chapter calls for portrayals of human conflict with greater complexity and ethical awareness.

DOI: 10.4018/979-8-3373-7589-2.ch009

INTRODUCTION

War has consistently been one of the most prevalent and definitive themes in storytelling in human storytelling. From ancient epics inscribed on stone to political rhetoric and modern cinematic extravaganzas, war narratives have shaped people's perceptions of themselves their enemies abroad, and the cosmos around them. It represents an organized state of conflict that can last for extended periods. States non-state actors, or groups with ideological motivations wage these battles (Clausewitz, 1976). Extreme violence social upheaval, and political turmoil mark these conflicts. War goes beyond battlefield confrontations; it shows how humans act when driven by power, belief systems, fear, or personal identity (Wright, 1942). Throughout history, wars occur from small tribal fights to global conflicts, and human societies use them to dominate others or resist oppression.

The concept of war means more than just military strategies; it runs deep in the realm of cultural narratives, religious doctrines, and national identities. Philosophers and researches have shown contradicting opinions on the concept of war, with Thomas Hobbes (1651) suggesting war was how humans naturally behaved, with ideas of 'war of all against all', whereas Immanuel Kant (1795) believed that using human intelligence and national cooperation could lead to lasting peace if countries worked together and followed international rules. These divergent ideas about war shape how societies decide whether wars are acceptable.

Romanticism and Ideology in War

In many cultures, people often paint war in a romantic light and turn it into myths, often viewing it as a site for Heroism, sacrifice, and national pride (Elshtain, 1998). This happens even though war leads to deep human suffering and tragedies. While, some view it as an unfortunate necessity, others see it as a tragic consequence of human conflicts. Either way, it depicts the complex psychological and moral battles that human society faces, with a stage to explore these dilemmas. This romantic narrative of war, are embedded into human psychology and elicit emotional responses and an attachment towards such ideas.

Psychological Needs and Cultural Projections

The archetypes of Hero and Villain emerge from the cultural and psychological framing of war. When war is seen as noble, it demands noble figures such as, Heroes who embody courage, loyalty, and sacrifice, and Villains symbolizing chaos, evil, or moral decay. These roles simplify the complex realities of war, offering emotional clarity and reinforcing the belief that war is justifiable. The binary does not just simplify realities of war and nation-building, but it also runs as a construct of emotional clarity and moral resolution.

The Hero and the Villain are not just simply antithetical figure, but rather extraordinary and diverse individuals who shaped by their choices and the paths they choose to follow. The Hero is conventionally shown as arising from adversity, characterised by empathy and an inherent motivation to assist others, despite their own personal challenges. Their strength comes not from victory, but from standing firm and with head up in the face of defeat, from the quiet resolve and ethical resolve to do the right thing. The Villain too, experiences struggle, but their psychological disposition is distorted by fear, with conviction that they have been wronged and deserve power for self-preservation. Villains are not motivated purely by malice but rather by uncharted psychological conflict. They seek control and power in order to fill internal empty and hollow space without awareness that they're only creating more internal fragmentation and resulting void. This typically leads them to treat others as objects as they justify their violence. Heroes search for purpose through service to other people while villains are searching for purpose through fulfilling their own selfish desires. Their conflict exposes the complex spectrum of human behavior and potential, shaped through belief, environmental influences, and choices.

Thematic Conflicts in Epics Narratives

Among diverse literary literatures, from ancient epics such as *The Iliad* from Greek mythology, *The Epic of Gilgamesh* from ancient Mesopotamia, *The Aeneid* by Virgil, *Beowulf* from Anglo-Saxon England, *The Song of Roland* from medieval France, *The Nibelungenlie'* from Germanic tradition, *The Poem of the Cid'* from Spain, *The Mahabharata* and *Ramayana* from religious Hindu texts; all of them explore the themes of Heroism, battle, and the consequences of war within their respective cultures. These all have pervasive commonality, centric war and conflicts, with the defeat of evil over good. Every story has fundamental structural elements of Hero and Villain. The Hero's role is to function as the saviour, and is represented as the personification of all human virtue and greatness, whereas the Villain's role is to instigate disorder, perpetrate harm, and embody destructive forces. To create a balance amongst these opposing forces and re-establish the equilibrium, the Hero goes through great lengths to deliver justice.

Archetypes in Conflict Narratives

These figures are referred to as archetypes, which have evolved over time but their function remains central to our understanding of the human experience within the context of conflict. This chapter seeks to deconstruct these archetypes in war narratives, exploring how the concepts of Heroism and Villainy have been portrayed and expanded in literature, film, and other forms of storytelling.

War and storytelling have always been closely linked throughout history. Storytelling does not merely record war it also helps interpret, justify and mythologize it. Through stories societies can come to terms with the trauma they have suffered, build their national identity, and understand the loss and violence. Storytelling consists of many forms including oral histories, epic poems, propaganda movies, and novels, shaping the collective memory and influencing how future generations will remember understand war. These narratives frequently cast war in such dualities such as heroes and villains, good and evil, providing emotional closure to morally ambiguous situations. Consequently, the act of storytelling is not only an echo of war's influence but also a tool that enables its history to be formed and continued.

Specifically, in the World War I and II are explored and analysed from the perspectives of Hero and Villain archetypes. An examination of the psychological impacts these archetypes have on soldiers and civilians, followed by a discussion of their evolution throughout history. Their roles in global conflicts are explored, alongside how nationalism shapes these archetypes, with the emergence of the Anti-Hero in war stories, before reframing the Villain archetype by considering their psychological motivations and victimization. Furthermore, gender dynamics in these archetypes are analysed, as well as how postmodern narratives challenge traditional portrayals of war. Additionally, it will look at the globalization of conflict narratives and their effects on international relations to understand the dynamic in comparisons.

In sum, the chapter questions these archetypes and critically examines their roles in war situations, arguing for representations that reflect human complexity, challenging ethical concerns and dominant ideological constructs.

JUNGIAN PSYCHOLOGY AND THE ARCHETYPES OF WAR NARRATIVES

Where war narratives are unpicked at a depth-psychological level, it is through the lens of Jung's psychology that one gains the most profound understanding of the libidinal symbolic structures that underlie them. The Hero and Shadow archetypes are important components of collective unconscious, playing significant roles in determining how societies construct meaning regarding conflict, identity, and morality (Jung, 1969). In war-time, these archetypes not only serve as narrative function but also a crucial psychological function, serving as a frame through which we can process violence, heroism, and the human condition.

Carl Jung's theory of archetypes, introduced in 1919, explains how universal symbols like the Hero and the Shadow (Villain) reside within the collective unconscious, shaping human experiences across cultures and time periods. In war narratives, the Hero embarks on a journey to overcome life's challenges, while the Shadow represents repressed negative forces like chaos, destruction, and evil. The struggle between these figures reflects both external battles and the internal psychological, moral, emotional, and ethical conflicts individuals face in war, influenced by duty, trauma, identity, social pressures, cultural values, and personal beliefs. This section examines how these archetypes influence the construction of war stories, exploring the Hero's internal battles with fear, guilt, and ethics, and how the Villain embodies not just an external enemy but also the darker aspects of individual personality.

Jung (1969) suggested that there exists another region beyond our personal consciousness, that of the collective unconscious, which is the subsurface of the conscious mind and common to all humanity. This unconscious includes 'archetypes' which are patterned symbols, such as the Hero, the Shadow, the Mother, and the Wise Old Man, manifesting in myths, dreams, literature, and religious and cultural tradition. When confronted with overwhelming circumstances, such as Nazi Germany or Stalinist Russia, people often justify their actions as righteous. This is particularly evident in wartime, where adversaries are often symbolically cast as pure evil or as externalizations of collective repressed fears and base instincts, with some individuals even rationalizing harmful acts as expressions of justice or peace.

The Hero Archetype

The Hero archetype typically embodies qualities of bravery, determination, and selflessness, often demonstrating self-sacrifice for a perceived greater good. Such figures serve as protagonists in stories and myths, and recognised for their courage, selflessness, altruism, achieve a righteous objective and desire to help others. While a character of this archetype typically appears to be ordinary, but they possess a remarkable ability that sets them apart. A classical Hero uses their tremendous skill to defeat a foe or for a righteous purpose. Typically, the battle they fight on the surface coincides with an internal battle, a struggle with identity, a powerful temptation. In classical epics, such as Homer's *Iliad* and *Odyssey*, Heroes like Achilles and Odysseus are depicted as strong, formidable warriors who go on perilous journeys, overcome obstacles, and fight for a noble cause. In these works, the Hero represents the idealized human condition: one who rises above the chaos of war to achieve greatness.

Allison and Goethals (2011) found that Heroic figures across cultures are associated with self-sacrifice, resilience, and moral integrity, reinforcing Jung's idea that the Hero is a symbolic carrier of the collective good. They also identified the "Great Eight" traits that are frequently attributed to Heroic individuals: intelligence, strength, selflessness, compassion, charisma, resilience, dependability, and the ability to inspire others. They further argue that societal perceptions of Heroism are shaped by diverse mental models, encompassing archetypal paths like the underdog's triumph against overwhelming odds (e.g., the biblical David and Goliath) or the individual's moral renewal through self-transformation.

Heroism is a multifaceted and intrinsically subjective concept. What one individual perceives as heroism, another may see as defiance or even recklessness. We, humans, tend to forget that the Heros are also humans at the end of the day and that they are not as flawless as idealised. Heroism, in its most authentic form, is characterized by both strength and fragility, universality and individuality; it is a profound reflection of the noblest and most imperfect attributes of humanity.

The Villain Archetype

Contrary to the Hero, the Villain embodies chaos, destruction, moral corruption and embodiment of darkness. The Villain or Shadow is projected onto the enemy, becoming the personification of danger, immorality, or alien (otherness) (Jung, 1959). This projection not only justifies aggression but also maintains a sense of moral superiority. As Jung wrote, "The Shadow is a moral problem that challenges the whole ego-personality" (Jung, 1959, p. 20). Functioning as the narrative's antagonist, the Villain opposes the Hero. While Heroes are typically guided by principles and duty, Villains are driven by egoistic desires, animosity, or a lust for power. In war narratives, the Villain is often the dehumanized and demonized embodiment of the enemy, simplifying the complex realities of conflict into a stark dichotomy of good versus evil. However, this simplistic binary overlooks the inherent spectrum of human morality, where absolute good and pure evil are rare. For example, consider a school setting, an introverted or atypical student is mistaken as arrogant or menacing. Peers might label them as the social 'villain', not for their actual behavior, but for challenging group expectations and norms. This projection allows the group to evade confronting its own biases, revealing it insecurities, fear of difference, and collective exclusion disguised under the cover of moral high ground directed at the student.

Clasen (2017) suggests that the dark and threatening negative characters often seen in films are an expression of culturally conditioned fears, such as of being betrayed, of the unknown, or of forces beyond our control. Their eventual annihilation or destruction offers a sense of reassurance, helping society feel restored and safe again. Archetypal villain types, seen in stories like traitors, supernatural entities, or uncontrollable originate from both societal and biological roots. These characters often externalize cultural anxieties, related to xenophobia, distrust, or moral instability. The defeat or eradication of such characters are perceived as a restoration of order and a reassertion of societal norms which is subjective to each society. This symbolic annihilation contributes to societal reintegration through 'moral cleansing,' reinforcing the concept of a 'good society' against a dangerous 'other'.

The War-Time Shadow model, demonstrating how societies, particularly during conflict, project their own undesirable qualities onto adversaries, effectively casting them as incarnations of darkness (Samuels,1985). Such an archetypical dynamic fuels dehumanization, fragmentation and vilification in propaganda. It highlights how nations, just like individuals, frequently divide their psyche between good and evil, idealizing their own side while demonizing the other. This reflects Jung's concept of the Ego and the Shadow

Ultimately, the Villain's journey epitomises the intricacies of the human psyche, driven by the pursuit of power, the apprehension of loss, the yearning for visibility, and an intrinsic quest for acknowledgement and importance within the wider human narrative.

THE ORIGINS OF HEROES AND VILLAINS

The Hero and Villain archetypes did not emerge randomly; rather, they arose from humanity's earliest attempts to comprehend morality, identity, and survival through storytelling. Early societies viewed these archetypes to be mythic representations of cultural ideals along with existential fears, transcending mere characters. They symbolized values, fears, and aspirations of their originating cultures, representing the dualities in life such as light and dark, order and chaos, and virtue and vice. This section investigates mythological, religious, and psychological foundations of these figures.

Heroes as Cultural Ideals

In ancient myths, Heroes personified the paramount values with their founding societies. Idols like in Gilgamesh (from Mesopotamia), Achilles (from Greece), and Rama (from India) were revered for their moral and spiritual importance and for their strength and valour. These Heroes were often depicted as semi-divine or were chosen by gods, and linked their earthly deeds to a divine purpose.

For instance, Gilgamesh does not pursue just physical conquest within *The Epic of Gilgames*, instead he comes to terms with mortality then seeks wisdom (Sandars, 1960). Achilles in relation to *The Iliad* represents both bravery and also wrath. He feels inner chaos because honour and additionally the death shows larger human conflicts (Homer, 2007). These Heroes reflected cultural ideals: masculine strength, loyalty to kin or nation, and the pursuit of an enduring legacy.

Religious texts similarly contributed to the development of the Hero archetype. Hindu stories feature Arjuna from the *Mahabharata* as an ideal Hero. His significance derives not merely from martial prowess but from his profound moral dilemmas. His talk with Krishna in the *Bhagavad Gita* highlights dharma (duty) ethical behavior, and spiritual obligations (Easwaran, 1989). Moses, David, and Jesus in Jewish and Christian stories act as Heroes who complete godly tasks often facing huge challenges.

These narratives establish the fundamental framework for the Hero's journey: an individual who confronts challenges, adheres to their values, and ultimately imparts justice or insight to the world.

Villains as Cultural Fears

Villains were often used to point out cultural fears. In old myths, Villains mostly were represented as monsters or foreigners like Grendel in *Beowulf*, Ravana in the *Ramayana*, and Medusa, in Greek mythology. These images were the archetype of disorder, evil, or the destruction of social norms. Grendel is the type of character described as being the dweller in darkness and the grudge-bearer of the joy of others, thus embodying social and spiritual isolation. The figure of Beowulf has demonstrated not just victory in physical combat but also a restoration of societal order. Similarly, Ravana, though portrayed as a powerful and learned ruler is often characterized by arrogance and lust, serving as a cautionary figure against unchecked ambition (Brockington, 1998). Villains were often the personification of the unfamiliar both in physical appearance and moral character. They were creators of the "others", a notion that was further used for the protection and reinforcement of cultural boundaries.

In religious books, Villains are people who are not only evil but also a moral failure transformed into human body. Lucifer (from Christian theology), Cain (from the Book of Genesis in the Hebrew Bible/Old Testament), Judas Iscariot (from the New Testament of the Christian Bible), and Duryodhana (from the Indian epic *Mahabharata*) are not only opposers, but also persons who symbolize betrayal, envy, or greed against God. These narratives serve as effective instruments for enhancing the moral character of various communities. Upon transgressing or engaging in malevolence, the characters elucidate the importance of contravening sacred obligations or embracing the virtuous course of existence. In religious legends, villains were frequently associated with theological transgressions. These people serve not merely as adversaries or antagonists but also as cautionary figures.

Duality and the Emergence of Moral Polarization

The opposition between a Hero and a Villain transcends beyond a narrative device; it is intrinsically embedded within human cognitive frameworks and belief systems. Such dualities are reflected in ancient religions, mythologies, and cultural philosophies, often linked to a broader moral universe. This dichotomous thinking simplifies complex human experiences leading to bifurcated conceptualization significantly influences the construction and reception of war narratives.

Zoroastrianism originating in ancient Persia around the 6th century B.C.E., represents one of the earliest and most explicit instances of moral dualism. In the Zoroastrian worldview, the battle between Ahura Mazda, the god of light, truth, and order, and Angra Mainyu (or Ahriman) the spirit of darkness, deceit, and chaos, is eternal (Boyce, 2001). These opposing forces of power continue to fight in perpetual conflict, with human choices determining the ultimate outcome. This system of divine laws has served as the basis of the dualistic moral systems that were to come, giving a clear moral responsibility referent.

In Christian doctrine, similar to Zoroastrianism, articulates a cosmic fight featuring God as the ultimate protagonist and Satan as the principal antagonist. Heaven and Hell serve as the spiritual and moral correlates of the world of good and evil. Just as Angra Mainyu embodies a violation of divine order, Satan represents a prime mover of discord; despite his apparent dominance, he remains ultimately subordinate. The dualistic structure outlines the battlefield between the realm of spirits and parallels the archetypal war basis, where each single soul is among the participants. According to Pagels (1995), as the emerging Christian religion and its boundaries or heretics were made perfectly clear, the quite insignificant figure of Satan was transformed so as to become a diabolical cosmic Villain. It is this spiritual and moral polarity which is translated into the language of war, where God's will be called upon to justify killing and the enemy is portrayed as morally wicked. Religious wars through the ages have used this binary to enshrine Heroes as 'soldiers of God' and Villains as 'agents of the devil', driving up the emotional cost of battle.

This duality between good and evil in mythical stories can also be seen clearly. In the mythos of Greece, gods such as Zeus and Apollo are the symbols of order and civilizations, for they represent them while creatures such as Typhon and Eris serve as metaphors for the chaos and destruction. In the Mahabharata, Krishna and the Pandavas depict dharma (cosmic order) while Duryodhana and the Kauravas are the symbols of adharma (unrighteousness). These tales are not morally neutral, they are designed to convey clear ethical messages about the nature of good and evil, often reinforcing societal norms and spiritual doctrines.

Joseph Campbell (1968), in *The Hero with a Thousand Faces*, explores this universal structure through the concept of the monomyth, or the Hero's Journey. He posits:

A Hero ventures forth from the world of common day into a region of supernatural wonder: fabulous forces are there encountered and a decisive victory is won: the Hero comes back from this mysterious adventure with the power to bestow boons on his fellow man.

(Campbell, 1968, p. 30)

According to Campbell, every culture's Hero must face trials, encounter evil forces, and ultimately triumph to restore balance. The Hero takes many risks and experiences each kind of tragedy to come through the whole journey successively. Crucially, the Hero must confront the narrative's Villain archetype, as these confrontations, while often symbolic rather than literal, facilitate a necessary engagement with difficult truths, leading to positive transformation. The drama of the Hero and Villain stage is an essential part of the story that provides individuals a sense of clarity and the feeling of control amid the chaotic world of existence.

Boyer (2001) argues that supernatural ideas like 'moralizing gods' helped underpin pro-social behaviour in early human societies, with serving as the necessary antithesis to divine virtue. Religious belief is not a unique mental faculty or specialized evolutionary adaptation, but rather a byproduct of regular cognitive processes like agency detection, memory, and theory of mind. Our brains are programmed and predisposed to detect intentionality and pattern, leading to natural development of the ideas concerning supernatural agents even where there isn't any basis or evidence in reality.

PSYCHOLOGICAL THEORIES AND CONCEPTS UNDERPINNING THE HERO AND VILLAIN ARCHETYPES

Beyond their function as narrative devices, the Hero and Villain archetypes are deeply ingrained in the psyche. They assist in the interpretation of emotions like fear, pride, aggression, guilt, and the search for identity on both personal and societal levels. This part examines the psychological theories that account for the profound existence of these archetypes in war stories.

Freudian Psychology

Sigmund Freud's work on aggression, particularly in *Civilization and Its Discontents* (1930), provides another lens to understand the presence of Villains in human storytelling. To Freud, every human carries a death drive (Thanatos) somewhere within the subconscious, which manifests as a primal instinctual drive toward aggression, serving to redirect destructive impulses onto external foes. External antagonists, embodying culturally constructed metaphors of evil serve a society's need to disperse collective aggressive energy. Contemporary societies have unparalleled tools to create and distribute representations of a specified adversary. The necessity to forego ethical considerations to unleash primitive drives for violence offers society equivalent satisfaction and catharsis to that experienced in warfare. Adversaries instigate moral justification and euphemism, while dehumanisation diminishes constraints on violent impulses (Freud, 1930).

In addition, the "Uncanny" (Das Unheimliche) is a very crucial factor in understanding the construction of a frightful Villain. The concept of "Uncanny" refers to something both familiar and strange, evoking deep terror and dread. Villains frequently embody this concept by being imperfect human replicas, subtly distorted humanoid figures, or familiar archetypes transformed into grotesque entities. This pervasive oddity evokes repressed worries and anxieties, rendering the Villain more unsettling by directly addressing matters that people subconsciously avoid contemplating (Freud, 1953/1919).

Moral Disengagement and Dehumanization

Theory of moral disengagement posits that the execution of immoral behaviours requires a disconnection from moral principles, enabling individuals to perform acts that would typically elicit guilt, such as killing or engaging in warfare (Bandura, 1999). Moral justification, euphemistic labelling, and dehumanization are but a few of the strategies needed in constructing the archetype of the Villain devoid of humanity who has committed acts so evil their representation in any artwork from any period in history would be deemed constrained.

Both soldiers and civilians no longer perceive the opponent as human, owing to pervasive efforts to depict the adversary as a horrific and subhuman menace. This framing fosters a culture where any form of violence may be construed as courageous rather than shameful. Minimising the significance of the adversary through meticulously chosen language or imagery significantly exacerbates the escalation of violence.

Social Identity Theory

According to Henri Tajfel's social identity theory, self-concept is derived from group dynamics, where there is an in-group favouritism and discrimination against out-group (Tajfel & Turner, 2004). In warfare, this escalates into glorification of one's own soldiers as Heroes and demonization of the enemy as Villains. Emphasis on the national identity and shared values, particularly in opposition to a common adversary, strengthens in-group cohesion. The out-group Villain serves as essential antagonist to cohesion and justification of aggression.

Delegitimization, defined as the systematic denial of an outgroup's legitimacy and is manifested through its depiction as violation of the fundamental societal norms and values. The actions that are often implemented using such methods as dehumanizing language, stereotypes, negative attributions and historical distortion. These views are reinforced by societal and cultural institutions, including education, media and politics. It serves to excuse hostile actions, impede empathy and disrupt the resolution process of conflicts, thereby perpetuating violence. (Bar-Tal, 1990).

Cognitive Dissonance

Conflict between morality and violence enactment creates dissonance at war for moral self in the case of a soldier. Cognitive dissonance elucidates how individuals reconcile conflicting beliefs and behaviours. Heroic narratives mitigate and contextualise morally ambiguous violence as essential sacrifices (Festinger, 1957). Psychological discomfort is reduced when the enemy is villainized, for viewing them as purely evil allows one to justify harm without experiencing guilt or shame. Dehumanization also resolves soldiers' cognitive dissonance by removing the enemy's human essence, which eliminates the moral inconsistency in killing them. In addition, since legitimate authority figures commit violence, the soldier can abdicate personal moral responsibility through legitimation. Once targets are entirely dehumanized, and violence is authorized by official authorities, a soldier can engage in 'violent acts without a restraint on morality' while preserving the appearance of morality (Kelman, 1973).

Just World Theory

The Just World Theory (Lerner, 1980) rests on the idea that people need to believe the world is inherently fair, individuals receive what they deserve and deserve what they get. This belief in a morally ordered society serves as a psychological defense mechanism, helping individuals keep a feeling of control, predictability and security in an often-chaotic world. In situations of conflict and injustice, it shapes the way in which people interpret the actions and fates. victims are blamed for their suffering; heroes are thought to be rewarded because they are virtuous or make sacrifices. Villains, too, as defined by their moral failing, are thought to deserve punishment because of their essence; this can justify wartime dehumanization of civilians in enemy territories, the disregard of refugee suffering, or the rationalization of collateral damage. These beliefs are particularly dangerous when activated in military or political decisions.

Virtuous violence (Fiske & Rai, 2015) posits that individuals use violence to uphold moral order rather than to subvert it, in accordance with social roles, relationships, and cultural norms. What one culture sees as unjust violence, another may see as morally necessary (e.g., ritual punishment, war, honour defense). This suggests that cultural relativity shapes moral codes followed by the society. In war discourses, this contributes to the substantiation of moral dualism in the construction of Heroes and Villains, whereby both parties see their conduct as morally justifiable. Consistent with Just World Theory, this framework likewise challenges rigid dichotomies and elucidates how good and evil are formed through interpretation and context.

Narrative psychology

Narrative psychology, which was developed by Dan McAdams and other theorists, posits that people make sense of their lives owing to the existence of stories. These stories serve the purpose of organizing an individual's memory, emotion, and identity (McAdams & McLean, 2013). Within the context of wartime, people are able to make sense of suffering, trauma, and loss due to the existence of Heroic tales. Consider combat veterans who, as part of their self-narratives, may depict themselves as guardians or, at worst, sacrificial victims. This fosters psychologic recuperation skewed towards the Heroic identity.

Those who, in contrast, are unable to construct such narratives tend to suffer from moral injury, a profound and pervasive sense of being wronged or guilt which comes from perpetrating actions that contradict one’s belief system and creates a rupture in the self-story (Litz et al., 2009).

THE ROLE OF PROPAGANDA IN REINFORCING ARCHETYPES

Propaganda has always been key in moulding how society sees war through the creation and reinforcement of stories about Heroes and Villains. This idea comes from Carl Jung's theory of archetypes. Governments, the military, and political groups use emotional images, symbols, and tales to sway public opinion, make war seem acceptable, and build national identity.

The Reinforcement of Hero Archetypes

U.S. propaganda consistently portrays its soldiers as guardians of democracy and freedom. Movies like *American Sniper* (2014) and *Saving Private Ryan* (1998) paint American soldiers as Heroes who make big sacrifices. These stories tap into the 'citizen-soldier' ideal within American culture, reinforcing ideological agendas and influence societal understandings of terrorism and security. (Boggs & Pollard, 2006). Similarly, in Cold War Soviet and Chinese propaganda depicted their soldiers as loyal defenders of socialist ideas. Russian soldiers, both in World War II (which they call the Great Patriotic War) and in the current Ukraine conflict, are shown as protectors of the Rodina (motherland). Films like The *Cranes Are Flying* (1957) and *Ballad of a Soldier* (1959) show young men going to war with emotional sacrifice and moral clarity, reflecting Soviet ideals of bravery and duty. More recently, *T-34* (2018) romanticize the escape of a Russian tank crew from Nazi captivity, utilising action sequences to glorify courage and resilience. These stories focus on duty, toughness, and national unity.

In war media is a powerful tool to shape public opinion, build national identity and justify military action. In socialist countries, media is state controlled and used to promote government ideology and national unity. In China, outlets like *People's Daily* or *PLA Daily* are mouthpieces for the ruling party to shape public opinion, through carefully crafted content that emphasizes loyalty, discipline and collective identity. Censorship is common and dissenting views are suppressed. In capitalist countries media is independent or privately owned and there is more diversity of views. While they may still be influenced by corporate interests or political bias there is more space for investigative journalism, public criticism and media pluralism. Freedom of the press is more protected but not absolute.

Other countries are replicating these approaches. Indian film and media, including state-sponsored pictures such as *Uri: The Surgical Strike* (2019), portray soldiers as symbols of national honour and solidarity. Likewise, Israel's military identity is often framed in existential terms as defenders of a threatened nation. Historically, Britain narratives, from colonization down to the Falklands War, military personnel are equated with the execution of global order and civilizational duty (Mackenzie, 1984). This is reflected in films like *Zulu* (1964), which portrays British soldiers heroically defending against overwhelming Zulu forces, reinforcing ideas of discipline, bravery, and imperial responsibility and *The Falklands Play* (2002) which dramatizes Britain's political and military involvement in the Falklands conflict, presenting British leadership and soldiers as protectors of national integrity and international law.

However, other revolutionary forces and movements such as the Cuban guerrillas and the Viet Cong, deploy 'freedom fighter' archetype, recasting resistance against imperialism as Heroic struggle.

The Construction of Villain Archetypes

Propaganda also constructs evil stereotypes to portray the enemy as inhuman and deserving of hostilities. During World War II, German troops were portrayed by the Allies not just as human, but bestial and sadistic, often conflating all German with the SS (Schutzstaffel or Protection Squadron) mass murderer collective. Cold War rhetoric depicted Soviet communists as a homogenous 'red menace', an enemy to Western way of life (Medhurst et al., 1997). After the 9/11 attacks, the American media and political discourse framed Middle Eastern terrorists as irrational, fanatical, and subhuman, a narrative that enabled widespread public support for the War on Terror (Jackson, 2005). North Korean soldiers continue to be framed in Western media as brainwashed automatons, evoking fear, and disdain.

Simultaneously, anti-Western propaganda performs this function as well. In Vietnamese, Palestinian, and Pakistani tales, American, Israeli, and Indian soldiers are depicted as ruthless conquerors and violators of human rights. In jihadist propaganda, Western soldiers are portrayed as modern crusaders, embodying not only geopolitical roles but also theological and existential dimensions. These instances illuminate how archetypal conceptions are mirrored and repurposed throughout ideological conflicts, with each faction employing the Hero and the Villain to preserve internal cohesion and vilify the opposition.

Modern Archetypal Dynamics

Nevertheless, modern movies and media progressively provide opportunities to defy these tropes as well. Films like *Jarhead* (2005) and *The Hurt Locker* (2008) depict American soldiers grappling with psychological distress and ethical quandaries. These depictions complicate the conventional Hero's journey by exploring the psychological ramifications and moral ambiguity of warfare. The movie *Zero Dark Thirty* (2012) presents a more complex picture of Taliban fighters, hinting that one man's Villain might be highly influenced by social, political, and historical determinants. This trend reflects a broader shift in cultural narratives that question the black-and-white moral framing of conflict and instead highlight its human cost (Alford, 2016).

Besides media depictions, cultural institutions are powerful forces of archetypal repetition. For instance, school curriculums are known to provide sanitized or biased representations of national history, focusing on military victories and courageous individuals and downplaying atrocities or civilian hardships. These systems also inculcate citizens into believing war narratives of their rulers and to identify themselves with national myths. Cultural context, including religious beliefs and historical grievances, as well as collective trauma, also profoundly shape how archetypes are experienced personally, and advanced on the level of shared human language.

These conceptions serve predominantly political functions. According to Lippmann (1922), public opinion is shaped not by direct experience but by the 'pictures in our heads' provided by powerful institutions. Propaganda utilises these visuals to portray, while simultaneously obscuring, the progression of events and the identity of actors in manners that provoke specific reactions such as pride, anger, and fear. It does so by whitewashing the moral complexity of war, turning complicated stories of competing values into simple tales of Good Guys vs. Evil Villains. This reductionism makes it easy for people to give their consent to violence, to silence those who speak up and perpetuates cycles of conflict.

MEDIA FRAMING OF WAR

At times of conflict, the narratives in the media can oversimplify conflicts of such geopolitical nature into metaphoric battles in black and white of good versus evil. Heroes and Villains are constantly used in media narratives to construct an engaging story. This mythic fiction is where Heroes embodies the purest virtues of the nation, imbued with moral courage, and Villains represent concentrated malignancy and oppose the collective values of society. This reduction of conflicts is more comprehensible and emotionally relatable for the public.

National Cinemas and Heroic Narratives

As explored in the preceding section, nations frequently leverage film to shape perceptions of their conflicts and appeal to audiences emotionally. For example, while the German movie *Das Boot* (1981) is famous for its graphic and claustrophobic portrayal of U-boat warfare, it also emphasizes themes of duty, bonding, and and internal conflict among German crew, effectively humanizing their experience with wartime.

Similarly, Japan's *Letters from Iwo Jima* (2006) portrays the soldiers as resolute and brave defenders of their home against powerful enemies. In a comparable effort to reclaim national perspective, Italy's *El Alamein – The Line of Fire* (2002) shifts the focus to the North African campaign and offers an Italian viewpoint on the war, illustrating the struggle and suffering endured by Italian soldiers which is often overlooked.

South Korea also confronts the traumatic legacy of war through films like *Tae Guk Gi: The Brotherhood of War* (2004), which depicts the brutality of the Korean War through the bond of two brothers conscripted into the South Korean army, emphasizing sacrifice, loyalty, and the emotional toll of internal conflict. This film bolsters the overarching story of a nation's resilience and the indomitable spirit of an individual battling through grave struggles.

Despite commonly portraying the destructive consequences of war, many films, such as *Eagle Wings* (2021) from Nigeria, attempts to showcase their military's counter insurgency abilities by collaborating with the Nigerian military to ensure authenticity and patriotically Heroize their forces. These disparate national cinemas are constantly seeking to reinforce the perception of their soldiers as Heroic figures who personify the concepts of militarism, national identity, and valorous sacrifice.

The "Good German" Myth

During World War II, Allied media showed German Field Marshal Erwin Rommel as the 'good German', emphasizing his strategic battle smarts, chivalry, and honour, particularly his tactical brilliance in North Africa, while concurrently minimizing his affiliations with the Nazi regime (Samuels, 2017). This story helped make the enemy seem more human and paved the way for facilitating post-war reconciliation. The Rommel myth exemplifies the media's capacity to generate and perpetuate simplified narratives that diverge significantly from the intricate truths of history. It contests simplistic historical narratives that depict wartime media as a mere passive reflector of events; rather, it elucidates media as a proactive entity that constructs meaning, potentially incites militarism, and endorses enduring political objectives such as post-war stability and the reintegration of defeated adversaries into the global community.

Framing Terrorism Post-9/11

Another case is that after the 9/11 attacks, U.S. media and politicians' discourse frequently framed Osama bin Laden as pure evil, terming him the 'face of terror'. This framing made it easier to ignore the complex social, political, historical, and economic factors that contribute to the actions of al-Qaeda and similar extremist groups. By focusing on Osama bin Laden as a single, clearly identifiable villain, the narrative simplified a multifaceted issue into a straightforward battle of good versus evil, obscuring the deeper causes of terrorism. It simplified multiple U.S. foreign relations issues such as policies regarding the Middle East, American military bases in Saudi Arabia, the Israeli-Palestinian conflict, and even power struggles within Arab nations. Individuals who endeavoured to contrast the proactive policies of the

U.S. with regional economic disparities or the presence of tyrannical regimes were swiftly branded as anti-American or sympathisers of the opposition. Scholars like Noam Chomsky and Edward S. Herman argued in *Manufacturing Consent* (1988) how media operates within what they referred to as a 'propaganda model', wherein, during times of crisis, capturing dissenting narratives becomes nearly impossible while dominant ones get amplified without any challenge. This dynamic proved vital for justifying the enduring military campaigns, the usefulness and far-reaching consequences of which had become deeply contested. This reductionist perspective, albeit emotionally persuasive, simplified the intricate global concerns behind terrorism, garnering public support for military interventions in Afghanistan and Iraq. By echoing official stories without question, the media helped create an uncritical, us-versus-them view of the world. This marginalized alternative perspectives and justifying prolonged military engagements.

The Vietnam War's Moral Ambiguity

The Vietnam War, akin to the Korean War, was initially framed by Americans as an effort to contain and fight communist aggression (the "Villain"). America intervened to ostensibly protect democracy (the "Hero"), not out of ideological conviction, but out of duty and a sense of unity morphing into a graceless conflict. American soldiers in the My Lai Massacre in March 1968 brutally executed defenceless Vietnamese civilians, exemplifying this dehumanizing transformation. These civilians were reframed as enemies within a propagandistic framework, where the justification of American 'Heroes' seemingly necessitated the obliteration of any potential enemies. This metamorphosis was frequently rationalized by the military culture and media framing that the Vietnamese, especially those located in areas of presumed Viet Cong sentiment, were nothing but an undifferentiated menace, enabling soldiers to commit war crimes by adorning as "True Heroes" (Hersh, 1970). Chris Hedges explains this through his book *War Is a Force That Gives Us Meaning* (2002). His critique details the anti-war arguments and the moral detachment from reality which takes place during wars, which are too different to be encapsulated by one phrase or blanket statement.

Iraq War's Unintended Consequences

The defeat of Saddam Hussein in the 2003 Iraq War was hailed as the downfall of a tyrannical Villain, but the power vacuum that followed, the dissolution of the army, and the inability to build good governance created the conditions that led directly to the kind of sectarian violence and, finally, the formation of ISIS (Islamic State of Iraq and Syria). This exacerbated by the controversial decision to dissolve the

Iraqi army and to de-Ba'athify (referring to the Ba'ath party, meaning "resurrection" or "renaissance" in Arabic) the government, created a fertile ground for instability and insurgency (Dodge, 2005). For many Iraqis, the perceived 'end' of one war of the conflict merely marked the beginning of prolonged and severe turbulence. John Mearsheimer's *The Tragedy of Great Power Politics* (2001) and other realist-based international relations theories emphasise how the unipolar moment after the Cold War and interventions on liberal grounds often ignore the persistence of power politics and the unintended consequences of regime change. In such scenarios, the Villain's doesn't eradicate power dynamics; it merely reframes them.

GENDERED REPRESENTATIONS OF HEROES AND VILLAINS

The narrative tradition has consistently made use of frameworks that are distinctly gendered. These portrayals are not just reflections of sexualised gender norms and expectations but rather are actively reproduced.

Heroic Gender Archetypes

Male Heroes typically embody traits of hegemonic masculinity, such as strength, stoicism, rationality, and physical prowess (Miller et al., 2016). A male Hero's dominance and accomplishment are quite often based on his ability to take decisive action, which often involves violence to defend or protect social order, frequently from the primary agents of violent conflict resolution. This archetypal masculinity is deeply rooted in notions of guardianship and ultimate sacrifice.

In modern media, female heroes often exhibit heightened sexualization and face significant under-representation. Typically, they are portrayed as emotional beings driven by maternal instincts or depicted as needing guidance and support from male characters. They are often shown as emotional, attractive, prone to worry, and excited in a crisis (Baker & Raney, 2007). Their agency is predominantly less focused on action-driven plots in nearly every situation. They are more inclined to express it through their emotional range or nurturing disposition. Although entertainment media have introduced more resilient female protagonists in recent years, they remain constrained by narratives that diminish their roles as mere plot devices, so inhibiting character development, as their suffering predominantly serves to further male characters' arcs. During World War I and II, especially Western countries placed strong emphasis on specific gender roles. It was common to see women featured on recruitment posters and presented in a domestic setting, such as the one which urged, "Women of Britain Say 'Go'!" (Imperial War Museums, 1915). The character of a woman was expected to enforce the idea that a man's Heroism required him to abandon everything and fight for the country and his home. In the Soviet Union, even when celebrating a woman's combat Heroism, suppose as a sniper or pilot, the narrative sometimes navigated a delicate balance, not straying too far from traditional practices of femininity and infant care.

Villainous Gender Archetypes

Representations of gender also reach our Villains. Male Villains are usually rendered as essentially evil, driven by ambition, greed, or a lust for power, and often with limited penetration of their complex motivations. They pose a direct, destructive, and active threat to the established order.

On the other hand, female Villains have historically been typecast as seductive femme fatales, emotionally unstable, or jealous and vengeful, and their villainy often stems from personal betrayal or an unconventional quest for power that challenges the patriarchal status quo. A notable real-life illustration of this is Ilse Koch, infamously known as "The Witch of Buchenwald" during World War II. The most extreme depiction of her irrational villainy is not framed around a strategic argument; instead, sensational narratives of sadism, depravity, and cruelty are tortured, that is, alleged lampshades made of human skins burnished by one who is considered to be the epitome of traditional feminine roles (Drumbl & Mouthaan, 2021). This portrayal reinforced stereotypical ideals of women who hold power destructively, hence personalizing Nazi evil through a distinctly gendered lens rather than focusing on systemic atrocities (Jardim, 2023).

For the same reason, as barbarian figures with animalistic features, the Japanese were often cartooned characteristically with faces like a savage being. This helped Allies justify their own extremely violent treatment of an enemy whom they could be demonize as being 'less than human'. Again, in a Nazi propaganda poster, Jews were depicted as unattractive characters who would manipulate others and threaten racial purity, elements which fall right into a stereotype of 'inherently evil' designed to justify persecution or genocide. Soviet women spy in Western propaganda during the Cold War were portrayed as two-faced, deceitful temptresses and femme fatales who abuse their femininity for disloyalties These images reflect a fear that female power can become subversive, suggesting that the true threat lies not only in blatant violence but also in the potential for deception and manipulation. This fear implies that traumatizing women could cause more harm. During the period of war propaganda, even when women served as soldiers, their roles were rarely acknowledged. When a female figure was portrayed as an enemy, she was often depicted as morally corrupt or mentally unstable, typically exploiting feminine traits instead of demonstrating strategic or physical prowess.

MORAL REVERSAL AND DILEMMA

Moral Transformation of Antagonist Figures

Redemption arcs make great narrative devices, tracing a character's painful journey from selfishness and amorality, or simply being amoral, to repentance and atonement. This voyage commonly leads to an act of great self-sacrifice or a continued demonstration in some new moral behaviour that is fundamentally different from what has gone before. Redemptive narratives provide significant dramatic potency as they transform people from mere representations of malevolence, imbuing them with humanity and complexity. This shift is often a catalyst, triggered by a moment of deep understanding, a massive personal loss, or some crucial confrontation with an overpowering moral force. This approach allows the society to empathize with the Villain 's internal struggles (McElroy, 2011). Although this concept is used mostly in entertainment and media, it holds high real-life implications as well.

But genuine and true redemption requires genuine contrition and an explicit admission of past wrongs. It also calls for an honest commitment to change, often accompanied by personal sacrifice or intensive measures to redress wrongs done and harms inflicted on others. In war films, a notable example is Sergeant Elias Grodin in Platoon (Stone, 1986). Though portrayed with greater moral integrity than his brutal counterpart Sergeant Barnes, Elias's attempt to save his men, despite being mortally wounded, becomes his redemptive act, offering a poignant contrast to the moral decay surrounding him.

Oskar Schindler, a prominent German entrepreneur during World War II and member of the Nazi Party, is another fascinating case study. In the beginning, Schindler used Jewish workers in his factories purely for economic and personal gain. But then, observing the atrocities of the Holocaust – including the demise of the Kraków Ghetto, where Jews from the city were pursued and incarcerated – Schindler spent his entire fortune on bribing Nazi officials to run his factories as a haven, providing a powerful case of real-world redemption. Although there was also his use of wealth, power, and influence to ultimately help save over 1200 Jews from extermination, all of which represents an enormous change in moral direction. He cannot be called a good man, as various articles and historical writings indicate his nature to be greedy, opportunistic and unethical, but the fact that he still did some good suggests a moral reversal.

Victims as Villains

The victims are Villains trope is a particularly odious and problematic narrative framing where the very same individual who has experienced harm or injustice is portrayed in a negative light, blamed and seen as culpable for their plight, and may even be depicted as an antagonist. This fundamental inversion of narrative is not uncommon and may be based on our cognitive biases, notably the just-world hypothesis. When a victim experiences misfortune, this bias can cause the victim to be seen in a different light, as if they somehow invited or deserved the harm that has happened to them – taking the blame away from the perpetrator. This shift in narrative is an efficient mechanism for managing cognitive dissonance, allowing the observers or perpetrators to have a way to rationalize each act of harm with a sense of justice.

In this trope, dehumanization of the victim, the 'undeserving' term is contradictory. Victims are Villainized if they do not exemplify the societal approved 'ideal victim', someone who perfectly embodies innocence, helplessness, and victimization by a third party that is seen as completely unavoidable (Christie, 1986). For example, in the exploitable images of victims when real war is occurring, the occupied or ethnically cleansed population is sometimes drawn upon as an apparently existentially dangerous subject using propaganda from the aggressor's state, thus dehumanizing them in the process.

One historical example is the Hutu Power regime in Rwanda, which employed dehumanizing language (the Radio Télévision Libre des Mille Collines, a station which referred to the enemy as "cockroaches"), and they essentially redefined Tutsi people as 'criminals' or 'Villains' via their media, allowing the various perpetrators to justify their genocidal atrocities (Strauss et al., 2000). This dehumanization from victim to Villain eased the moral burden of those committing the atrocities. This occurred in a somewhat similar manner to the Nazi propaganda that used media and subsequently victimized the Jewish people portraying them as a parasitic body, perhaps manipulative, evil, and a threat to the German nation, to Villainize the Jewish people from victims of persecution to Villain to justify the Holocaust.

Another example of redefining a victim of antisemitism is Alfred Dreyfus, a Jewish French army officer who was accused and convicted of treason in 1884. Dreyfus was wrongfully convicted in an espionage case based on flimsy evidence due to the metastasizing impact of the antisemitism Dreyfus had experienced in the French military during that time. Dreyfus eventually served time on Devil's Island and then experienced a public shaming. But eventually the truth unfolded that another officer, Major Esterhazy, was actually the real traitor. Other people such as Émile Zola directed public attention and exposed the injustice instead of the legitimacy of the original case against Dreyfus, leading to his exoneration in 1906 (Landsman & Bredin, 1987).

These cases illustrate the ease with which an individual can be portrayed as an adversary or villain, resulting in disastrous consequences for the victim and irrevocably altering their lives.

The Paradox of Heroic Villainy

This idea of characters who perform and commit traditionally 'Villainous' actions, such as (but not limited to) violence, deceit or rebelling against accepted conventions, and yet are portrayed in a sympathetic or fascinating light, muddying the Hero and Villain binary. This paradox emerges when characters engage in morally questionable actions motivated by a sense of justice, or, conversely, when their malevolent deeds are a response to a more significant, often systemic injustice. The society is encouraged to wrestle with this moral ambiguity, debating over whether the means are justified by the ends. Nowhere is this more evident than in anti-Hero stories, where Heroes and Heroines do morally questionable things to obtain some sort greater good, or where their Villainy is ultimately a necessary evil.

In actual warfare, Che Guevara, the Argentine Marxist revolutionary, embodies the dilemma of morality i.e., right and wrong. To several individuals in the revolutionary and anti-imperialist movements, he is seen as a hero who devoted his life to combating repressive regimes and aspired for the day when global disparities could be dismantled through armed conflict (Brincat, 2023). For his victims and others who opposed communist revolutions while denouncing his tactics of guerilla warfare and revolutionary brutality, he is only a villain. His actions were brutal, and he felt that to realise his objective of a socialist revolution, he utilised his authority to dismantle existing structures, hence generating a pronounced dichotomy in viewpoints. Likewise, amongst at activist figures, Nelson Mandela, a South African anti-apartheid revolutionary and future politician, might have been viewed at certain times in his early activism through this paradoxical perspective, and indeed he was, by the apartheid regime that he opposed. Now celebrated as a founding Hero of peace and racial justice worldwide, Mandela was once labelled a terrorist and a saboteur by the apartheid government for leading the ANC's (African National Congress) armed wing, Umkhonto we Sizwe (Spear of the Nation) (Mandela, 2008). This wing's committed acts of sabotage against state infrastructure and symbols of apartheid, which were legally defined by the regime as acts of terrorism and treason, justifying his long imprisonment (Ellis, 2015). This led to him being labelled as Villain and an enemy to the state.

Mahatma Gandhi stood as an Indian anti-colonial nationalist who demonstrated the complex moral dilemma associated with warfare. Gandhi earned his Heroic status through his *ahimsa* and non-violent resistance philosophy which led India to independence with civil disobedience. Both British authorities and some Allied leaders denounced Gandhi's "Quit India Movement" as a threat to the Allied war effort against fascism because they believed it obstructed their fight against a more significant global menace. Some people labelled him as a Villain because he refused to sanction violence for eliminating oppression. His steadfast adherence to ahimsa brings attention to crucial moral and justice-related issues.

We must evaluate the intentions and outcomes of their legacies because history's evaluation does not offer straightforward answers and the “greater good” remains a controversial notion.

DILEMMA IN WAR: BEYOND GOOD VS. EVIL

The real human experiences extend beyond just good vs evil, we are diverse, rich, and amazing beings shaped by a mosaic of emotions, experiences, and choices. The philosophical framing of war through these archetypes raises questions about the nature of good and evil. Is the Hero always doing the right thing? Is the Villain always the worst being? This moral ambiguity is often glossed over in media portrayals, as the focus remains on celebrating the Hero’s victories and vilifying the enemy. These narratives explore the moral grey areas of war, questioning the righteousness of the Hero’s actions and humanizing the enemy. The glorification of Heroes can obscure the traumatic and destructive nature of war, framing it as a noble endeavour rather than a human tragedy. The moral compass is a compass which needs to be delved deeper into.

But what if this narrative is dismantled? What if the Hero is not always right, and the Villain isn't purely evil? There is a glorification of wartime Heroes that often obscures the psychic, moral and moralistic costs of conflict. Heroes do the hard things, even when such things result in the destruction of innocent lives. Their successes are often built with the bodies of civilians, dissolved promises, and lifelong guilt. Can someone still be considered a Hero if they do terrible things in the name of the ‘greater good’? Who defines that ‘good’? And what if the enemy soldier, too, believes they are defending their homeland, their people, their truth?

It is a painful, but necessary, part of recognizing the enemy's humanity. The soldiers on the two sides could have the same fears, hopes, vulnerabilities. The temptation to monochromatize must be resisted. All individuals are entitled to articulate their perspectives and justify their acts, regardless of their success or failure. They might be drafted youth, traumatized survivors or people corralled into decisions they had never hoped to make. Humanising the 'other' does not justify their conduct; rather, it demands empathy and nuance in our comprehension of them. The moral compass during wartime does not consistently point in a single way. It revolves around politics, propaganda, trauma, and survival. For some, the 'correct action' may involve refraining from combat, defying directives, or exhibiting empathy towards the enemy. These acts of moral courage are often overshadowed by narratives of triumph. Comprehending the war necessitates the acceptance of its moral ambiguity.

CONCLUSION

The Hero and Villain archetypes, deeply ingrained in cultural, religious, and psychological frameworks, wield immense power in war narratives. While they often reduce complex struggles to simplistic moralistic binaries, these roles are fundamentally products of perspective, ideology, and narrative manipulation. Through critical examination, these archetypes are deconstructed and problematized, fostering a fuller awareness of the profound moral ambiguity inherent in warfare and revealing the nuanced humanity that exists beneath the surface categories of 'hero' and 'villain'.

REFERENCES

Alford, M. (2016). The Political Impact of the Department of Defense on Hollywood Cinema. *Quarterly Review of Film and Video, 33*(4), 1–16. DOI: 10.1080/10509208.2015.1086614

Allison, S. T., & Goethals, G. R. (2011). *Heroes: What they do and why we need them*. Oxford University Press.

Anderson, Benedict. (2006). Imagined communities: reflections on the origin and spread of nationalism.

Baker, K., & Raney, A. A. (2007). Equally super?: Gender-role stereotyping of superHeroes in children's animated programs. *Mass Communication & Society, 10*(1), 25–41. DOI: 10.1080/15205430709337003

Bandura, A. (1999). Moral disengagement in the perpetration of inhumanities. *Personality and Social Psychology Review, 3*(3), 193–209. DOI: 10.1207/s15327957pspr0303_3 PMID: 15661671

Bar-Tal, D. (1990). Causes and consequences of delegitimization: Models of conflict and ethnocentrism. *The Journal of Social Issues, 46*(1), 65–81. DOI: 10.1111/j.1540-4560.1990.tb00272.x

Bigelow, K. (Director). (2008). The hurt locker [Film]. Summit Entertainment.

Bigelow, K. (Director). (2012). Zero dark thirty [Film]. Columbia Pictures.

Boggs, C., & Pollard, T. (2006). Hollywood and the Spectacle of Terrorism. *New Political Science, 28*(3), 335–351. DOI: 10.1080/07393140600856151

Boyce, M. (2001). *Zoroastrians: Their Religious Beliefs and Practices*. Psychology Press.

Boyer, Pascal. (2001). Religion Explained: The Evolutionary Origins of Religious Thought.

Brincat, S. (2023). Introduction: Che Guevara and world politics. *Globalizations, 20*(8), 1426–1446. DOI: 10.1080/14747731.2023.2274649

Brockington, J. L. (1998). *The Sanskrit Epics*. BRILL. DOI: 10.1163/9789004492677

Campbell, J. (1968). *The Hero with a thousand faces*. Princeton University Press.

Campbell, J. (2008). *The Hero with a thousand faces*. New World Library.

Christie, N. (1986). The ideal victim. In Palgrave Macmillan UK eBooks (pp. 17–30). https://doi.org/DOI: 10.1007/978-1-349-08305-3_2

Clasen, M. (2017). *Why horror seduces*. Oxford University Press. DOI: 10.1093/oso/9780190666507.001.0001

Clifford, J., & Said, E. W. (1980). Orientalism. *History and Theory*, *19*(2), 204–223. DOI: 10.2307/2504800

Dodge, Toby. (2005). Iraq's future: the aftermath of regime change.

Drumbl, M., & Mouthaan, S. (2021). 'A Hussy Who Rode on Horseback in Sexy Underwear in Front of the Prisoners': The Trials of Buchenwald's Ilse Koch. *International Criminal Law Review*, *21*(2), 280–312. DOI: 10.1163/15718123-bja10047

Eastwood, C. (Director). (2014). American sniper [Film]. Warner Bros. Pictures.

Easwaran, E. (1989). Classics of Indian spirituality.

Ellis, S. (2015). Nelson Mandela, the South African Communist Party and the origins of Umkhonto we Sizwe. *Cold War History*, *16*(1), 1–18. DOI: 10.1080/14682745.2015.1078315

Elshtain, J. B. (1998). Women and War: Ten Years On. [suspicious link removed]. *Review of International Studies*, *24*(4), 447–460. DOI: 10.1017/S0260210598004471

Festinger, L. (1957). *A theory of cognitive dissonance*. Stanford University Press. DOI: 10.1515/9781503620766

Fiske, A. P., & Rai, T. S. (2015). *Virtuous violence: Hurting and killing to create, sustain, end, and honour social relationships*. Cambridge University Press.

Freud, S. (1930). *Civilization and its discontents*. Hogarth.

Freud, S. (1953). The uncanny (J. Strachey, Ed. & Trans.). In J. Strachey (Ed. & Trans.), The standard edition of the complete psychological works of Sigmund Freud (Vol. 17, pp. 217–252). London: The Hogarth Press. (Original work published 1919)

Gaufman, E. (2016). Security threats and public perception. In New security challenges. https://doi.org/DOI: 10.1007/978-3-319-43201-4

Hedges, C. (2002). *War is a force that gives us meaning*. PublicAffairs.

Hen, K., Tsui, H., & Lam, D. (Directors). (2021). The battle at Lake Changjin [Film]. Bona Film Group.

Herman, E. S., & Chomsky, N. (1988). *Manufacturing consent: The political economy of the mass media*. Pantheon Books.

Hersh, S. M. (1970). *My Lai 4: A report on the massacre and its aftermath*. Random House.

Hobbes, T. (1651). *Leviathan*. Penguin Books.

Hobbes, T. (2018). LEVIATHAN (Complete Edition). e-artnow.

Homer, . (2007). *The Iliad*. University of Michigan Press.

Howard, S. (Director). (2005). Jarhead [Film]. Universal Pictures.

Imperial War Museums. (1915). Women of Britain say 'Go!' Retrieved February 1, 2023, from https://www.iwm.org.uk/collections/item/object/205020521

Jackson, R. (2005). *Writing the War on Terrorism: Language*. Politics and Counter-terrorism.

Jardim, T. (2023). *Ilse Koch on Trial: Making the "Bitch of Buchenwald"*. Harvard University Press., DOI: 10.2307/jj.362392

JUNG, C. G. (1959). Collected Works of C.G. Jung, Volume 9 (Part 2): Aion: Researches into the Phenomenology of the Self (G. ADLER & R. F. C. HULL, Eds.). Princeton University Press. [suspicious link removed]

JUNG, C. G. (1969). Collected Works of C.G. Jung, Volume 9 (Part 1): Archetypes and the Collective Unconscious (G. ADLER & R. F. C. HULL, Eds.). Princeton University Press. [suspicious link removed]

Kant, I. (1795). *Perpetual Peace: A Philosophical Sketch*. Essays on Kant's Cosmopolitan Ideal.

Kelman, H. C. (1973). Violence without moral restraint: Reflections on the dehumanization of victims and victimizers. *The Journal of Social Issues*, *29*(4), 25–61. DOI: 10.1111/j.1540-4560.1973.tb00102.x

Landsman, S., & Bredin, J. (1987). The triumph of justice. *Michigan Law Review*, *85*(5/6), 1095. DOI: 10.2307/1289035

Lerner, M. J. (1980). The Belief in a Just World. In *The Belief in a Just World. Perspectives in Social Psychology*. Springer., DOI: 10.1007/978-1-4899-0448-5_2

Lippmann, W. (1922). *Public opinion*. Harcourt, Brace.

Litz, B. T., Stein, N., Delaney, E., Lebowitz, L., Nash, W. P., Silva, C., & Maguen, S. (2009). Moral injury and moral repair in war veterans: A preliminary model and intervention strategy. *Clinical Psychology Review*, *29*(8), 695–706. DOI: 10.1016/j.cpr.2009.07.003 PMID: 19683376

MacKenzie, J. M. (1984). *Propaganda and empire: The manipulation of British public opinion, 1880-1960* (Rev. ed.). Manchester University Press.

Mandela, N. (2008). *Long walk to freedom: The Autobiography of Nelson Mandela.* Little, Brown.

McAdams, D. P., & McLean, K. C. (2013). Narrative Identity. Current Directions in Psychological Science, 22(3), 233-238. https://doi.org/ (Original work published 2013)DOI: 10.1177/0963721413475622

McElroy, P. C. (2011). On redemption, the value of putting bad things to good use. Stanford Digital Repository. https://purl.stanford.edu/gq597rc8146

Mearsheimer, J. J. (2001). *The tragedy of great power politics.* W. W. Norton & Company.

Medhurst, M. J., Ivie, R. L., Wander, P., & Scott, R. L. (1997). *Cold War Rhetoric: Strategy, Metaphor, and Ideology.* Michigan State University Press. [suspicious link removed]

Miller, M. K., Clark, J. D., & Jehle, A. (2015). *Cognitive Dissonance Theory (Festinger).* The Blackwell Encyclopedia of Sociology., DOI: 10.1002/9781405165518.wbeosc058.pub2

Miller, M. K., Rauch, J. A., & Kaplan, T. (2016) Gender Differences in Movie SuperHeroes' Roles, Appearances, and Violence. Ada: A Journal of Gender, New Media, and Technology, No. 10. DOI: 10.7264/N3HX19ZK

Monteleone, E. (Director). (2002). El Alamein: The line of fire [Film]. Fiorella Film, RAI Cinema.

Nabavi, T. (2012). *Razieh & Bijandi, Mohammad.* Bandura's Social Learning Theory & Social Cognitive Learning Theory.

Pagels, E. (1995). *The Origin of Satan: How Christians Demonized Jews, Pagans, and Heretics.* Vintage Books.

Peace, P. (1917). *A Philosophical Essay, translated with Introduction and Notes by M. Campbell Smith, with a Preface by L. Latta.* George Allen and Unwin.

Samuels, A. (1985). *Jung and the post-Jungians* (1st ed.). Routledge., DOI: 10.4324/9780203359297

Samuels, M. (2017). Erwin Rommel and German Military Doctrine, 1912–1940. War in History, 24(3), 308-335. https://doi.org/ (Original work published 2017) DOI: 10.1177/0968344515623982

Sandars, N. K. (1960). The epic of Gilgamesh. http://www.aina.org/books/eog/eog.pdf

Spielberg, S. (Director). (1998). Saving private ryan [Film]. DreamWorks Pictures; Paramount Pictures.Stone, O. (Director). (1986). Platoon [Film]. Hemdale Film Corporation.

Strauss, S., Gourevitch, P., & Forges, A. D. (2000). Genocide in Rwanda. *African Studies Review*, *43*(2), 126. DOI: 10.2307/524988

Tajfel, H., & Turner, J. C. (2004). The Social Identity Theory of Intergroup Behavior. In Jost, J. T., & Sidanius, J. (Eds.), *Political psychology: Key readings* (pp. 276–293). Psychology Press., DOI: 10.4324/9780203505984-16

Traesar, L., & Saktiningrum, N. (2024). THE PORTRAYAL OF FEMALE VILLAINS: A REPRESENTATION STUDY ON CRUELLA. *Rubikon Journal of Transnational American Studies*, *11*(2), 327. DOI: 10.22146/rubikon.v11i2.94900

WEST. R. (1984). On War (M. HOWARD & P. PARET, Eds.). Princeton University Press. https://doi.org/DOI: 10.2307/j.ctt7svzz

Wright, Q. (1942). *A study of war* (Vols. 1–2). University of Chicago Press.

ADDITIONAL READINGS

Assmann, J. (2011). *Cultural memory and early civilization*. https://doi.org/DOI: 10.1017/CBO9780511996306

Franco, Z. E., Blau, K., & Zimbardo, P. G. (2011). Heroism: A conceptual analysis and differentiation between heroic action and altruism. *Review of General Psychology*, *15*(2), 99–113. DOI: 10.1037/a0022672

Fussell, P. (2000). *The Great War and modern Memory*. Oxford University Press.

Hoskins, Andrew & O'Loughlin, Ben. (2010). War and Media: The Emergence of Diffused War.

Hourihan, M. (1997). *Deconstructing the hero: Literary Theory and Children's Literature*. Psychology Press.

Jesperson, S. (2013). The Media at War: Communication and Conflict in the Twentieth Century, 2nd edn. by Susan Carruthers. Journal of Peace Research. 50. DOI: 10.2307/23441191

Salamucha, A. (1970). Philip G. Zimbardo. The Lucifer Effect: Understanding How Good People Turn Evil. *Forum Philosophicum.*, *14*(1), 166–168. DOI: 10.35765/forphil.2009.1401.14

Chapter 10
Voices and Void:
Trauma and Testimony

Sandhya Devi N. K.
https://orcid.org/0000-0001-5028-1777
RV University, Bangalore, India

ABSTRACT

The goal of this study to investigate the atrocities against the Yazidi community, especially the women, during ISIS-led genocide. Using discourse analysis, this chapter examines Nadia Murad's memoir The Last Girl: My Story of Captivity, and My Fight against the Islamic State (2018) as a site of testimonial articulation that transcends personal trauma to represent the collective trauma. Furthermore, alongside the memoir, the chapter incorporates media appearances to analyse Murad's self-advocacy to navigate and challenge institutional mediation and inaction. The study uses testimonial theory and gender theory to understand testimony and the sexual slavery of Yazidi women as they were captured, traded and sexually violated. The study found that sexual enslavement decimated their individuality and connection to their culture while also denied their autonomy, narrative agency and expression. However, Murad's testimony served both personal and collective as it not only documented genocidal trauma, but demanded for recognition, accountability and justice.

INTRODUCTION

Testimony aims to "communicate" or "convey information" to an audience (Sosa, 1994, p. 219). It is shaped by conviction, intention, purposeful address, and resilience. These components are rooted in real-life experiences, truth-telling, and beliefs. Language plays a predominant role in constructing and elucidating credible

DOI: 10.4018/979-8-3373-7589-2.ch010

assertions while also accentuating its crucial function of witnessing (Harris & Koenig 2006, p. 505). While speakers of testimonials relay messages, they also encompass written and non-verbal communication (Harris & Lane, 2014; Robinson et al., 2013). This includes several forms of expressions such as religious proclamations, political statements, and personal narratives based on the narrator's first-hand experience. Furthermore, the characteristics of testimony involve memory, perception, trust, witnessing, and intervention. Although testimony is a source of knowledge, it is also a channel through which the narrating voice communicates, recounts, and reflects on past traumatic experiences. Besides, it functions as a literary documentation that is not merely metaphorical or abstract but rather a form of proof that embodies the experiences of true events, oppression, and suffering (Detue & Lacoste, 2021). Further, testimonies are not only personal, but function as historical texts that have witnessed true events. Simultaneously, they advocate for justice, making testimony an impactful tool for reclaiming agency and personal healing.

Genocide may not be an immediate destruction of a nation. Instead, it can be a slow, churning-out process, that amounts to a calculated attempt at erasure. The term "genocide" means the killing (-cide) of a race or a tribe (genos) coined by Raphael Lemkin, a Polish Jewish jurist (Simon, 1996). Genocide was defined for the first time in Article II of the Convention on the Prevention and Punishment of the Crime of Genocide. It is an act committed to intentionally destroy a particular religious, national, ethnic, and racial groups. This includes the killing of entire groups, causing serious bodily/mental harm, destruction of life conditions, imposition to prevent new births, and the transference of children from one group to another (Cetorelli et al., 2017).

The Yazidis, an ethnic and religious minority group, primarily lived in the Mount Sinjar region of northern Iraq. Because their faith differs from the mainstream Islamic beliefs, they have long been marginalized and contemptuously labelled as "devil-worshipers", making them targets of extreme violence (Murad, 2018). This led to persecutions, systematic murders, forced conversions, and psychological and physical assaults. One of the most horrific events was the genocide of the Yazidi community in Iraq, during which they were hunted, captured, and killed by the Islamic State of Iraq and Syria (ISIS), in 2014. Yazidis were forced to undergo dehumanizing treatment which included torture and forcible religious conversion. While Yazidi women were kidnapped, sold, traded, and raped, younger boys were separated, indoctrinated and placed among ISIS fighters (Kizilhan, 2025). Cetorelli et al. (2017) posit that 3000 Yazidis were killed, while 7000 were kidnapped.

Although the United Nations (UN) offered humanitarian support, legal assistance, and documented Yazidi genocide and domestic courts of Iraq held trials with few convictions, full accountability remains incomplete. As a consequence, the justice sought by Yazidi men and women remains obscure and distant. However, testifying to such atrocities is essential for representing and supporting the Yazidi community, while advocating for justice and accountability. For this purpose, this chapter uses Nadia Murad's memoir to bring out Yazidi genocide, rape, captivity and the call for justice. An eminent activist and a Nobel Peace Prize laureate, Nadia Murad frequently appears in international media and humanitarian forums to represent Yazidis and showcase their suffering. Using discourse analysis, this chapter examines how her memoir *The Last Girl: My Story of Captivity, and My Fight against the Islamic State* (2018) functions as testimony that transcends personal trauma to represent the collective Yazidi suffering. Furthermore, alongside the memoir, the chapter also incorporates speeches and media appearances to analyze Murad's self-advocacy to navigate and challenge institutional mediation.

LITERATURE REVIEW

Assmann (2006) emphasizes the relationship between memory and history, and that both function as reliable sources of knowledge. Saul Friedländer in his book *Nazi Germany and the Jews* (2014) explores this connection between testimonial experience, its reliability, and factual history. Oliver (2001) discusses the struggles of oppressed communities for recognition and representation, while also accentuating how subjectivity is innately structured through witnessing in duality: being an eyewitness and bearing witness to the unwitnessed experiences of the self. Consequently, testimony includes the struggle for self-representations and "soldering" the self, as the experiences of soldiers are used to analyze effacement, healing, fragmentation, and reflexivity (Chouliaraki, 2016). Also, witnessing and testifying are literary practices that conjoin ethical, political, and aesthetic factors. This conjointly documents victimized experiences and resists erasure (Detue & Lacoste, 2021).

Testimonial studies further discuss the political and interventionist potential through forms such as testimonio, human rights discussions, subaltern studies, and the difficulties in fully representing traumatic experiences through trauma and holocaust studies (Cubilié & Good, 2003; Dutro, 2008; Popkin, 2003). However, testimony can be problematic, as it can be both informative and performative, raising questions about the authenticity of the testified messages (Delaperrière, 2014). Jones (2019) discusses the relationship between text (produced by witnessing) and audience. He further argues that the formulation of resolution can occur only through understanding oppression and experiences. This consequently emphasizes the need for acknowledgement, empathy, as well as trusting the text, the victim, and understanding the perpetrator's intention. Testimony, although contested within academic literary genres, it legitimizes and helps process traumatic memories, cultural resistance, and ethical recuperation of a community that has been marginalized (Levine,

2006; Strejilevich, 2006; Vice, 2008). Further, testimony allows reclamation and emphasizes the responsibility of the narrator and interviewer; as well as the reader's role in experiencing the othered trauma (Bernard-Donals, 2001; Levine, 2006).

Furthermore, the limitations of language, the performative nature of testimony, and the employment of catharsis are critical in comprehending the multiple layers of testimony (Popkin, 2003; Robinson, 2011; Shirm, 2023). Further, the relationality between the audience and the narrator is significant in negotiating author agency in testimonial texts (Woods, 2021). Although testimony is narrated by the narrator, a collaborative or supporting approach from the community may help in articulation, providing emotional and physical security, and address the skepticism regarding victim's testimony (Knowles, 2023).

Testimonial narratives and self-advocacy not only document trauma but enable national and international recognition, humanitarian networks, and wider audience reach (Fassin, 2011). This further allows for narrative authenticity, control, appeals for justice, prosecutions, and inaction. As a result, such testimonies can lead to militaristic, humanistic, economic, and political interventions (Cubilié & Good, 2003). Ticktin (2011) reveals that compassion is a product of humanitarian support that can help provide medical services and asylum. However, Forsythe (2005) discusses the confusion between nationalism and cosmopolitanism. This showcases cyclic tensions between national interest and humanistic interest; state and human safety, legal state action and the moral necessity to protect human dignity (Woods, 2021). As a result of such complexity, international support may fall short in providing aid, legal solutions, and asylums. Further, institutional mediation may victimize survivors, or reframe testimonial voices (Delaperrière, 2014). This questions the role of national and international mediation in recognizing, acknowledging, prosecuting, and reforming legal and political issues of the disadvantaged individuals (Fricker, 2008; Sellars, 2010).

ISIS employed systemic violence to brutally cleanse the Yazidi community (Mohammed, 2022). This included sexual slavery, mass executions, indoctrinations, forced conversions, and marriages (Cetorelli et al., 2017). As a consequence, the survivors underwent extreme psychological instability and disorders due to enduring loss and trauma (Cheterian, 2021). Many survivors also experienced psychosocial stress as a consequence of trauma, and residing in refugee camps (Kizilhan et al., 2023). Yazidi girls and younger women were abducted and subjected to sexual assault and slavery. As a result, these women suffered chronic pains, stress, guilt, and mental illness (Goodman et al., 2020). Further, forced marriages, conversions, domestic roles, and sexual enslavement complicate and blur the definition of rape (San José, 2020). The extermination of Yazidi community, which is also called as "jihad", was carried out to humiliate them (Jaffal, 2020). These women underwent trauma due to perceived social rejection. Further, the aftermath of rape often led to social rejection and mental illness, similar to war-related psychological disorders and illnesses (Ibrahim et al., 2018). Akhavan et al., (2020) examine the complicated nature of justice for Yazidis, given the several layers of discrimination, trauma, and genocide (Potot-Warren, 2020; Kizilhan, 2025). Although UN Investigative Team to Promote Accountability for Crimes Committed by ISIS (UNITAD) was established to seek justice, its effectiveness is debatable and questionable due to the unpredictability of Iraqi politics (Van Schaack, 2018). El-Masri (2018) discusses the brutality of ISIS-induced Yazidi genocide and foregrounds the need of investigations, reforms, and justice through truth commissions and local and global bodies.

THEORETICAL BACKGROUND

This study uses testimony as a theoretical background using *Testimonio: On the Politics of Truth* by John Beverly (2004) and *Testimony: Crises of Witnessing in Literature, Psychoanalysis, and History* by Shoshana Felman and Dori Laub (1992). Beverly (2004) conceptualizes that testimony is a first-person narrative that is born out of societal, cultural, political subjugation, multi-leveled inequality, trauma, and forced silencing. As Jara (1984) writes in the context of Chilian authoritarian regimes and the consequent atrocities, notes that testimony is an urgent narrative since it involves an ongoing issue either through memory or an ongoing situation. Further, Sobel & Kushnir (2013) illustrate that testimony is evaluated through previous knowledge and contextual experience to avoid wrongful representation. They proclaim that by exposing truths, testimony brings out the silenced voices which consequently represent the subaltern, while also avoiding misrepresentation, fictionalization, erasure, and misappropriation. However, the mediation of hegemonic institutions, political powers, and beliefs endangers testimony due to the over-simplification, editing or catering to these structures. Further, this theory emphasizes truth-telling and the authenticity of the marginalized experiences. By centering authenticity, it also expands through cultural, societal and historical experiences. Consequently, through such outreaches, testimony becomes a form of political intervention by actively participating in the struggles of the marginalized. By addressing controversial, questionable, unjust circumstances, testimony reveals and conquers ambiguities, silences, and erasures. However, testimony theory also showcases the contradictory nature of healing and reliving of traumatic experiences through intrusive and continuous reminders, triggers, and memories. This makes the traumatized victim a symptom of history, unable to heal or reclaim the self in its entirety (Sternbach, 1991). Felman and Laub (1992) conceptualize testimony as a form of witnessing which includes the crisis inherent in witnessing. They highlight the inability and impossibility of fully articulating the traumatic experiences

(Caruth, 2016). Also, testimony transcends genre as it involves taking moral responsibility to both testify and represent, and speak to the audience, while also allowing them to trust the narrator (Marusic, 2015) (who conceptualizes trust in testimony as a moralistic and ethical stance founded in relational belief). Further, testimony demonstrates the dilemma between resistance and vulnerability, as it embraces and challenges political and existential aspects of testimony.

Simone De Beauvior (2010) posits that society is constructed around male-perspectives, resulting in a male-centered world. Although men may experience occasional forms of oppression, women are systematically marginalized, and forced into gender submission. Due to this, women are located in spaces in "less power, less prestige, less pay or no pay" (Lindsey, 2020, p. 03). This inevitably showcases that men occupy the space of power through roles of husbands, brothers, fathers, bosses, and superiors. Such dominance is substantiated through domestic spatial domination, physical and sexual assault, verbal abuse, private and public spaces (Kelly, 2013).

This emphasizes the political form of intimidation which is witnessed as it is enforced upon the society and its citizens (Brownmiller, 2005). Rape occurs when a man induces non-consensual sexual intercourse. Through this act, men display power over women. This illustrates sexual supremacy through which sexual dominance, gender dominance, and virility are exhibited (Blackstone, 2003). Further, sexual violence conjoined with physical assault and aggressive behaviors control women, while also causing the disintegration of their psychological and physical autonomy.

One of the main functions of testimony is its ability to represent the subaltern by providing a platform that otherwise is non-existent. The oppressed are also profoundly silenced and suppressed, as they are often transformed into consumable objects through capitalistic and patriarchal structures. Subalternity is comprised of socio-economic oppression, while also enduring invisibility (Spivak, 2023). As a result, although the oppressed women attempt to speak, they are controlled by the hegemonic and epistemological structures unless they are dismantled. Subsequently, the limitations of testimony and representation through subaltern experiences emphasize the linguistic, socio-cultural, political, and the representational blockades.

In summary, the theoretical background sections combine testimonial theory, feminist theory and subaltern underpinnings to investigate the layered silencing and gendered violence experienced by Yazidi women. Testimonial theory conceptualizes first-person testimony characterized by continuity, urgency, and the endeavor to speak under hegemonic subjugation. This theory aims to expose truths while also showcasing the inability to completely articulate trauma. Gendered violence, specifically, sexual violence illustrates male dominance over women's spaces, autonomy, and bodies (Brownmiller, 2005; de Beauvoir, 2010). Subsequently, subalternity demonstrates the erasure of disadvantaged voices and restrains of representability through patriarchal contexts (Spivak, 2023). Altogether, these frameworks act as a base to critically analyze Yazidi women's gendered violence, and enforced silencing, while also addressing and representing Yazidi collective testimony by highlighting hegemonic inequalities, across gender, power structures, psychological and physical spaces and language.

BEYOND PERSONAL TESTIMONY

The Yazidi community is a staunch religious group rooted within Yazidism, which is intertwined within the fabric of their livelihood. While most of them speak Kurdish, some are considered as ethnic Kurds. However, many identify as Yazidis and possess a distinct ethnic and religious identity. Since Yazidism does not follow mainstream Islamic traditions, Yazidis have long been targets for mainstream religions, including Sunni Arabs, Sunni Kurds, and Baathists (Murad, 2018; Kizilhan et al., 2023). Further, the secluded settlements in Sinjar mountains and lack of support added to their vulnerabilities. They were called as "kuffars" meaning "unbelievers worthy of killing" (Murad, 2018, p. 04). Nadia, through her memoir, recounts the unravelling events as ISIS invaded and captured several parts of Iraq, including Mosul and Kocho, while simultaneously and specifically targeting the minority Yazidi population (Nadia's place of residence before invasion).

Murad (2018) begins her memoir by illustrating the escalating anxiety and terror as ISIS drew nearer. Trauma scholars call this phenomenon a "proleptic experience", where danger and violence are lived before the actual occurrence, highlighting the psychological uneasiness (DeRosa, 2014). This creates hyper-awareness and the consequent psychological responses, highlighting the collective communal fear, unrest and terror (Caruth, 2016; Van Der Kolk, 2014). This anxiety worsens as they are betrayed by Sunni Arabs who were their neighbors, friends, and teachers. Their denial of help and collaboration with ISIS causes social annihilation, since the dismantling of trusted relationships leaves an existential and psychological wound. This showcases how Yazidis suffered a social death as their identities were systematically erased and left unacknowledged by their society (Patterson, 2018).

On the onset of ISIS invasion, Yazidis reaches a peaking point highlighting the rejection of Yazidi identity, religion, and culture. Laub's (1991) observation that the completion of trauma is through witnessing and listening. However, the rejection of their calls for help is ignored, treated with indifference, and coercively and indirectly silenced. Further, the abandonment by the Peshmerga (internal forces of Kurdistan) and the American army leaves Yazidis with nothing but a sliver of hope. This showcases the fracturing of the dialogic structure of testimony, where their listener refuses to listen, causing meaninglessness and absurdity (Das & Sarangi, 2017). Murad's consistent and persuasive testimony emphasizes the unwillingness of listeners, as the Peshmerga refer to their withdrawal from Kocho as "tactical", showcasing the failure to uphold their assurances to protect the Yazidi community. This illustrates the dehumanization of the Yazidi community, where their situation was considered reductive and futile. This led to the dismissive manner in which their plight was received by both national and international structures. The departure at such a critical moment causes the community to experience a deep sense of helplessness, fragility, vulnerability, and abandonment.

The lack of institutional response indicates the bureaucratization of violence, where decisions such as withdrawal are made through administrative aspects that deny moral notions and responsibilities (Bauman, 2023). This further illustrates the "moral distancing" since this strategy is considered as an amoral and technical decision (Bloom et al., 2020). The displacement and dissemination of responsibility, and the subsequent structural and administrative invisibility, make their security unviable and their lives expendable (Bartels, 2008). This indicates that violence is not confined to the act itself, but rather includes the periods/events before and after the event. This destroys everyday functioning and psychological stability which is influenced by betrayal, voicelessness, silence, and survival (Das, 2006).

The sudden and panicked flight of the Yazidis to the Sinjar mountains indicates a geographical displacement. Fleeing caused them to be trapped in conditions with no water, food, and shelter and in precarious temperatures, while in many cases, these individuals suffered untimely, illogical, and pointless deaths (Kizilhan, 2025). This illustrates how certain voices, cultures, and identities are rendered irrelevant and unintelligible. This is because the process of assigning hierarchies happens within, and by the hegemonic institutions, since these structures define and constitute meaning-making systems and response-making systems (Spivak, 2023). This leads to the construction of a testimonial field that is ruptured by betrayals, abandonment, invalidation, and brutality, disrupting the preconditions for witnessing. Their experiences remain suspended, neglected, and dismissed within the dialogic structures, causing a consequent second wound, which is called as testimonial injustice (Carel & Kidd, 2023). Testimonial injustice degrades credibility due to institutional intolerance, discrimination, and opacity.

Within this fragmented state, Nadia's testimony is placed at the margins of national and international discourses, which in turn highlights the fraught act of testifying from the position as a subaltern. Nadia, as a woman and as a Yazidi, is rarely allowed to articulate, and is rarely considered credible (Davidson & Pierson, 2001). Her testimony, in this circumstance, however, reveals an attempt to speak within these conflicting structures rather than escape from them. This showcases the necessity to connect with her agency and portray her hermeneutical, communicative, and testimonial capabilities (Kidd et al., 2017). Also, Murad's testimony unravels from community suffering and political urgency, making her testimony "representational" (Sperber et al., 2010). This demonstrates the plurality of pain and suffering making her testimony a plea to be heard, while also challenging and questioning the creation of hegemonic hierarchies.

Although Nadia Murad recounts her personal experiences, Murad concurrently broadens the experiences to the Yazidi community. The initial phase of ISIS's entrance does not entail ravage or invasion. Rather, invading Kocho and by extension, Yazidi community is systematic, specific, and meticulous, to implement a slow and tactical process of isolation from the international community (El-Masri, 2018). Religious conversion, severed communication, and denial of basic necessities such as food and electricity, and restricted routes were demanded and enforced. The trauma already undergone and the trauma underway disintegrates both the individual's and the community's context of reference, causing heightened emotional reactions (Edy, 2006). They experienced unprecedented anxiety and gripping fear, causing instability and psychological terror indicated by mass weeping, sleeplessness, restlessness, and loss of appetite. Such reactions are responses to the collapse of safety, inescapability, anticipation of discrimination, violence, and death, and the unknown future. Gerhart et al., (2014) posit that the impairment of functional disruption describes the psychological paralysis induced due to trauma. "Every second with ISIS was part of a slow, painful death-- of the body and the soul" (Murad, 2018, p. 119).

The personal and collective violence amplifies into gendered and genocidal voice. The killing of men, older women, separation of young women, the enslavement, the subsequent sexual exploitation, and indoctrination of the young boys showcase the systematic and strategic nature of achieving biological and cultural elimination of the Yazidi lineage through genocidal violence (Werner et al., 2021; Wessells & Strang, 2006). Furthermore, Murad's testimony records structural and gendered violence, and the historical trauma of the Yazidis (Beverley, 2004).

Nadia recounts the enforced passivity and the silencing of the female victims, as they were made to listen to the shooting of their male family members. These women were both the witnesses and passive participants, describing the subaltern witnessing of historical trauma, as they were excluded from responding and representing such heinous violence (LaCapra, 2014). Even though their trauma surpasses the intensity of the event, their witnessing is rendered irrelevant due to their inferiorized gender, passive victimization, coerced witnessing, and the imposition of silence. Testimony is nullified at this present moment, as they are allowed to hear, but not respond; allowed to witness, but not testify, experience pain, but not cry out. This showcases the prohibition of narrative, and linguistic representation, and action in the position of a subaltern woman (Morris, 2010). They are turned into unwilling witnesses to such violence and to the death of their family members, while also enduring the denial of basic human responses. Further, this showcases how the inability to communicate dismantles relational meaning-making, which is one of the core tenets of testimony (Frank, 1995). Instead, these women are perilously placed where trauma is both an experience and an erasure, without agency and voice.

Murad (2018) recounts the collective unrest, fear, and the consequent collapse of the Yazidi women in the slave market. Their screams, calls for help, and liberation, whether in Kurdish or Arabic, are unacknowledged and unanswered. This depicts the treatment of Yazidi voices with brutal indifference, the suppression of their voices, and the complete erasure of their autonomy, leading to the purposeful construction of silence (Felman & Laub, 1992; Laub, 1991). In these centers, women are forced to beautify themselves, and examined for virginity since youth, non-married women, and beauty produced high demand, while also being sold for higher prices (Murad, 2018). Although they protected each other from being sold, their resistance is met with retaliation as they are forced into compliance through verbal and physical assault. The elimination of the collective voice was apparent, as they are transformed into commodified, voiceless subjects devoid of rights and autonomy. This depicts the patterns employed for masculinist and cultural domination and objectification of these women (Weine, 2006).

The militarized traffic is followed by economic and sexual trafficking. The entrapment, displacement, trade, coercion, enslavement, and sexual exploitation reveal how these women were victims of both sex trafficking and sexual slavery (Mohammed, 2022). The systematic transportation, exhibition, trade, exchange between multiple owners, maintenance of trade records, conversion and marriage, and the subsequent rape illustrate the detailed planning and systematic nature of enslavement and rape (Brownmiller, 2005; Sellars, 2010). Thus, rape here is a meticulous and systematic strategy to subjugate, demoralize, and petrify women through domestic and sexual slavery (Sellars, 2010). Also, the forced nature of prostitution showcases the exercise of any or all rights concerned with ownership or proprietorship by ISIS members over the female Yazidi community (McDougall, 1998).

The layered brutality is observed as Nadia testifies about her terror as she was initially bought by a hugely built man. Her desperation is seen as she begs a non-threatening, thinner-appearing man to buy her. This showcases the desperation of Yazidi women for lesser forms of violence and survival, displaying the collapse of moral, ethical, social and bodily integrity (McJunkin, 2023). As she is transported to the Hajji Salman's house, she is initially not raped due to her menstruation. However, she is subjected to invasive touching and groping as her physical integrity is stripped during menstruation. Further, Hajji Salman's cruelty is observed as she is raped irrespective of illness and battered body displaying the inconsideration towards her bodily needs and physical well-being while her dignity is also shattered (McJunkin, 2023).

Another aspect of gendered violence is seen through targeting young girls. Murad testifies that these girls who were far too young for sexual intercourse, were nonetheless subjected to touching, sexual favors and exploitation. This foregrounds the prevalent gendered violence that surpasses age and physical and psychological maturity (Nelson, 2022). The sexualization of their bodies further illustrates the devaluation and degradation of their body. This causes a depreciated self which is comprised of shame, guilt and psychological rejection of their bodies (Bartky, 2015; Brownmiller, 2005). Further, this indicates how consent is gendered and mapped by masculinist frameworks that hierarchize whose consent is valid, whose dignity is prioritized, and what practices are healthy and acceptable.

Forced conversion and the subsequent forced marriages functioned as strategies that were justified by religious and political doctrines (Cetorelli et al., 2017). However, under the ISIS rule, these marriages were reconstructed and therefore, failed the norms of marital institutions, transforming marital-intimate bonds into meticulous system of rape, ownership, and religious indoctrination. These marriages also termed as "conjugal slavery" highlight the intersection with sexual slavery, coerced domestic labor, and marital duties (El-Masri, 2018). Further, Murad's testimony highlights how Yazidi women underwent continuous discrimination, demeaning detentions, verbal abuse, and rape, making them rootless, and spiritually unmoored.

Islamic religio-cultural structures under ISIS were employed to justify and legalize rape, as these women were categorized as "sabiyya", "infidels" and "pretend wives" (Murad, 2018) These designations made them "rapable" and "available", stripping them of consent and reducing them to mere objects (Bartky, 2015). Subsequently, religious doctrines and enslaved status collectively worked to silence them and erase their ability to refuse, situating them in a perpetual flux of sexual availability. As a result, the rapable status was naturalized, exhibiting it as fixed and unalterable. Oppressive sexual objectification is apparent, as these women are coercively tasked with female-related roles such as beautification, stylization, and sexual availability, which transformed them into the object of desire to their abusers (Bartky, 2015).

Coercion is further observed through bodily control and shaming. The constant humiliation regarding their hygiene, coupled with unwavering physical assault and aggressive behaviors to beautify themselves with cosmetics, and pose smiling in photographs with their captors reveals the complete lack of bodily autonomy (Taylor, 2018). This highlights the persevering and consistent male-dominated standards of beauty that manipulate femininity, culture, and glamorized gender expectations. The trauma traversed beyond Yazidi identity to female physic as their appearances mirrored the cultural invasion. Since Yazidi identity was aversive, beautification was employed to reshape Yazidi physical traits and fit into the image that was deemed attractive to their captors. This reveals that ethnic and female beauty standards are insufficiently seen as unfeminine, unattractive, and unhygienic which produced the "intimations of inferiority" (Marković, 2017). This confirms the notion that they were forced to manufacture acceptable "beautiful" bodies, while also correcting their lack of beauty standards according to their captors. This warped their self and self-representation and thereby the legitimacy of their testimony (Beverly, 2004). Further, Murad showcases her anger at having to cater to ISIS standards and individualized-self alienation, even as they hated the image they were forced to become.

Rebellion resulted in physical abuse, as resistance was perceived as an intolerable threat to patriarchal authority. This is because physical aggression, wit, and intellectuality are categorized as non-feminine gender traits, which aggravate gender power imbalances (Darke, 1990; Sandhya Devi, 2025). Verbal humiliation and targeted insults were deployed to shame them for their non-virginal status and to emphasize their exclusion from Yazidi cultural norms. This specifically targets the expectations of chastity, marital, and reproductive purity, religious fidelity, and same-religion marriages since these women symbolized familial honor, integrity, modesty, and purity (Taylor, 2018). Due to this, sexual assault led them to perceive themselves as a burden to their family and society, and thereby becoming pariahs (Kizilhan et al., 2023). These tactics of manipulation were aimed to lower their self-esteem, normalize sexual exploitation, and sever their ties with Yazidi identity (Ibrahim et al., 2018).

Murad discusses the anguish of perceiving themselves as having denigrated their social standing in the Yazidi community due to their transgressor status, and loss of Yazidi-communal identity (Felman & Laub, 1992). This consequently adds to the loss of testimonial position, and therefore the inability to speak. This further illustrates that their voice was made inessential to both ISIS and their own community. Through public and private shaming, and enforced cultivation of spousal roles associated with performing domestic chores; and simultaneously playing the contradictory roles of sexual slave, they internalized their "ruination". This unravels how subaltern women are doubly silenced echoing the masculinist-imperialist control over these women and cultural erasure (Morris, 2010; Spivak, 2023). Further, such epistemic silencing bolster historical erasure ascertaining that their marginalization remains unrecorded in official memory and missing in political reforms. As a result, they are unable to produce legible testimony. In totality, unavoidable and deliberate sexual assault conjoined with a range of verbal, physical, and behavioral abuses, feelings of shame and humiliation destroyed their fundamental value, highlighting their shared victimhood, alienation, and anguish (Somasundaram, 2007).

Sexual abuse is crucial to erase Yazidi voice and agency since coerced sexual activities against one's consent symbolize power and control. This disintegrates one's narrative authority, which is an essential component for narrative agency (Carel & Kidd, 2023). The inability to resist these activities, while being completely powerless causes excessive humiliation to the Yazidi women (Darke, 1990). Rape, in this case, is used as a tool for humiliation (Jaffal, 2020). It is deployed as a weapon of war to invade both women and their cultural identity (Brownmiller, 2005).

Additionally, the employment of rape is tactical for testimonial annihilation. This illustrates that rape not only destroys the physical and psychological well-being, but also decimates the victim's ability to assert authorship. However, since their script is already written by the perpetrator, it supersedes the victim's voice with narrative domination. The imposition of shame, cultural denigration, and severance renders the victim unspeakable within private and communal spatialities.

Beyond the physical assault, they were humiliated with ideological denunciations and degradation of their worth as women. Although they are blamed for their rape, it causes internalization of blame which consequently fortified the notion that they are complicit in their distress and responsible for their sexual assault. This self-imposed blame obstructs coherence in their testimony as it weakens the factors necessary for bearing witness (Popkin, 2003). This leads to a loss of their social recognition and legitimacy, which is essential for logical articulation. The intersection of ethnic-religious hostility is made visible through gender violence, where the female body becomes a predominant space of genocidal techniques.

Further, they were forced to wear revealing and demeaning dresses and perform demeaning jobs such as licking honey from the perpetrator's feet, causing extreme shame, and embarrassment. This elaborates on how ISIS terrorists regarded Yazidi women as commodified objects or trophies whose bodies were licensed to be controlled, assaulted, and exchanged at will, aligning with their desires and fantasies (Darke, 1990). Further, dependency, passivity, frailty, and emotionality are used to abdicate personal, gender, and social agency which causes unprecedented shame and disempowerment (Proulx et al., 1994).

Nadia's rescue is marked with precarity, as her survival depended on placing trust on those she encountered during her escape. Her vulnerability is magnified since her rescuers may return her due to the material reward or elevated status (Murad, 2018). Further, the hesitation among Yazidi sympathizers is due to the risks associated with hiding and aiding sexual slaves, which would further endanger their family. This showcases how testimonial subjectivity continues to remain at the periphery even during escape (Sperber et al. 2010). However, Nasser and his family's help in hiding and rescuing Nadia illustrates solidarity and the dismantling of sectarian and ethnic divide. This showcases that the crux of meaning systems is interconnected and formed during crisis, where individual and collective trauma converge (Somasundaram, 2007). Nadia and Nasser are situated within intersecting power hierarchies, gender oppression, vulnerable status to state and non-state violence, and their collective poverty. Their shared meaning through testimony is found in the calamity around them, compounded by fear, and resistance. This showcases that meaning and survivor testimony are relational, and are formulated acts of solidarity and resistance (Hirschberger, 2018). This also reveals the necessity of inter-subjective recognition and listeners to understand testimonial theory.

Upon reunion with her brothers, they are enveloped in shared grief and the liberation of emotional repression. Her ensuing encounter with an older Yazidi woman who resembled her mother evokes visceral response, symbolizing the psychic fragmentation caused by prolonged captivity, trauma, and unknown status of her family members. Such response is deep-seated in the subsequent process of sense-making amid social identity crisis, and individual and collective degradation (Canetti et al., 2018). These parallel responses of grief echo the older women's loss of her three sons and her consequential vow to remain unclean until the recovery of their remains. This highlights the unresolved lamentation that pervades post-conflict Yazidi life (Shnabel & Nadler, 2008). Further, this illustrates how the aftermath of trauma confines them within their community and families (Somasundaram, 2014). However, this showcases the limits of expression, which consequently delays the unspeakable articulation of trauma. Yet Murad's eventual personal and collective testimony materializes as a courageous breach of this silence. She becomes a conduit for the unthinkable and inexpressible trauma, as she translates internalized suffering into an external, political voice that challenges silencing and confinement.

The influence on generational loss is crucial to her testimony. The disappearance of Murad's mother reflects the generational rupture through the elimination of the elderly Yazidi community. This demonstrates the shared victimhood that creates avenues for correlative recognition, yet in this circumstance, it indicates the impossibility of complete recuperation and restoration of justice. Further, Nadia's survivor's guilt over the continued captivity of her sisters reveals how trauma is a protracted event extending beyond a singular event (Shnabel & Nadler, 2008). This illustrates that trauma is not only based on destruction, but it is an irreplaceable component in the creation of meaning through its persistence and irreparability. Further, this reveals the encumbrance of bearing witness to the unresolved circumstances and the unsaid testimonies of the Yazidis still in captivity.

The Yazidi cultural identity further challenges the process of meaning-making. The socio-cultural elements of Yazidism form the crux of Yazidi identity which influences the construction of truth. These truths are primordial, forming the fundamental assumptions and belief systems of the self, others, and the community. Yazidi experiences lead to the construction of a collective Yazidi meaning systems. This "collectiveness" of Yazidi experience creates an inter-dependent system that sustains their community and religion. When such collective structures are destroyed through major upheavals such as genocide and sexual slavery, the resulting disintegration overwhelms the society's recuperative capacity (Audergon, 2004). While individual trauma, as in the Murad's case, has an immediate psychological impact due to her ties to Yazidism, collective trauma impacts the entirety of the community (Gouni & Verny, 2023). When such truths are violated, their disintegration and the attempted reconstruction target the society's restorative capacity by dismantling the social and cultural practices. Therefore, their trauma causes existential and absurd ruptures that fragment the reference systems that individuals and communities use to navigate and construct relative meaning systems. This is observed as the survivors live in a "limbo state".

Although inhabited by survivors, the house was a place of misery. My brothers who had been so full of life, were like empty bodies, staying awake in the day only because impossible to sleep all the time. (Murad, 2018, p. 277)

This reveals common symptoms such as withdrawal and isolation, and in some cases, identifying enemies or surrendering to external forces (Audergon, 2004). The endured trauma disintegrates the context of reference, comprised of the self, members of a community, and the world. This demonstrates how collective trauma is present and active in their day-to-day life, crippling their sense of togetherness and the socio-cultural aspects of the Yazidi community (Somasundaram, 2007). Further, Murad discusses the harrowing lives in refugee camps and half-built houses, accompanied by extreme poverty, deaths, and a lack of basic necessities that contribute to their suffering.

When Nadia and the female members of her family are rescued, they choose to remain silent, despite their family's acceptance. While their silence is partly rooted in shame and the intolerable weight of their memories, it is appended with both the unbearable and unthinkable nature of memories. Many of the Yazidis still live as internally displaced people (IDP) in camps and continue to suffer informally with perpetual anguish. This is because reliving such memories and knowing what happened to them is both unbearable and unthinkable. The limitations of narration are illustrated through collective trauma as it does not lend itself to narration; rather, it guides what can be perceived, recollected, and responded to (Hirschberger, 2018; Kizilhan, 2025). Further, witnessing itself is traumatic, as the witnesses may choose to distance themselves from the self and the event. The part of the self that is capable of narrating and responding does not want to witness the trauma as it unravels (Levine, 2006). However, only through repeated attempts to witness the frozen moment of oppression can the "self" begin to testify and heal.

The lack of choice during the recollection of past trauma illustrates the lack of control over the original traumatic event such as undergoing rape, enslavement, and collective death of her community (Somasundaram, 2014). However, Murad chooses to testify by disintegrating the identity of a victim and thus becomes an advocate through activism, memoir writing, media appearances, and speeches in both national and international venues. Further, the Yazidi community's attempts to recuperate by structuring their trauma and assigning meaning as they cumulatively fight for recognition and justice for the violence against them. This is performed by providing mutual protection and support, as Murad's family and community come together as they face threats, pain, and suffering (Kizilhan et al., 2023). This also reveals the employment of varied coping strategies through feelings of responsibility and guilt (Norris et al., 2008).

The communal trauma is aggravated by survivors' guilt, collective fragmentation, and displacement. However, migrants as in the case of Nadia experience guilt for seeking asylum in foreign countries, including Germany or Turkey, while surviving family members remain in refugee camps (such as the Kurdistan Region of Iraq or Syria) (Kizilhan, 2025). The displacement causes a harrowing and conflicting circumstance, disrupting their timescape with an agonising past and an uncertain future (Das, 2006). This entanglement of grief with memory and guilt demonstrates the psychological and emotional aspects of geographic displacement (Wessells & Strang, 2006). It also includes Murad's attempt to reconcile with survivors' guilt, collective fragmentation, and suffering. This illustrates the impossible inseparableness between individual pain, communal pain and remembrance. Murad's guilt is a strong reminder of the Yazidi genocide and her displacement to Germany. As a result, Murad's memory represents multidirectional memory, since her individual recollections borrow, translate, recognize, echo, and shape her testimony. The convergence further amplifies her need to represent and remember Yazidi trauma (Rothberg, 2014).

Further, Nadia Murad's collective testimony confronts systematic marginalization by refusing to remain voiceless, as her testimony becomes a counter-narrative (Mora, 2014). This demonstrates Nadia's responsibility to challenge the atrocities inflicted by her perpetrators, extremist propaganda, and the ideological infrastructure that sustains such brutality (Bamberg, 2008; Marusic, 2015). This enables her to overpowering the induced shame, and reclaim both agency and voice. This transforms trauma into testimony, achieving positive transformation, and taking control of time, space, and experiences. As a result, her testimony disrupts the intended impact of shame, containment, and obstruction of voicing, thereby asserting the political and historical significance of Yazidi female exploitation.

Murad's testimony transcends individual trauma and extends to a powerful expression of collective Yazidi trauma, suffering, and survival. Her testimony illustrates both individual and broader disintegration of Yazidi socio-cultural structures, religious customs, traditions, and intergenerational community. Her suffering and grief are not individualistic but shared. Although her experiences are extremely personal, her testimony represents the silence experienced by Yazidis. Her testimony acts both as an archival tool of perseverance and trauma, and as a site where memories of survival are both mobilized, mourned, and employed for resistance.

Challenging the Ignored Representation

With a stark contrast between historical connections and her experiences, Murad unravels Yazidi genocide and systemic rape (Doja, 2019). Her exploration of her own rape, the surrender of her physical and psychological agency, and multiple attempts at escapes showcases that her rape is not a moral debate, but a struggle, a living-death, and a crime. Through this, she emphasizes that her experiences, her victimization, personal loss, and loss of agency are beyond the personal, making her narration logical, reliable, and authentic (Bernard-Donals, 2001).

The use of testimony is an imperative tool to include war-related conflicts and warring to interrogate the ideological discourses of nationhood, belonging, and religio-cultural freedom, to invoke human rights discussions through politics of recognition. Scherr (2015) describes the politics of recognition as countering the unjust systems that discriminate them by establishing and recognizing "humanness" within the specific context. Since imprisonments, exterminations, civil unrest, and conflict deny "humanness", testimony, self-advocacy, and first-person narratives resist and reclaim individual and collective voice (Sandford, 2024). Subsequently, Nadia Murad's advocacy is based on her memoir, national, and international speeches, interviews, and media appearances. This section will analyze the memoir initially and then move to media appearances to self-advocacy.

Nadia's memories through testimony, are reliable, consistent, and authentic (Sperber et al., 2010). While her testimony is at times personal, her invocation of Yazidi experiences is consistent across various platforms. In addition, her corroboration with Yazidi survivors bolsters her credibility and concreteness of her testimony (Felman & Laub, 1992; Human Rights Watch, 2017). Further, her credibility is bolstered by prolonged, systematic, and calculated nature of her abduction, forced conversion, marriage, and rape. Her testimony withstands accusations of emotionality, hysteria, and invalidity due to her gender, which are common strategies used to discredit women's speech (Butler, 2004). Within patriarchal societies and rape cultures, the ongoing speculation, lack of conviction, and the fabling of coerced sexual assault assist the perception of rape victims as illogical and tainted witness. This illustrates Fricker (2008)'s testimonial injustice where although women have been testifying and advocating, they have been purposefully undermined.

Nadia's resistance operates on both narrative and political levels, as her lived experiences affirm truth, legitimize testimony and resist epistemic erasure (Serisier, 2018). The recognition by UNITAD of Yazidi trauma as "genocide" further strengthens her testimony (International Criminal Court, 2015). As a result, Murad's testimony attempts to induce a humanistic response and hold the terrorists accountable and induce humane response.

My story, told honestly and matter-of-factly, is the best weapon I have against terrorism, and I plan on using it until those terrorists are put on trait. There is still so much that needs to be done. World leaders and particularly Muslim religious leaders need to stand up and protect the oppressed. (Murad, 2018, p. 306)

The employment of memoirs, story-telling, and media appearances enables the use of narrative approaches to amplify and represent Yazidi suffering (Shafto et al., 2012). Although these approaches may be prohibitive, they can nonetheless mediate humanitarian support and political assistance.

Murad's self-advocacy facilitates the creation of a meta-history as she combines facts and self-representation. This leads to the effective engagement with the audience who are communal, national, and international (Assmann, 2006). This allows healing to be transformed into a therapeutic discourse as private suffering is turned into a public statement (Puvimanasinghe & Price, 2016). While her self-advocacy necessitates a moral response from her audience, it also elicits action from humanitarian global agencies. She advocates for welfare service provisions, resources, as well as the persecution of the ISIS, while also seeking acknowledgment for her suffering and loss. Her persistent calls to humanitarian agencies also include their ability to civilize and save the Third World (Mutua, 2002). Further, her employment of text and dialogic narrative allows her to create immediacy where an unfolding present is unraveled rather than a completed past event. This parallels 'a uncensored, unmediated, intimate and immediate view of the reality of conflict' (Burden, 2006, p. 05).

In her memoir, Nadia narrates how Nasser and his family helped her escape ISIS-occupied territory to Kurdistan, a region in Iraq. During the Yazidi persecution, many Yazidis tried to escape to Kurdistan for safety. Because of their shared histories and linguistic familiarity, those who properly identified were allowed entry without interrogation. However, when Nasser and Nadia reached the Kurdish borders, Nadia temporarily refuses to speak Kurdish and continued to pretend like a Sunni Muslim as per her false identification. The refusal to speak symbolises anger and resentment over Kurdish abandonment and the lack of protection. This reveals institutional inaction and neglect, particularly by regional authorities who were expected to protect them. Further, her refusal is also an act of resistance against Kurdish authorities who failed to acknowledge their lingual commonality, shared heritage, and mutual oppression from the dominant Iraq institutions. These actions not only expose institutional inaction but also highlight political interference and the lack of accountability.

Murad's repeated use of terms such as "remember" and "never forget" reflects her rebellion and her strong will to face her trauma. Further, her bold stances against her perpetrators and loud activism emphasize that she is not indifferent, passive, or merely a victim. Although, her testimonies carry political insinuations, call for actions, and references, her language also renders injustice, suffering and appeals for empathy.

By bringing the past to the present, she includes textuality, dialogicality, and temporality to culminate advocacy, genocide, and representation (Chouliaraki, 2016). She reminds Yazidi genocide, exile, and displacement for the readers, humanitarian forces, policymakers, and stakeholders through her memoir and human-rights activism. This reflects her struggle to apprehend ISIS and hold them accountable to their suffering by embodying authority and knowledge, while also integrating her voice with feminist voices. Further, her testimonies, appearances, and discussions in BBC news (2024), The Spectrum Conversation (2024), and Seton University Address (2024) extend beyond personal expression. By vigorously testifying through international forums, she not only creates relatable social awareness, but also employs advocacy as a strategic instrument to include a reception of societal structures and political aspects, advancement goals and activism (Avieson et al., 2017). Subsequently, testifying through personal truths, bearing witness to collective trauma, and transforming them into evidence, Murad establishes herself as an authoritative voice (Sobel & Kushnir, 2013).

Survivor testimony specifically in the case of Nadia Murad, is influenced not only by the content of the narrative but also by the media through which it is distributed and received. Institutional platforms such as mainstream news outlets, United Nations, and human rights campaign structure survivor voices within familiar humanitarian dialogues to highlight victimhood, psychological suffering, and liberation. However, these frameworks can restrict narrative complexities which in turn reduces testimony to comprehensible fragments coordinated with global agendas. Simultaneously, digital platforms like YouTube, Twitter, and Instagram allow for reach, visibility, quick transmission, and global support. However, these platforms essentialize content that is shorter, emotionally powerful, and visually pleasing to succeed. As a result, the content becomes more audience-centric while also ignoring the role of critiquing hegemonic structures. Murad's Instagram account resonates these dynamics: although Murad continually strives to advocate for accountability, women's rights and broader injustices, posts are aesthetically curated with captions, images, and short videos. Although social media testimony is authentic, personal, and collective, it can also be tailored through data prediction logics, audience expectations, and current trends. Collectively, this illustrates how narratives as in the case of Murad, may be commodified, substantiated, or silenced.

Her activism results in tangible institutional outcomes. The establishment of the UN Investigative Team to Promote Accountability for Crimes Committed by ISIS (UNITAD) is a resultant of her advocacy through her Instagram account, media, "Nadia's Initiative" (a non-profit organization for community rebuilding and advocacy for Yazidi survivors), collaboration with Amal Clooney, and lobbying for rights.

This showcases Murad's persistence for representation, call for action, and justice. Further, the absence of international tribunal for Yazidi rape victims, the distorted, rushed, and ambiguous Iraq trails, the lack of ISIS prosecutions and limited international efforts enable the boundary crossing from personal to political testimony (Diedrich, 2021; Smith & Watson, 2010). Nadia's political goal also aligns with the concept of "social biography" as she opens her personal wounds/experiences and places the collective experiences in the historical context (Avieson et al., 2017, p.03).

Although many ISIS members were prosecuted for terrorism, the Iraqi domestic courts failed to prosecute crimes such as sexual slavery and abduction. While genocide, systematic rape, and war crimes are recognized as international crimes, they were not criminalized or incorporated into the Iraq Penal Code (Human Rights Watch, 2017). Furthermore, Iraqi courts and the Kurdistan Regional Government (KRD) prosecuted ISIS members through broader counterterrorism laws, under which they were convicted for affiliation rather than individual acts of violence. This reveals Iraq's epistemic ignorance and disregard for Yazidi oppression. This showcases the multi-level discrimination and ill-treatment due to their minority status, religion, and gender bias. Further, the reliance on counterterrorism that led to legal proceedings ignored the oppression experienced by Yazidi girls and women (Bamberg, 2008). The consequent accountability gap is thus portrayed by Nadia as she emphasizes national injustice and failure.

Even though United Nations Investigative Team to Promote Accountability for Crimes Committed by Daesh/ISIL (UNITAD) was able to gather information and construct a comprehensive assessment, still specific prosecutions remain untouched. Further, UNITAD's effectiveness was eroded due to Iraqi politics showcasing the challenges associated with holding terrorists accountable for such heinous crimes. This illustrates that mediation with such complicated and sensitive issues requires careful, meticulous, and strong legal frameworks, investigations, explorations, and international collaborations (Akhavan et al., 2020).

Murad's testimony is moulded by resistance rather than compliance. It includes witnessing and retelling of her experiences and the sexual assault she endured. Therefore, her testimony is in accordance with survivor's terms and not through victimhood which is constructed according to the perpetrator's terms. By doing this, her testimony functions as a counter testimony to the gender, religious, national and international spatial structures. The affirmation of the victim's truth is essential since many times legal systems and agencies often fails (Fernández-Morales, 2024). This ensures that her trauma is not neglected, not fabled and not revised as per the prevailing institutional structures (Traynor, 2021). This emphasizes Nadia's defiance and unwavering commitment against patriarchy, socio-cultural structures, identity elimination, and sexual assault. Although her trauma is unspeakable and difficult to recall, her assertiveness in truth-telling illustrates the necessary nature of testimony and the continuous nature of trauma thereby defying the very logic of genocide.

Murad's resistance to assimilate forced religious identity, marriage, sexual enslavement and trade emphasizes her perseverance in upholding Yazidi religious identity, lineage, and cultural practices during and after genocide (Stefanovic & Loizides, 2011). Although forced marriages are recognized as a crime against humanity, their tactical usage as a tool for genocide, particularly through forced conversions and rape remains largely unaddressed. This reveals the failure of national, and international agencies to prosecute such acts as genocidal crimes (Cheterian, 2021; Revkin & Wood, 2021).

Further, her advocacy illustrates the difference between recognition and reform. Subsequently, her advocacy demands redefinitions to include specific experiences. In 2024, UNITAD recognized that mass marriages, systematic enslavement, and sexual slavery as war crimes. Despite this recognition, prosecutions remain limited. Also, ICC (International Criminal Court) recognized gendered-based war crimes, yet the rape and subjugation of Yazidi girls and women remain neglected, showcasing the gap between recognition and actual reform, readdress, and official solutions (International Criminal Court, 2015). The weaponization of forced marriages functions as a tool of ethnic cleansing, gendered domination, and ownership which causes significant mental and physical trauma to the community. Further, the institutional inaction has accorded to the ongoing erasure of Yazidi culture and religion. This has complicated repatriation, given the pervading regional instability, lack of fundamental amenities, and infrastructure, and continued absence of legal reform and justice. In her Nobel lecture, she advocates "Justice is the only way to achieve peace and co-existence" to amplify that her culture is preserved while also allowing re-occupation of their homeland and co-existence.

Murad's reclamation of Yazidi identity, unhinged defiance, repeated calls for legal action, accountability appeals, and media engagement serve as an impactful tool for self-advocacy. In her speech, Nadia Murad pleads at the United Nations General Assembly:

If beheadings, sexual enslavement, and child rape, and the displacement of millions do not force you to move, when will you move? Life was not created solely for you and your families. We also want life and deserve to live it.

This showcases how her testimony assumes moral responsibility to represent and speak to the audience that demands them to trust the narrator (Marusic, 2015). The culmination of Murad's reclamation alongside Yazidi women further, emphasizes the transformation from personal resistance to political resistance. Even though her claims and appeals have reached international sectors, the genocidal strategy of forced marriages demands exploration, advocacy, policy changes, and scrutiny (Richey & Ponte, 2011). Inscribing forced marriages under sexual violence rather than a tactical tool for cultural erasure, risks the reduction of Yazidi injustices and suffering.

Murad's advocacy reveals a paradigm shift as it transfers culpability from the victim to the perpetrator and challenge hegemonic institutions that naturalizes genocide. While Murad advocates for international recognition of Yazidi brutality, her testimony contributes in situating her within epistemological feminist practices (Brabeck, 2003). Further, Murad mediates injustices through reconstruction and recalling the past, despite her nightmarish memories and continued psychological anguish.

In addition, her speeches at international forums, news platforms, and alternate forums expand her advocacy and testimony, directed at global powers and world leaders. These speeches shift toward public performance that seeks justice, acknowledgments, reparations, and social change. Her presence in these platforms jointly represents that of a survivor and an activist thereby, transforming personal testimony into a political tool.

Murad's speech at the United Nations is personal, collective, and political simultaneously. Her address traverses the norms, beliefs, and conventions of international diplomacy, transforming her individualistic and collective trauma into a catalyst for societal change (Tufte, 2017). Nadia, while asserting her victimhood, does not allow herself to be reduced to her identity of a victim and a refugee. However, she uses her speech to become an activist and a survivor to claim that sexual slavery is a deliberate tactic of war. Through this, she challenges humanitarian agencies that reduce rape to personal harm and thereby, ignore its political dimensions. In doing so, she establishes the conflicting gendered and structural hegemonic institutions.

Her testimony occupies a dialogic space. Rather than being caught in suffering and social isolation, she employs "truth-telling" and "responsive agency" with her spectators (Mackinnon, 2017). Her advocacy asserts the acknowledgment of genocide through a call for political action instead of mere remembrance of trauma. Her call for responsible actions from the global community leads her to state: "We are called brave, but the courage we really want to see is from leaders in a position to do something." This denotes the interpersonal strand of her testimony as she urges her listeners to understand Yazidi victimization through trust and a change in societal norms (Faulkner, 2011). Further, she demands international protection under UN supervision since their minority status would only be targeted for increased crimes and violence.

Murad's advocacy expands beyond Yazidi women to include all sexual violence survivors, thereby moving beyond community-based advocacy. This universalized frame avoids identity-based limits and reframes rape as central to conflict and demanding targeted and transformative justice (Engle, 2005).

The paucity of total Yazidi autonomy, safe spaces, reclamation of homeland, and ISIS persecution leads to Murad's invocation of UN and UNITAD as she asks for international jurisdiction "What next?", "Send the case to the International Criminal Court or establish a hybrid court by treaty to prosecute ISIS crimes… other nations should follow Germany's example". Further, Murad's employment of institutional language and legal records tactically asserts a survivor-led approach to justice and make the invisible visible (Mulgan, 2000; Unerman & O'Dwyer, 2006).

Soliciting concrete reparations such as tangible healthcare, shelter, food are used to denounce abstract symbolic or material reparations to address systemic gendered inequalities while also demanding action on gender-sensitive and collective issues (Rubio-Marín, 2009). Murad includes the rehabilitation of Yazidi ancestral land which is largely agricultural is a core factor to her advocacy. Although she advocates, through Nadia's Initiative, efforts for land restoration are being taken place through water access, and supporting sustainable techniques to make their land livable again. This will also lead to gaining rootedness for the displaced community. Further, her persistent speeches, initiatives, and calls for action are directed to these agencies because their global power, resources, and legality can influence social change, resonate gender equality and provide justice (Bailey-Fakhoury, 2017).

Further, she focuses on survivor participation in her testimony which is integral in her call for action. By insisting to create spaces for survivor's opinions, knowledge and agency, leading to radically decentering global inaction, and justice responses. And moreover, this also showcases the reimagining of international justice as participatory and not paternalistic (Cohen, 2001). Therefore, healing includes participation, ethical engagement, acknowledgement and recognition (Oliver, 2001).

In her speech at UN, she says "Survivors know best what they need to heal and recover - so why not ask them?". Murad affirmatively asks that survivors be given a voice since their first-hand experience and their prolonged bodily and mental healing illustrates what Oliver (2001) calls as the capacity of response. This further shows how public testimony should be grounded within survivor's articulation of healing, integrity, trauma and justice.

The inaction and superficiality of media coverage also led to the need for survivor-led testimonies since this causes invisibility and erasure of survivor testimony (Rubio-Marín, 2009). These disadvantaged subaltern individuals should be listened to and interacted for ethical considerations and decisions, and therefore unlearn publicity-induced artificial knowledge (McFarlane, 2006). Murad's launch of Murad's Code is a direct resolution and intervention to the thievery of victim's pain (Madlingozi, 2010). This code employs ethical guidelines for testimony and advocacy to appeal against the commodification of trauma narratives. It also helps address policy-making, welfare service provision, and the creation of authentic and reliable discourses (Hudson, 2002; Oliver, 2001). Further, such testimonies contribute to post-conflict recovery rooted in personal integrity, consent and according to the survivor's terms.

This further epitomizes a "post humanitarian" narrative where voice is given greater significance than spectacle (Chouliaraki, 2013). Murad's code symbolizes her rebellion against exclusion in dominant discourse and social, economic, and political marginalization (Spivak, 2023). Further, Murad's Code indicates the shift from relying on international and national organizations to self-produced organization. It also denotes a shift from external to internal/survivor-led ethical advocacy, thereby, centralizing survivor experiences.

Murad's emphasis on integrity focuses on internal issues within individuals and communities, concentrating on the self. Even as a victim, her advocacy rooted in ethical and moral conviction, represents Yazidi loss, trauma, and need for justice (Lewis & Madon, 2004). Murad's assertion of "accountability as solidarity" portrays the demand for physical, psychological, and economic recuperation. This is an attempt to emphasize the need of collective healing, not limited to psychological healing or media-broadcasted support. Her partnership with the Yazidi community illustrates the need for community building, ethnic protection, and safe environments. Further, the need for solidarity and survivor-led testimony also tackles poverty and patriarchy (Muoghalu, 2012). Through this, Murad's advocacy demonstrates her perpetual fight for social change and social justice (Wilkins, 2020).

In her Sky News interview with Yalda Hakim, Murad discusses the rejection of children born of rape, even though the rape survivors were reintegrated into the Yazidi community. This is because Yazidi religion accepted only those who are born to Yazidi parents as Yazidi (San José, 2020). Rape survivors and their children face the worst of the controversies, because their children were not only born out of rape but they were fathered by non-Yazidi individuals. As a result, these women were continually subjugated by their own communities. Consequently, Murad highlights their anguish, spacelessness, and societal rejection as they faced the dilemma of choosing between reintegration into their community and retaining custody of their children. Further, Murad's collaboration with the female victims enables better articulation, assertion, and communication. This consequently allows integration of all sexual violence victims and improved prospects for external support.

Although, ISIS was defeated in 2017, Yazidi rape victims remain chained to patriarchal, institutional, and national structures, facing minimal representation, voicelessness, and incomplete social acceptance. Further, the Yazidi trauma endures through inadequate justice and reforms, reintegration challenges, and unaddressed cultural and generational rupture. As a result, Murad's advocacy gives voice to the voiceless and represents the disadvantaged. Moreover, her persistent advocacy attempts to gain better international consciousness, deploy humanitarian efforts, and governing global discourses on genocide and gendered violence. Through this she exposes the continued effects of trauma, remembered and relived through personal and collective memory and public testimony.

FINDINGS

The study found that Murad's memoir served as both a personal and collective testimony that documented and represented the trauma suffered by the Yazidi community. International abandonment, religious and sectarian differences, internal conflicts led to genocide and sexual slavery. This fragmented the fabric of the Yazidi community, which was centred around social and familial relationships, religion, and communal identity. Further, the study found that the community was symbolically, structurally, geographically and psychologically displaced. This led to social death, as their communities disintegrated, and they were rejected, invalidated, and rendered meaningless.

The experiences of Yazidi women were extremely traumatizing as they were subjected to enslavement, trade, objectification, rape, and assault. Conjugal slavery, forced conversions, and humiliating punishments destroyed their individuality and connection to their culture. This caused unprecedented prolonged psychological and physical disorders. This showcases the interplay of patriarchal, fundamentalist structures that promoted gendered submission and violence. Although the Yazidi women were accepted, the rejection of their children illustrated the oppression within their community. This demonstrates how Yazidi women were commodified and were denied narrative agency and expression.

The ignorance by national and international bodies, also called as testimonial injustice led to a second trauma. However, Nadia Murad's testimony and advocacy have been impactful as she has reached international, national, and wider audiences through her memoir, speeches, and media appearances. Her transformation from victim to survivor illustrates the resilience of her community. The study reveals that survivor-led testimonies, accounts, and stories should be recognized as significant and valid and thereby, employed for reparations and reforms.

This chapter goes beyond hegemonic and state-of-the-art literature on sexual violence in armed conflict scenarios. This is performed through accepting rape as a weapon of war by featuring how Yazidi women's experiences are moulded by intersecting communal, religious and cultural dynamics, which are frequently disregarded in mainstream literature. While dominant theories describe rape as a weapon of war employed to violate bodies and national identity, this chapter extends that scope by illustrating and uncovering how violence endures itself through testimonial silencing, shame, and repudiation of survivors who are particularly silenced by their communities. This chapter also challenges the conceptualization that healing occurs from the time of rescue by revealing that testimonial injustice and cultural erasure at institutional, state, and communal levels continue to distress survivors.

The chapter also contributes to post-conflict theory and testimonial injustice-related theories by highlighting the boundaries of community reintegration directed by purity and legitimacy, which is observed through selective acceptance of women and rejection of their children. The paradoxical nature of survivor and advocate demonstrates the emotional labour and retelling required of the victims within post-humanitarian discourses. Furthermore, the enforced beautification imposed by ISIS adds to feminist theories of objectification, gender performativity, body and visual politics by illustrating the aesthetic domination within wartime sexual abuse, revealing the demotion of ethnic and cultural identity, contrasting with imposed beautification standards. Through these interventions, this chapter deepens culturally footed and intersectional comprehension of sexual assault that adjuncts and critiques mainstream theories. These discernments call for an approach highlighting the intersectional, cultural perception of sexual abuse and slavery, and their aftermath

CONCLUSION

In conclusion, Murad's speech serves as an impactful act of testimonial resistance that surpasses personal grief to seek structural transformation. Through aspects like religious, linguistic, and ethnic minority-related, sectarian-related and gender-related vulnerabilities, she exposes the oppression against the Yazidi community. Nadia Murad's memoir functions as a powerful example of self-advocacy, where she not only testifies her personal and collective trauma, but also engages with world leaders and global structures regarding prolonged silences and inaction. She uses advocacy to demand recognition, accountability, and prosecution of crimes such as sexual slavery, forced marriages, and religious conversion. Murad reframes and confirms that survivorhood is not a passive experience but rather expands into political authority. By imploring both national and international bodies, she reclaims agency after being suppressed and marginalized. She asserts that justice should include logically defined and survivor-centric resolutions.

She transcends victimhood and reclaims her agency and personhood through her testimony while also critiquing wrongful representation, invisibility, and erasure. Her memoir functions both as testimony and a political tool to resist systemic marginalization and the silencing of the Yazidi community. Through this, she calls for structural change, culpability, reparations and justice through her testimony and advocacy. Her advocacy challenges communal, institutional, patriarchal, and extremist structures while also seeking social welfare and restorative support aimed at rebuilding and reclaiming their original homes. She particularly focuses at national

REFERENCES

Akhavan, P., Ashraph, S., Barzani, B., & Matyas, D. (2020). What justice for the Yazidi Genocide?: Voices from below. *Human Rights Quarterly*, *42*(1), 1–47. DOI: 10.1353/hrq.2020.0000

Assmann, A. (2006). History, memory, and the genre of testimony. *Poetics Today*, *27*(2), 261–273. DOI: 10.1215/03335372-2005-003

Audergon, A. (2004). Collective trauma: The nightmare of history. *Psychotherapy and Politics International*, *2*(1), 16–31. DOI: 10.1002/ppi.67

Avieson, B., Giles, F., & Joseph, S. (Eds.). (2017). *Mediating memory: Tracing the limits of memoir*. Routledge., DOI: 10.4324/9781315107349

Bailey-Fakhoury, C. (2017). Learning through collective testimony: African American motherwork, womanism, and praxis. *Journal of the Motherhood Initiative for Research and Community Involvement*.

Bamberg, M. (2008). Considering counter narratives. In M. Bamberg, M. Andrews & J. Benjamins (Eds.), *Considering counter-narratives: Narrating, resisting, making sense* (pp. 351-371). John Benjamins Publishing Company. https://doi.org/DOI: 10.1075/sin.4.43bam

Bartels, D. M. (2008). Principled moral sentiment and the flexibility of moral judgment and decision making. *Cognition*, *108*(2), 381–417. DOI: 10.1016/j.cognition.2008.03.001

Bartky, S. L. (2015). *Femininity and domination: Studies in the phenomenology of oppression*. Routledge., DOI: 10.4324/9780203825259

Bauman, Z. (2023). Modernity and the Holocaust. In W. Longhofer & D. Winchester (Eds.), *Social Theory Re-Wired* (pp. 300-318). Routledge. https://doi.org/DOI: 10.4324/9781003320609-38

Bernard-Donals, M. (2001). Beyond the question of authenticity: Witness and testimony in the Fragments controversy. *PMLA*, *116*(5), 1302–1315. DOI: 10.1632/pmla.2001.116.5.1302

Beverley, J. (2004). *Testimonio: On the politics of truth*. University of Minnesota Press.

Blackstone, A. M. (2003). *Gender roles and society*. ABC-CLIO.

Bloom, P. B. N., Kimhi, S., Fachter, S., Shamai, M., & Canetti, D. (2020). Coping with moral threat: Moral judgment amid war on terror. *The Journal of Conflict Resolution*, *64*(2-3), 231–260. DOI: 10.1177/0022002719854209

Brabeck, K. (2003). IV. Testimonio: A strategy for collective resistance, cultural survival and building solidarity. *Feminism & Psychology*, *13*(2), 252–258. DOI: 10.1177/0959353503013002009

Brownmiller, S. (2005). *Against Our will: Men, women and rape*. Fawcett.

Burden, M. C. (2006). *The blog of war: Front-line dispatches from soldiers in Iraq and Afghanistan*. Simon & Schuster.

Butler, J. (2004). *Precarious life: The powers of mourning and violence*. Verso.

Carel, H., & Kidd, I. J. (2023). Institutional opacity, epistemic vulnerability, and institutional testimonial justice. In Altanian, M., & Baghramian, M. (Eds.), *Testimonial Injustice and Trust* (pp. 273–296). Routledge., DOI: 10.4324/9781003396789-20

Caruth, C. (2016). *Unclaimed experience: Trauma, narrative, and history*. JHU press., DOI: 10.56021/9781421421650

Cetorelli, V., Sasson, I., Shabila, N., & Burnham, G. (2017). Mortality and kidnapping estimates for the Yazidi population in the area of Mount Sinjar, Iraq, in August 2014: A retrospective household survey. *PLoS Medicine*, *14*(5), e1002297. DOI: 10.1371/journal.pmed.1002297

Cheterian, V. (2021). ISIS genocide against the Yazidis and mass violence in the Middle East. *British Journal of Middle Eastern Studies*, *48*(4), 629–641. DOI: 10.1080/13530194.2019.1683718

Chouliaraki, L. (2013). *The ironic spectator: Solidarity in the age of post-humanitarianism*. John Wiley & Sons.

Chouliaraki, L. (2016). Authoring the self: Media, voice and testimony in soldiers' memoirs. *Media, War & Conflict*, *9*(1), 58–75. DOI: 10.1177/1750635216636509

Cohen, J. (2001). Questions of credibility: Omissions, discrepancies and errors of recall in the testimony of asylum seekers. *International Journal of Refugee Law*, *13*(3), 293–303. DOI: 10.1093/ijrl/13.3.293

Cubilié, A., & Good, C. (2003). Introduction: The future of testimony. *Discourse (Berkeley, Calif.)*, *25*(1), 4–18. DOI: 10.1353/dis.2004.0003

Darke, J. L. (1990). Sexual aggression: Achieving power through humiliation. In W.L. Marshall., D.R. Laws, & H. E. Barbaree (Eds.), *Handbook of sexual assault: Issues, theories, and treatment of the offender* (pp. 55-72). Springer US. https://doi.org/DOI: 10.1007/978-1-4899-0915-2_4

Das, B., & Sarangi, I. (2017). Voiceless victims of war: An absurd truth. *Rupkatha Journal on Interdisciplinary Studies in Humanities, 9*(1). Advance online publication. DOI: 10.21659/rupkatha.v9n1.22

Das, V. (2006). *Life and words: Violence and the descent into the ordinary*. University of California Press. DOI: 10.1525/9780520939530

Davidson, T., & Pierson, R. R. (2001). Voices from the margins: Subaltern women speak... and rewrite history. *Journal of Women's History, 13*(2), 169–179. DOI: 10.1353/jowh.2001.0045

De Beauvoir, S. (2010). *The Second Sex*. Knopf.

Delaperrière, M. (2014). Testimony as a literary problem. *Teksty Drugie, 2*, 42–54.

DeRosa, A. (2014). The end of futurity: Proleptic nostalgia and the war on terror. *Literature, Interpretation, Theory, 25*(2), 88–107. DOI: 10.1080/10436928.2014.904709

Detue, F., & Lacoste, C. (2021). What testimony does to literature. *Synthesis: An Anglophone Journal of Comparative Literary Studies, 13*(13), 22–36. DOI: 10.12681/syn.27559

Doja, A. (2019). Politics of mass rapes in ethnic conflict: A morphodynamics of raw madness and cooked evil. *Crime, Law, and Social Change, 71*(5), 541–580. DOI: 10.1007/s10611-018-9800-0

Dutro, E. (2008). 'That's why I was crying on this book': Trauma as testimony in responses to literature. *Changing English, 15*(4), 423–434. DOI: 10.1080/13586840802493076

Edy, J. (2006). *Troubled pasts: News and the collective memory of social unrest*. Temple University Press.

El-Masri, S. (2018). Prosecuting ISIS for the sexual slavery of the Yazidi women and girls. *International Journal of Human Rights, 22*(8), 1047–1066. DOI: 10.1080/13642987.2018.1495195

Engle, K. (2005). Feminism and its (dis) contents: Criminalizing wartime rape in Bosnia and Herzegovina. *The American Journal of International Law, 99*(4), 778–816. DOI: 10.2307/3396669

Fassin, D. (2011). *Humanitarian reason: a moral history of the present.* University of California Press.

Faulkner, P. (2011). *Knowledge on trust.* Oxford University Press., DOI: 10.1093/acprof:oso/9780199589784.001.0001

Felman, S., & Laub, D. (1992). *Testimony: Crises of witnessing in literature, psychoanalysis, and history*. Taylor & Francis.

Fernández-Morales, M. (2024). Survivor memory and rape memoir: Chanel Miller's Know my name. *Life Writing*, *21*(2), 401–418. DOI: 10.1080/14484528.2023.2196367

Forsythe, D. P. (2005). International humanitarianism in the contemporary world: Forms and issues. *Human Rights & Human Welfare*, *5*(1), 57. DOI: 10.56902/HRHW.2005.5.1.57

Frank, A. (1995). The wounded storyteller: Body, illness. *Ethics*. Advance online publication. DOI: 10.7208/chicago/9780226260037.001.0001

Fricker, M. (2008). Forum on Miranda Fricker's "Epistemic injustice: Power and the ethics of knowing". *Theoria*, *23*(1), 69–71. DOI: 10.1387/theoria.7

Friedlander, S. (2014). *Nazi Germany and the Jews: The years of persecution: 1933-1939*. Hachette UK.

Gerhart, J. I., Hall, B. J., Russ, E. U., Canetti, D., & Hobfoll, S. E. (2014). Sleep disturbances predict later trauma-related distress: Cross-panel investigation amidst violent turmoil. *Health Psychology*, *33*(4), 365–372. DOI: 10.1037/a0032572

Goodman, A., Bergbower, H., Perrotte, V., & Chaudhary, A. (2020). Survival after sexual violence and genocide: Trauma and healing for Yazidi women in Northern Iraq. *Health (Irvine, Calif.)*, *12*(06), 612–628. DOI: 10.4236/health.2020.126046

Gouni, O., & Verny, T. (2023). The human toll of collective trauma: The ravages of war and persecution. *International Journal of Prenatal & Life Sciences*, 1–40. https://doi.org/DOI: 10.24946/IJPLS/20231010

Harris, P. L., & Koenig, M. A. (2006). Trust in testimony: How children learn about science and religion. *Child Development*, *77*(3), 505–524. DOI: 10.1111/j.1467-8624.2006.00886.x

Harris, P. L., & Lane, J. D. (2014). Infants understand how testimony works. *Topoi*, *33*(2), 443–458. DOI: 10.1007/s11245-013-9180-0

Hirschberger, G. (2018). Collective trauma and the social construction of meaning. *Frontiers in Psychology*, *9*, 1441. DOI: 10.3389/fpsyg.2018.01441

Hudson, A. (2002). Advocacy by UK-based development NGOs. *Nonprofit and Voluntary Sector Quarterly, 31*(3), 402–418. DOI: 10.1177/0899764002313005

Human Rights Watch. (2017). *Flawed justice: Accountability for ISIS crimes in Iraq.* https://www.hrw.org/report/2017/12/05/flawed-justice/accountability-isis-crimes-iraq

Ibrahim, H., Ertl, V., Catani, C., Ismail, A. A., & Neuner, F. (2018). Trauma and perceived social rejection among Yazidi women and girls who survived enslavement and genocide. *BMC Medicine, 16*(1), 1–11. DOI: 10.1186/s12916-018-1140-5

International Criminal Court. (2015). *Statement of the prosecutor of the International Criminal Court, Fatou Bensouda, on the alleged crimes committed by ISIS.* https://www.icc-cpi.int/news/statement-prosecutor-international-criminal-court-fatou-bensouda-alleged-crimes-committed-isis

Jaffal, Z. (2020). Rape as genocide crime in International Criminal Law-The case of Yazidi women in Iraq. *International Journal of Criminal Justice Sciences, 15*(2), 230–247.

Jara, J. (1984). *An unfinished song: the life of Victor Jara.* Ticknor and Fields.

Jones, S. (2019). Testimony through culture: Towards a theoretical framework. *Rethinking History, 23*(3), 257–278. DOI: 10.1080/13642529.2019.1620909

Kelly, L. (2013). *Surviving sexual violence.* John Wiley & Sons.

Kidd, I. J., Medina, J., & Pohlhaus, G.Jr., (Eds.). (2017). *The routledge handbook of epistemic injustice* (1st ed.). Routledge., DOI: 10.4324/9781315212043

Kizilhan, J. I. (2025). Mental health and collective trauma among Yazidi Genocide Survivors. *Journal of Mental Health Disorders, 5*(1), 47–49. DOI: 10.33696/mentalhealth.5.036

Kizilhan, J. I., Berger, T., Sennhauser, L., & Wenzel, T. (2023). The psychological impact of genocide on the Yazidis. *Frontiers in Psychology, 14*, 1074283. DOI: 10.3389/fpsyg.2023.1074283

Knowles, C. (2023). Articulating understanding: a phenomenological approach to testimony on gendered violence. In M, Altanian & M. Baghramian (Eds.), *Testimonial Injustice and Trust* (pp. 173-197). Routledge. https://doi.org/DOI: 10.4324/9781003396789-14

LaCapra, D. (2014). *Writing history, writing trauma.* John Hopkins University Press., DOI: 10.56021/9781421414003

Laub, D. (1991). Truth and testimony the process and the struggle. *The American Imago*, *48*(1), 75–91.

Levine, M. G. (2006). *The belated witness: Literature, testimony, and the question of Holocaust survival.* Stanford University Press., DOI: 10.1515/9781503616820

Lewis, D., & Madon, S. (2004). Information systems and nongovernmental development organizations: Advocacy, organizational learning, and accountability. *The Information Society*, *20*(2), 117–126. DOI: 10.1080/01972240490423049

Lindsey, L. L. (2020). *Gender: Sociological perspectives*. Routledge., DOI: 10.4324/9781315102023

MacKinnon, C. A. (2017). Rape, genocide, and women's human rights. In Lattimer, M. (Ed.), *Genocide and Human Rights* (pp. 133–144). Routledge., DOI: 10.4324/9781351157568-4

Madlingozi, T. (2010). On transitional justice entrepreneurs and the production of victims. *Journal of Human Rights Practice*, *2*(2), 208–228. DOI: 10.1093/jhuman/huq005

Marković, S., & Bulut, T. (2017). Attractiveness of the female body: Preference for the average or the supernormal? *Psihologija (Beograd)*, *50*(3), 403–426. DOI: 10.2298/PSI1703403M

Marusic, B. (2015). *Evidence and agency: Norms of belief for promising and resolving*. OUP Oxford., DOI: 10.1093/acprof:oso/9780198714040.001.0001

McJunkin, B. A. (2023). Rape as indignity. *SSRN*, *109*(385). Advance online publication. DOI: 10.2139/ssrn.4347325

Mohammed, H. A. (2022). The last genocide against the Yazidi People. *Review of Middle East Studies*, *56*(1), 108–122. DOI: 10.1017/rms.2023.1

Mora, R. A. (2014). Counter-narrative. *Qualitative Inquiry*, *8*(1), 23–44.

Morris, R. C. (Ed.). (2010). *Can the subaltern speak?: Reflections on the history of an idea*. Columbia University Press.

Mulgan, R. (2000). 'Accountability': An ever-expanding concept? *Public Administration*, *78*(3), 555–573. DOI: 10.1111/1467-9299.00218

Muoghalu, C. O. (2012). Rape and women's sexual health in Nigeria: The stark realities of being female in a patriarchal world. *American Anthropologist*, *19*(1-2), 33–41.

Murad, N. (2018). *The last girl: my story of captivity, and my fight against the Islamic State*. Crown.

Mutua, M. (2002). *Human rights: A political and cultural critique*. University of Pennsylvania Press., DOI: 10.9783/9780812204155

Nelson, K. C. (2022). Malefaction and atrocities against women and girl child. *The Indian Journal of Law and Legal Research*, *4*(1).

Norris, F. H., Stevens, S. P., Pfefferbaum, B., Wyche, K. F., & Pfefferbaum, R. L. (2008). Community resilience as a metaphor, theory, set of capacities, and strategy for disaster readiness. *American Journal of Community Psychology*, *41*(1-2), 127–150. DOI: 10.1007/s10464-007-9156-6

Oliver, K. (2001). *Witnessing: beyond recognition*. University of Minnesota Press.

Patterson, O. (2018). *Slavery and social death: A comparative study, with a new preface*. Harvard University Press.

Popkin, J. D. (2003). Holocaust memories, historians' memoirs: First-person narrative and the memory of the Holocaust. *History & Memory*, *15*(1), 49–84. DOI: 10.2979/his.2003.15.1.49

Potot-Warren, J. (2020). Identifying Genocide: The Yazidi massacre in the context of the convention on the prevention and punishment of genocide 1948. *The Student Journal of Professional Practice and Academic Research*, *2*(1), 30–49. DOI: 10.19164/sjppar.v2i1.922

Proulx, J., Aubut, J., McKibben, A., & Côté, M. (1994). Penile responses of rapists and nonrapists to rape stimuli involving physical violence or humiliation. *Archives of Sexual Behavior*, *23*, 295–310. DOI: 10.1007/BF01541565

Puvimanasinghe, T. S., & Price, I. R. (2016). Healing through giving testimony: An empirical study with Sri Lankan torture survivors. *Transcultural Psychiatry*, *53*(5), 531–550. DOI: 10.1177/1363461516651361

Revkin, M. R., & Wood, E. J. (2021). The Islamic State's pattern of sexual violence: Ideology and institutions, policies and practices. *Journal of Global Security Studies*, *6*(2), ogaa038. Advance online publication. DOI: 10.1093/jogss/ogaa038

Richey, L. A., & Ponte, S. (2011). *Brand aid: Shopping well to save the world*. U of Minnesota Press.

Robinson, E. J., Einav, S., & Fox, A. (2013). Reading to learn: Prereaders' and early readers' trust in text as a source of knowledge. *Developmental Psychology*, *49*(3), 505–513. DOI: 10.1037/a0029494

Robinson, L. (2011). Soldiers' stories of the Falklands war: Recomposing trauma in memoir. *Contemporary British History*, *25*(4), 569–589. DOI: 10.1080/13619462.2011.623866

Rothberg, M. (2014). Multidirectional memory in migratory settings: The case of post-Holocaust Germany. In C. De Cesari & A. Rigney, *Transnational Memory: Circulation, Articulation, Scales*, 123-145. https://doi.org/DOI: 10.1515/9783110359107.123

Rubio-Marín, R. (Ed.). (2009). *The gender of reparations: unsettling sexual hierarchies while redressing human rights violations*. Cambridge University Press., DOI: 10.1017/CBO9780511596711

San José, P. C. (2020). The rapes committed against the Yazidi women: a genocide?: A Study of the crime of rape as a form of genocide in International Criminal Law. *Comillas Journal of International Relations*, (18), 50–71. DOI: 10.14422/cir.i18.y2020.003

Sandford, S. (2024). "On their own terms": Agency, advocacy and representation in refugee Webcomics. *Journal of Australian Studies*, *48*(2), 194–208. DOI: 10.1080/14443058.2024.2343827

Sandhya Devi, N. K. (2025). Unweaving the accustomed silence: A study of comfort women using How We Disappeared by Jing Jing Lee (2020). *Journal of Gender-Based Violence*, •••, 1–17. DOI: 10.1332/23986808Y2025D000000080

Scherr, R. (2015). Framing human rights: Comics form and the politics of recognition in Joe Sacco's Footnotes in Gaza. *Textual Practice*, *29*(1), 111–131. DOI: 10.1080/0950236X.2014.952771

Sellars, K. (2010). Imperfect justice at Nuremberg and Tokyo. *European Journal of International Law*, *21*(4), 1085–1102. DOI: 10.1093/ejil/chq070

Shafto, P., Eaves, B., Navarro, D. J., & Perfors, A. (2012). Epistemic trust: Modeling children's reasoning about others' knowledge and intent. *Developmental Science*, *15*(3), 436–447. DOI: 10.1111/j.1467-7687.2012.01135.x

Shirm, G. (2023). Testimony, reconstructed: the secondary witness and testimonial possession in Bart van Es' The cut out girl: A story of war and family, lost and found. *Textual Practice*, *37*(8), 1281–1301. DOI: 10.1080/0950236X.2022.2111702

Shnabel, N., & Nadler, A. (2008). A needs-based model of reconciliation: Satisfying the differential emotional needs of victim and perpetrator as a key to promoting reconciliation. *Journal of Personality and Social Psychology*, *94*(1), 116–132. DOI: 10.1037/0022-3514.94.1.116

Simon, T. W. (1996). Defining genocide. *Wisconsin International Law Journal, 15*, 243.

Sobel, D. M., & Kushnir, T. (2013). Knowledge matters: How children evaluate the reliability of testimony as a process of rational inference. *Psychological Review, 120*(4), 779–797. DOI: 10.1037/a0034191

Somasundaram, D. (2007). Collective trauma in northern Sri Lanka: A qualitative psychosocial-ecological study. *International Journal of Mental Health Systems, 1*(1), 1–27. DOI: 10.1186/1752-4458-1-5

Somasundaram, D. (2014). Addressing collective trauma: Conceptualisations and interventions. *Intervention (Amstelveen, Netherlands), 12*(4), 43–60. DOI: 10.1097/WTF.0000000000000068

Sosa, E. (1994). Testimony and coherence. In Matilal, B. K., & Chakrabarti, A. (Eds.), *Knowing from words: Western and Indian philosophical analysis of understanding and testimony* (pp. 59–67). Springer Netherlands., DOI: 10.1007/978-94-017-2018-2_5

Sperber, D., Clément, F., Heintz, C., Mascaro, O., Mercier, H., Origgi, G., & Wilson, D. (2010). Epistemic vigilance. *Mind & Language, 25*(4), 359–393. DOI: 10.1111/j.1468-0017.2010.01394.x

Spivak, G. C. (2023). Can the subaltern speak? In Cain, P. H., & Harrison, M. (Eds.), *Imperialism* (pp. 171–219). Routledge.

Stefanovic, D., & Loizides, N. (2011). The way home: Peaceful return of victims of ethnic cleansing. *Human Rights Quarterly, 33*(2), 408–430. DOI: 10.1353/hrq.2011.0029

Sternbach, H. (1991). The serotonin syndrome. *The American Journal of Psychiatry, 148*(6), 705–713. DOI: 10.1176/ajp.148.6.705

Strejilevich, N. (2006). Testimony: Beyond the language of truth. *Human Rights Quarterly, 28*(3), 701–713. DOI: 10.1353/hrq.2006.0038

Taylor, D. (2018). Humiliation as a harm of sexual violence: Feminist versus neo-liberal perspectives. *Hypatia, 33*(3), 434–450. DOI: 10.1111/hypa.12427

Ticktin, M. I. (2011). *Casualties of care: Immigration and the politics of humanitarianism in France*. Univ of California Press. DOI: 10.1525/9780520950535

Traynor, M. (2021). Beyond the limits of imagination: Abductive inferences from imagined phenomena. *Synthese, 199*(5), 14293–14315. DOI: 10.1007/s11229-021-03421-z

Tufte, T. (2017). *Communication and social change: A citizen perspective*. John Wiley & Sons.

Unerman, J., & O'Dwyer, B. (2006). Theorising accountability for NGO advocacy. *Accounting, Auditing & Accountability Journal, 19*(3), 349–376. DOI: 10.1108/09513570610670334

Van der Kolk, B. (2014). The body keeps the score: Brain, mind, and body in the healing of trauma. *New York, 3*, 14-211.

Van Schaack, B. (2018). The Iraq investigative team and prospects for justice for the Yazidi genocide. *Journal of International Criminal Justice, 16*(1), 113–139. DOI: 10.1093/jicj/mqy002

Vice, S. (2008). Holocaust poetry and testimony. *Critical Survey, 20*(2), 7–17. DOI: 10.3167/cs.2008.200202

Weine, S. (2006). *Testimony after catastrophe: Narrating the traumas of political violence*. Northwestern University Press.

Werner, G. G., Riemann, D., & Ehring, T. (2021). Fear of sleep and trauma-induced insomnia: A review and conceptual model. *Sleep Medicine Reviews, 55*, 101383. DOI: 10.1016/j.smrv.2020.101383

Wessells, M., & Strang, A. (2006). Religion as resource and risk. *A world turned upside down: Social ecological approaches to children in war zones*, In N. Boothby, A. Strang & M. Wessells (Eds.), *A World Turned Upside Down- Social Eclological Approaches to Children in War Zones* (199-222). Bloom-fied, CT: Kumarian Press. https://doi.org/DOI: 10.1515/9781565493179-012

Wilkins, K. G. (2020). Advocating with accountability for social justice. In Tacchi, J., & Tufte, T. (Eds.), *Communicating for Change: Concepts to Think With* (pp. 53–62)., DOI: 10.1007/978-3-030-42513-5_5

Woods, R. (2021). Testimony and its mediations in life writing. *Life Writing, 19*(3), 441–454. DOI: 10.1080/14484528.2021.1873718

ADDITIONAL READINGS

Cohen, D. K. (2016). *Rape during civil war*. Cornell University Press., DOI: 10.7591/9781501705984

Merry, S. E. (2009). *Human rights & gender violence: Translating international law into local justice*. University of Chicago Press.

Oliver, K. (2001). *Witnessing: beyond recognition*. University of Minnesota Press.

Orgad, S. (2014). *Media representation and the global imagination*. John Wiley & Sons.

Skjelsbæk, I. (2012). *The political psychology of war rape: Studies from Bosnia and Herzegovina*. Routledge., DOI: 10.4324/9780203695616

Chapter 11
2023 Serbia Mass Shootings:
Crisis Communication, Moral Outrage, and Social Impact

Marko Selakovic
https://orcid.org/0000-0002-6568-6627
SP Jain School of Global Management, UAE

Layal Hmaidi
https://orcid.org/0000-0003-3626-5847
American University in the Emirates, UAE

ABSTRACT

The aim of the research is to identify the crisis pathway and examine the impact of crisis responses on the mass shooting in Belgrade, that occurred in May 2023. Official crisis responses and media reporting have been analyzed within the context of Situational Crisis Communication Theory, social impact, moral outrage, and crisis risk mitigation. Based on the document analysis, the flow and timeline of the crisis as well as the official responses have been identified and structured. Moral outrage and social impact have been assessed through the quantitative survey of the citizens of the republic of Serbia (N=518).The findings indicate that the initial crisis responses produced negative social impact and, in turn, did not mitigate moral outrage. Despite concrete measures proposed by the state, inconsistent crisis responses enabled the tabloidization of the first mass shooting event, potentially causing a contagion effect, thereby contributing to the second mass shooting.

DOI: 10.4018/979-8-3373-7589-2.ch011

INTRODUCTION

"Tragedy has been converted to sensation and sensation is operationalized into viewership, Nielsen points, and market share, which is then materialized in advertising revenues.... The sensationalism of a rampage shooting can provide headlines for three days of news cycles" (Muschert, 2002)

In May 2023, two subsequent mass shooting events took place in Serbia: Vladislav Ribnikar school shooting, and Dubona and Malo Orašje mass shooting.

On May 3, 2023, a 13-year-old student opened fire at Vladislav Ribnikar Primary School in Belgrade. Armed with weapons legally owned by his father, the student initially killed nine people: eight students and a school security guard. He also wounded seven others, including six students and a teacher (one wounded student died from the injuries several days after the shooting). The attack was meticulously planned over several months, with a selection of specific classrooms and a compilation of a list of targeted students.

The following day, on May 4, 2023, a 21-year-old man carried out a shooting spree in the villages of Dubona and Malo Orašje. Armed with an automatic rifle, the shooter killed nine people and injured 12 others. Due to the random targeting of victims, the attack was classified as indiscriminate.

Such shocking events attracted the immediate interest of the media, with mass shootings at the absolute epicentre of both media reporting and political action. This event was salient as Serbia has never experienced school shooting incidents, let alone back-to-back mass shooting incidents, despite the history of wars in 1990s and the large concentration of weapons. Even previous cases with 5 or more casualties, such as Jabukovac (2007), Velika Ivanča (2013), Kanjiža (2015) or Žitište (2016) were shooting sprees without follow-up incidents and without large-scale protests.Due to specific nature of these mass shootings and lack of literature related to school shootings and back-to-back mass shootings in Serbia, the study aims to analyze the progression of the crisis, and the official responses elicited, with a particular focus on their influence on the crisis's trajectory, the moral outrage within the community, and the broader social repercussions.

Understanding the dynamics of such complex crisis situations is crucial within the context of crisis communication and media studies. Therefore, crisis response-based theories, such as Situational Crisis Communication Theory – SCCT (Coombs, 2007) offer valuable frameworks for examining how organizations and communities respond to crises. Moreover, the subsequent effects of the crisis responses on public perception and behavior have been examined.

THEORETICAL APPROACH AND FRAMEWORK

Although mass shootings are not among the most frequent crimes, especially in Europe, the number has increased across the countries in the last two decades (Lee et al., 2022; Lubsanov & Yantranova, 2022). Mass shootings in educational institutions are rare and distressing events, as they directly influence public perception of schools as safe havens for children and youth, impact public morale, and trigger a call for immediate action (Haner et al., 2019; Lee et al., 2022; Peterson & Densley, 2019). Societies with more exposure to school shootings, such as the United States, view such crimes as rare, but inevitable events. While the remaining societies perceive mass school shootings as an addressable social problem, that demands immediate remedial action (Lee et al., 2022). Despite the numerous classifications of mass murderers published in the scientific literature (Langman, 2015), limited research has identified specific psychological and behavioral characteristics useful for reliable distinguishment of possible school shooters from other students (Lubsanov & Yantranova, 2022).

School shootings tend to trigger intense emotions and provoke a demand for societal action. The sociological theory on framing and social movements suggests that attributional frame resonance (Benford & Snow, 2000) are possible explanations for the occurrence of public calls for action after events such as school shootings. Attributional frames are individualized and function as "seeds for mobilization, by shaping beliefs about the social problem's severity, and about the urgency, efficacy, and moral importance of taking remedial action" (Lee et al., 2022, p. 503). Within the context of school shootings, this posits that individuals perceive school shootings as a social problem that is necessary to address. Failure to appropriately address the challenges caused by mass shootings tends to erupt moral outrage, a phenomenon broadly defined as anger at the violation of a moral standard or principle (Batson et al., 2007; O`Mara et al., 2011), irrespective of the victim as "such principles apply universally and impartially (O'Mara et al., 2011, p. 41). Numerous literature sources indicate that moral outrage is capable of inspiring collective action, especially when an intentional and violent event going against the collective norms, such as a mass shooting, occurs (Cook, 2006; Coombs & Tachkova, 2023b; Mihailov et al., 2023). The Moral Foundations Theory (MFT) distinguishes between two types of triggers that evoke intuitive reactions and moral outrage; original triggers pertain to stimuli that evoke an evolutionary cognitive module, whereas current triggers pertain to the associated stimuli that prompt original triggers to come into consciousness (Haidt, 2012). Haidt (2012) conceptualized cognitive modules as adaptive challenges - original triggers, that universally intersect all cultures and time planes (O'Mara et al., 2011). The underlying backbone of adaptive challenges is moral intuitions which branch into five categories: Harm/Care, Fairness/Reciprocity, Authority/Respect,

Ingroup/Loyalty, and Purity/Sanctity (Haidt, 2012). A central adaptive A central adaptive challenge pertains to the moral virtue of caring for and protecting children. Consequently, any harm imposed on children such as school mass shootings will serve as a current trigger of the Harm/Care domain (Coombs & Tachkova, 2023b)

Current triggers can evoke moral outrage by the mere existence of media-powered depictions of school shootings. For, the Social Amplification Model of Moral Panics argues that the propensity of moral outrage tends to be mediated by social media amplification, whereby the virality of a school shooting in the media exasperates the perceived potential danger to a tangible societal threat (Puryear et al., 2024). This societal threat of school shootings heightens the public's feeling of distress due to the perceived rupture of societal security, and in turn, provokes a response of collective moral outrage that is manifested in the form of verbal outrage (via social media), silent marches, and communal protests (Skitka et al., 2005). Accordingly, individuals who are sympathetic and compassionate towards the suffering of children tend to demonstrate higher levels of moral virtue and by extension higher levels of moral outrage against policies – or the lack of - that do not impose preventative measures against school shootings. Cook (2006) indicated that murders have the power to trigger high levels of moral outrage in society, while Stanko (2000) reported the interdependence between high moral outrage and murders of schoolchildren. Hence, moral outrage caused by murders can potentially lead to further societal response and trigger protests (Lodewijkx et al., 2008; Miller et al., 2011).

A sociological cornerstone that elucidates the dynamics between school shootings, moral outrage, and punitive attitudes traces back to Durkheim's (1997[1893]) sociological theory, which proposes that moral outrage is evoked by deviant crimes that disintegrate the social fabric of a society by threatening its sacred values and violating its moral threshold (Burton et al., 2021; Silver, 2017). Upon shattering the "shared collective sentiment" of societal morality (Durkheim, 1997 [1893], p.58), deviant crimes subsequently undermine the authority's competency in enforcing retributive response that can re-claim society's moral law (Durkheim, 1997[1893]). At a psychological level, moral outrage tends to be molded through a bi-directional interplay between cognition and emotions. Kahneman's (2011) Dual Process Model argues that the initial implicit and reflexive emotion against social norm violations pertains to System 1 thinking as it occurs in an intuitive, automatic, and effortless manner. Whereas the secondary moral reasoning that cognitively informs moral actions to resolve conflict against norm violations pertains to System 2 thinking as it occurs more consciously and explicitly (Silver et al., 2019). In line with this argument, the Cognitive Appraisal theory of Emotion (Roseman, 1984) further posits that the underlying emotion of moral outrage pertains to anger against the transgressor due to the cognitive appraisal that moral standards have been violated (O'Mara et al., 2011). Whereas the underlying motivated action of moral outrage pertains to a retributive justice response against the transgressor due to the cognitive appraisal that moral standards are requisitioned to be re-established (O'Mara et al., 2011).

Albeit cognitive appraisals of moral decay evoke three emotions – contempt, anger, or disgust – anger specifically is the predominant moral emotion as it pertains to the harm of an individual's autonomy (Darley & Pittman, 2003; Rozin et al., 1999). That is, individuals who morally evaluate the transgressor as highly intentional with their crime, tend to emulate higher levels of anger. The moral emotion of anger evokes elevated moral - punitive beliefs, and by extension, mobilizes 'directed group actions' such as collective protests to restore justice through social change (Van Zomeren, 2006; Van Zomeren & Lodewijkx, 2005). More importantly, such cognitive–emotional correlates of moral anger and crime intentionality tend to manifest into augmented punitive attitudes against school shootings. Burton et al. (2021) reported that moral anger positively correlated with public attitudes against school shootings, whereby individuals endorsed policies such as gun control, school programs, and target hardening to curb school shootings. Therefore, moral outrage embodies a 'fight back' coping mechanism that emotionally, and cognitively, heightens the public's collective strain, and functions to re–establish societal stability by punishing the immoral transgressor to mitigate future threats (Ford & Feinberg, 2020; Henderson & Schnall, 2021; Silver, 2017).

Within the context of Situational Crisis Communication Theory (Coombs, 2017), moral outrage represents a crucial factor, in addition to the pre – defined crisis response factors and strategies, that stakeholders utilize to assess crisis threats. That is, moral outrage plays a role in the crisis triangle, which comprises of determinants of whether the situation is negative and evaluations of the crisis responsibility. Crises that trigger high emotional responses of moral outrage or "sticky crises" require specific crisis response as stake holders tend to engage in emotion – oriented processing of crisis – related events (Coombs & Tachkova, 2023b; Selakovic & Ljepava, 2023). Thus, sticky crises pose unique communicative demands from a political and managerial perspective as it creates deep uncertainty and evokes an extreme sense of urgency, thereby requiring systemic action (Yen & Salmon, 2017; Coombs et al., 2020; Coombs & Tachkova, 2023b). Going further, the decision-makers need to "re-assess their positions when enmeshed within an extreme moral outrage crisis and plot a new course from that point" (Coombe & Tachkova, 2023b, p. 2). Weak, confusing, or inappropriate responses to extreme events and sticky crises can cause a perception of deep value incongruence between the public and the decision-makers. This is an important aspect, as moral outrage-inducing crises have the potential to create greater perceptions of value incongruence than the average crises (Coombs & Tachkova, 2023b). The suggested response for an extreme moral outrage crisis should be "a form of atonement that reflects a combination of corrective actions (steps taken to prevent a repeat of the crisis), expression of concern, and a recognition of the moral violation" (Coombs & Tachkova, 2023b, p. 11).

The role of media in handling extreme crises needs to be considered as an additional catalyst to the Serbian mass school shooting. Sensationalistic responses to critical events and extreme publicity given to the mass murderer can cause a media contagion effect, that might result in copycat repetition of mass shootings (Helfgott, 2015; Dahmen et al., 2018; Kambam et al., 2020; Pescara-Kovach & Raleigh, 2017). The cultural indulgence in metastatic social media, reality TV, and '15 minutes of fame' has birthed the social media contagion effect, which denotes that the glorification of mass shootings using media coverage motivates a next perpetrator to commit a mass shooting due to their infatuation with sociocultural fame through news coverage and public attention (Pescara-Kovach & Raleigh, 2017; O'Tooleet al., 2014). Analogously, the media tends to sensationalize using headline-grabbing tactics of mass shooting reports to upsurge viewership, ad – revenue, and profit, as evident in major news outlets that publicize highly fatal shootings in the US (Silva & Capellan, 2019). Similar media patterns have been observed within the context of CEE countries, due to the positive correlation between the number of media sources and the total fatality rate of violent mass shooting incidents (Anisin, 2023). Accordingly, the contagion effect cultivates a copycat effect whereby the preparator consciously enacts and imitates prior behaviors of preceding notorious mass shooters to attain similar prominence. Albeit the rare prevalence of fame-seeking mass shootings within the US - 7% of 176 cases (Silva & Green- Colozzi, 2019; Petersen and Densley, 2021) and CEE context (Anisin, 2023), the atrocious nature of copycat shootings has been documented in the Sandy Hook shooting, Kerch College shooting, and current Serbian shooting (Anisin, 2023; Pejuskovic & Lecic-Tosevski, 2023).

Despite sociocultural differences between the North American and CEE context in terms of civilian gun ownership, firearm regulations (ie. stricter in the CEE region) and execution, it is evident that the contagion effect is an underlying precursor to mass shootings irrespective of culture-dependent attributes (Anisin, 2023). Accordingly, Lankford and Madfis (2017) have emphasized the pertinence of omitting the perpetrators' names from media reports, to reduce the magnitude of the unintended fame that motivates copycat mass shootings. For, sensationalized media coverage tends to fuel the psychological personality traits of narcissism and egocentrism found in fame-seeking mass shooters (Bushman, 2018) such as the current Serbian mass shooter who has been diagnosed by the media with a pathological disorder without regrets. The underlying theoretical foundation behind the contagion effect trickles down to the social cognitive theory (Bandura, 1977) which argues that behavior is learned through the stages of attention, retention, motivation, and potential. The socio-cultural context dictates the learned knowledge and behaviors behind mass shootings, that a perpetrator attends to and retains through observational learning of a model (ie. preceding mass shooter). Yet within the context of the contagion effect, the model is the sensationalized media coverage of the mass shooting (Pescara-Kovach & Raleigh, 2017) because it provides the 'know-how' of reproducing the necessary planning and execution of a mass shooting via observational learning (Pescara-Kovach & Raleigh, 2017). That is, the continuous replay of the former mass shooter's manifesto, life events, details of the event, and images of the victims, positively reinforces and rewards a copycat effect (Meindl & Ivy, 2017). Such practices call for the reinforcement of ethical standards and professional responsibility in journalism, particularly in digital, high-velocity communication environments, where the consequences can be amplified. These concerns are central to ongoing discussions on media ethics and professionalism in the digital age (Smokvina & Kovačič, 2021). Despite numerous initiatives related to media ethics by the international organizations such as UNESCO (Christodoulou, 2024), available literature indicates that Serbian media practices in reporting about violence are inclined towards sensationalism (Vulić & Pavlović, 2015).

Accordingly, media coverage should portray the perpetrator as cowardly to punish, and thereby curb copycat imitations of mass shootings. Fox et al. (2021) identified a strong correlation between sensationalistic reporting and media contagion effect with the occurrence of copycat mass shootings. Research by Garcia-Bernardo et al. (2018) assessed the impact of social media publicity and indicated that the probability of a school shooting doubles when tweets mentioning the words "school" and "shooting" increases five times. In addition to sensationalized mainstream media coverage, immediacy in spreading a multitude of sensationalized information across multiple digital channels has the power to trigger a spiral of media contagion that might lead to copycat mass shooting events (Pescara-Kovach & Raleigh, 2017). Nevertheless, literature has not robustly indicated whether media contagion, sensationalistic reporting, and increased social media attention related to school mass shootings could trigger copycat mass shootings outside the school setting. Moreover, there is no literature that has discussed the impact of government crisis communication pertaining to mass shootings on the development of the crisis.

Another fundamental precursor of mass school shooting that has been discussed within current literature pertains to the role of firearms within theories such as the 'Trinity of Violence', which asserts that a violent shooting is contingent upon the availability of firearms, random targets, and a motivated shooter (Ahmad et al., 2020). A positive correlation between the execution of mass shootings and easy access to firearms has been established as the incidence rate of mass shootings in the United States is four to ten times higher than in Asia or Europe (Lankford, 2022). On a parallel note, the United States has the highest number of gun ownership (90 guns per 100 people) universally and compared to Europe (Ahmad et al., 2020; Anisin, 2023; Richmond, 2020). A study examining mass shootings in the US, between 1982 and 2018, similarly reported that the number of firearms and the legal ownership of weapons (e.g. through their legal permit or the access of firearms from other family members with legal permits) predicted the total number of mass shooting victims (Yelderman et al., 2019). Within the Serbian context, Eagar (2023) and Engelbrecht (2023) similarly attributed the Serbian mass school shootings to the fact that weapons are embedded within the Serbian culture (Dumančić & Krolo,2017). Nevertheless, it is also important to note that Serbia does not have school shooting response trainings and simulations for students and teachers, which are important part of preparation (Strahler & Ziegert, 2015). Further to this, public sources analysis does not indicate that Serbia has rapid response strategy (Reeping et al., 2020) that would help the mitigation risks emerging from mass shooting incidents.

Nevertheless, firearm accessibility tends to be mediated by family dynamics and parental occupations, whereby the trajectory of violent behavior depends on the surrounding social strife in achieving manhood. That is, the generational background of mass shooters tends to entail parents who served in the military or law enforcement, as well as in educational institutions such as teachers or professors (Langman, 2017). Accordingly, it has been hypothesized that mass shooters who fail to attain familial expectations – be it by enlisting in the military or achieving academic excellence - tend to succumb to threats to their masculinity whereby the shooter's identity as a male is shattered by their inability to sustain familial military traditions (Langman, 2017; Kellner, 2012). A wounded masculine identity that is exposed to firearms at an early age, due to familial use and training in legal firearms, further exasperates the trajectory of a mass shooting (Langman, 2017). As noted by Kellner (2012), mass shooters undergo a "crisis in identity in which young men use guns and violence to create ultramasculine identities" (p. 497). Within the socio- political context of Serbia, the shooter's identity crisis may have been exasperated by witnessing a 'masculine' Balkan culture that has manifested into significant psychological and physical forms of male - dominated violence against the Serbian youth, due to the accumulation of post – war trauma, post-socialist conflict, ethno-nationalism, rigid gender roles, and overarching patriarchy (Babovic´ et al., 2010; Dumančić & Krolo,2017). However, no research has offered deeper insights into post-event moral outrage, social impact, call for action, and responses of Government, media, and society to the extreme crisis events.

RESEARCH QUESTIONS

1. Did crisis responses of the Republic of Serbia officials impact the flow of the crisis, and, if yes, in which way?

2. Did crisis responses of the Republic of Serbia officials trigger moral outrage of local population?
3. What was the overall social impact of the crisis on the Serbian community?

METHODOLOGY

Bearing in mind the complexity of the crisis, exploratory research has been deployed. Exploratory research enables a comprehensive understanding of a crisis flow and ensures a review of all critical steps in crisis communication handling.

In the initial stage of the research, a document analysis in the context of Situational Crisis Communication Theory (Coombs, 2017) was performed to understand the flow of the crisis and build both a crisis events timeline and a crisis response timeline. Identification of the responses has been performed by examining official statements of the institutions of the Republic of Serbia: President of the state, Prime Minister, Minister of Interior, Minister of Health, Minister of Education, Mayor of Belgrade, and Police. In addition to scrutiny of the official websites of the mentioned institutions, a systematic review of media outlets and digital news repositories has been employed to investigate communications of the Serbian institutions during the crisis. The list of media outlets and digital news repositories included the media in Serbian and English languages officially registered in the Republic of Serbia, as well as global media available in English language.

After the identification of the sources for document analysis, a comprehensive set of keywords related to the flow of the crisis has been set, in Serbian and English language. The set included the following keywords:

"Ubistvo" ("Murder"),
"Pucnjava" ("Shooting"),
"Ubica" ("Murderer")
"Masakr" ("Massacre"),
"Zlocin" ("Crime")
"Ubijeni" ("Murdered")
"Zrtve" ("Victims"),
"Protest" ("Protest")
"Okupljanje" ("Gathering")
"K.K." (initials of the underage shooter),
"Uros Blazic" (name of the second mass shooter),
"Aleksandar Vucic" (President of Serbia)
"Ana Brnabic" (Prime Minister of Serbia)
"Bratislav Gasic" (Minister of Interior of Serbia)
"Danica Grujicic" (Minister of Health of Serbia)

"Branko Ruzic" (Minister of Education of Serbia)
"Aleksandar Sapic" (Mayor of Belgrade)
"Veselin Milic" (Commander of Belgrade Police)
"Ribnikar" (name of the school where initial crisis event occurred),
"Dubona" (name of place where second mass shooting occurred),
"Malo Orasje" (name of place where second mass shooting continued),
The review has been conducted in three phases:

1. The first phase involved gathering data from government sources and media outlets to identify relevant events and key announcements. The content was further evaluated for its relevance based on the existence of keywords, returning a total of 5483 relevant findings incorporated in further analysis.
2. The second phase focused on analyzing the content to understand the chain of crisis events and associated crisis responses, including the events with social impact. This involved the linkage of the official government communications with the crisis events.
3. The third phase encompassed reviewing digital news repositories and social media to understand if any confounding government communications occurred. Once the existence of confounding events is eliminated, identified relevant content was systematically reviewed and analyzed to understand how the overall communication framework of the Serbian institutions could be contextualized within the crisis response strategies suggested by SCCT.

Following the review, a crisis timeline has been built, identifying all critical events and institutional response strategies.

Based on the crisis timeline, moral outrage and the social impact of different events included in the crisis timeline were assessed through the quantitative survey of the citizens of the Republic of Serbia (N=518). The survey responses have been collected through an online questionnaire posted on the Survey Monkey platform during June and July 2023. Survey respondents assessed their feelings about each event identified in the document analysis on a scale from 1 (extremely negative) to 5 (extremely positive). Snowball sampling has been used to identify the respondents in Serbia and disseminate information about the survey. Survey information has been circulated through emails and social media channels available in the Republic of Serbia. The suitability of the survey respondents has been identified through a set of eliminatory demographic questions related to the age group and nationality of the survey respondents. The survey responses are kept anonymous and have been analyzed using SPSS software. Only researchers had access to the response database.

Due to the specific and highly sensitive nature of the events included in the quantitative survey, additional ethical and risk mitigation measures have been deployed to reduce possible risks for survey respondents. The survey is set as fully compliant with the Australia *National Statement on Ethical Conduct in Human Research.* To reduce the distress that could be caused by the survey, the following measures have been additionally implemented:

- In the preamble of the survey questionnaire, it has been stated that the survey is dealing with tragic events, with the potential to cause distress or discomfort to the respondents. The preamble also contained an explanation that only those who want to proceed should participate in the survey, and that they can exit the survey at any moment in case they feel any kind of distress or discomfort.
- The survey questions are formulated in a manner to avoid any additional emphasis on the tragic events: they are focused only on a factual listing of the events that took place during the period that is in the focus of this research (from the school shooting occurrence on May 3rd, until post-crisis events on May 28th, 2023).

FINDINGS

The Crisis Timeline and Responses

Based on the findings of a document analysis, a detailed timeline of the crisis has been reconstructed. The timeline also included notable post-crisis events that had a social impact and that are associated with the crisis.

The findings of the document analysis indicate that this crisis situation by its nature can be defined as a mega-crisis. Extreme crises and mega-crises are "different" from traditional crises because they are more complex and challenging for crisis managers (Coombs et al., 2020, Tachkova & Coombs, 2022). Document analysis indicates that the initial crisis response of the officials was partly inconsistent, observing from Coombs's (2017) Situational Crisis Communication Theory principles: while the President of the state and Minister of Interior took a rebuilding stance towards the crisis in the initial crisis response, the Minister of Education opted to move with a victimage stance at the initial response. Observed from the angle of Situational Crisis Communication Theory (Coombs, 2017), partial inconsistency in the institutional responses is noticeable throughout the entire crisis period. Multiple crisis communication channels, confusing messages from various officials, and sensationalism in media reporting contributed to the spread of negative sentiment. In addition, this induced significant moral outrage amongst the citizens of the Republic of Serbia, deepening the crisis and creating an entire chain of subsequent events. The timeline of the crisis and chain of associated events, along with the officials` response type is shown in Table I.

Table 1.

Date / Time	Event	Officials' response type
03.05.2023. 8:40AM (trigger event)	Mass shooting – primary school	N/A
+4 hours	Press conference – ministers (interior, health, education) and police	Diminishment/Denial-Scapegoat/Victimage (blaming Western culture and media influences)
+9 hours	Press conference – President of the state – proposal of the measures (disarming, increased security)	Rebuilding (taking partial responsibility and proposal of the measures; facts about the act, murderer and his family; fear from copycat effect)
+11 hours	Citizen gathering near school, candle lighting, leaving flowers	N/A
+24 hours	Media speculations, tabloidization	N/A
+30 hours	Copycat event – knife attack in high school	N/A – no official response
+31 hours	Ministry of Interior – moratorium on weapons	Rebuilding
+35 hours	Citizen gathering near school, candle lighting, leaving flowers	N/A
+40 hours	Copycat event – 2nd mass shooting	N/A
+42 hours	Media statement – Minister of interior, announcement of 2nd mass shooting as terrorist act	Rebuilding (announcing implementation of measures, announcing 2nd murder as terrorism)
+50 hours	Press conference – President and Government	Rebuilding (capturing mass murderer, proposal of the measures)

The shift of the official crisis response stance from initial denial to rebuilding is especially noticeable throughout the crisis. However, due to the extremely high moral outrage caused by the initial mass shooting and protest gatherings initiated immediately (candle lighting and street protests), this shift was untimely and consequently ineffective. Moreover, when the post-crisis events got political context with the mass protests against violence holding public outburst and unrest (organized with support of the opposition political parties), the officials attempted to move the crisis to the political arena, as visible in BBC (2023) report. This move also did not yield solid results on post-crisis mitigation of the moral outrage, as the motivation for the mass protests was not predominantly political despite the involvement of the opposition – it has been more value-oriented. Hence, understanding the social impact of various crisis events and responses becomes even more important, taking into consideration findings associated with the timeline of the crisis and officials` crisis responses.

Social Impact and Moral Outrage

The research findings indicate that initial crisis responses did not help mitigate the moral outrage. Despite concrete measures proposed by the state and psychological support offered to the families of victims, the media reporting enabled the tabloidization of the first mass shooting event, possibly causing a copycat effect, and contributing to the second mass shooting to happen. The tabloidization was especially noticeable in the sensationalistic reports of the mainstream media about the mass shooters and their families (Beta, 2023; Telegraf, 2023). Moreover, survey results indicated that institutional statements and crisis communication had a negative social impact, without significant contribution to building community resilience and without mitigation of the moral outrage caused by the crisis: only the resignation of the Minister of Education and spontaneous gatherings of the citizens had a positive social impact. The results of the survey associated with the crisis events are shown in Table 2.

Table 2.

Date / Time	Event	Officials' response type
03.05.2023. 8:40AM (trigger event)	Mass shooting – primary school	N/A
+4 hours	Press conference – ministers (interior, health, education) and police	Diminishment/Denial-Scapegoat/Victimage (blaming Western culture and media influences)
+9 hours	Press conference – President of the state – proposal of the measures (disarming, increased security)	Rebuilding (taking partial responsibility and proposal of the measures; facts about the act, murderer and his family; fear from copycat effect)
+11 hours	Citizen gathering near school, candle lighting, leaving flowers	N/A
+24 hours	Media speculations, tabloidization	N/A
+30 hours	Copycat event – knife attack in high school	N/A – no official response
+31 hours	Ministry of Interior – moratorium on weapons	Rebuilding
+35 hours	Citizen gathering near school, candle lighting, leaving flowers	N/A
+40 hours	Copycat event – 2nd mass shooting	N/A
+42 hours	Media statement – Minister of interior, announcement of 2nd mass shooting as terrorist act	Rebuilding (announcing implementation of measures, announcing 2nd murder as terrorism)
+50 hours	Press conference – President and Government	Rebuilding (capturing mass murderer, proposal of the measures)

The findings of the sentiment analysis indicate extremely high levels of negative feelings associated with mass shootings. Interestingly, the satirical social media post of the Prime Minister triggered almost the same level of negative sentiments as mass shootings: this might be a solid explanation for why the number of attendees of protests against violence increased rapidly after the post. Moreover, it is important to note that numerous non-governmental organizations and civil society activists joined the protest as well. A possible reason could be that perceived value incongruence spread from mass shootings to the political decision-makers who did not show sufficient respect for the victims of violence and citizen who protested, as in the case of Instagram post of Serbian Prime Minister.

"There are many more of us than last time. Just a little more and we'll be on the Gazelle!" (central bridge in Belgrade) – text under Instagram post of the Prime Minister on May 12, 2023.

Additional statistical tests indicated that the population in Central Serbia, as well as the elderly population (older than 55 years of age), was somewhat less critical of the officials` communication than the population of other regions and the younger population. This correlates with the high level of political distrust noticed within the younger population of Serbia (Pantelic, 2022).

CONCLUSION

Based on the document analysis and findings of the survey, it can be concluded that the crisis has not been handled in an appropriate professional manner. Inappropriate handling of the crisis responses of the officials not only spread the risks and contributed to the appearance of copycat events, but also added to the spillover of moral outrage and the occurrence of numerous post-crisis events. Further, the crisis caused by mass shootings, as a mega-crisis, requires a specific and multifaceted approach to its handling. Thus, based on the findings of this research and insights from the available literature, the following recommendations for handling official crisis responses in case of mass shootings can be created:

- Respond timely and only through official channels
- Reduce publicity given to the mass shooter and focus publicity on the victims
- Minimize the possibility for tabloidization and different interpretations of the official responses
- Keep crisis responses consistent and coherent
- Reduce the number of official communication channels and sources
- Ensure that state officials refrain from posting any controversial content during the acute phase of the crisis

- Focus on mitigating moral outrage and initiating dialogue rather than focusing on polarization

In future research, the emphasis should be put on identifying social impacts after the acute stage of the crisis. Through the time series, perceptions and feelings about crisis responses can be compared. Comparative studies can be performed to compare the efficacy of crisis response in Serbia with the efficacy of the officials` crisis responses to mass shootings in other countries. Future research could also elaborate on the impact of cultural context on official crisis responses in cases of mega-crises. Moreover, initial recommendations for handling the official crisis responses in case of mega-crises may be further examined and expanded into the model, if applicable.

The research has certain limitations associated with usual survey biases. Biases and risks associated with survey research have been mitigated using a standard set of techniques for survey research (Ljepava & Selakovic, 2019). Further, in the document analysis, only sources in Serbian and English language have been included. In addition, recommendations for handling crisis responses in cases of mass shootings are grounded in the experience of mass shootings from Serbia and available literature sources, predominantly from the United States. Thus, the recommendations should be taken with certain reservations, as the cultural backgrounds and contexts of different countries may require adaptations of crisis response strategies (Coombs & Laufer, 2018). Overall, the research provides novel insights and contributions to the body of literature in terms of handling mega-crises and mass shootings, providing additional recommendations to the practitioners and Government communicators on managing crisis response strategies in crisis situations of extreme intensity.

REFERENCES

Ahmad, M., Mekala, H., Lone, J., Robinson, K., & Shah, K. (2020). Addressing mass shootings in a new light. *Cureus, 12*(7). Advance online publication. DOI: 10.7759/cureus.9298 PMID: 32832295

Anisin, A. (2023). Do mass shootings in central and Eastern Europe differ from US mass shootings? Insights from the MSCEE Data Set, *Journal of Mass Violence Research*, pp.1-13.

Babovic´, M., Katarina G., & Vukovic, O.(2010). Mapiranje porodicˇnog nasilja prema zˇenama u Centralnoj Srbiji [Mapping of Domestic Violence against Women in Central Serbia]. Ministry of Labor and Social Policy: Project against Sexual and Gender Based Violence. Belgrade, Serbia: SeConS

Bandura. (1977). *Social Learning Theory,* New York: Englewood Cliffs.

Batson, C. D., Kennedy, C. L., Nord, L. A., Stocks, E. L., Fleming, D. A., Marzette, C. M., Lishner, D. A., Hayes, R. E., Kolchinsky, L. M., & Zerger, T. (2007). Anger at unfairness: Is it moral outrage? *European Journal of Social Psychology, 37*(1), 1272–1285. DOI: 10.1002/ejsp.434

BBC. (2023). *Srbija i protesti: Desetine hiljada ljudi u šetnji i blokade mostova, opozicija najavila novi skup za 27. maj*. BBC. Retrieved from https://www.bbc.com/serbian/lat/srbija-65648580

Beta (2023). *Naslovne strane za četvrtak, 4. maj 2023. Godine*. Krstarica. Retrieved from https://www.krstarica.com/vesti/srbija/naslovne-strane-za-cetvrtak-4-maj-2023-godine/

Burton, A. L., Pickett, J. T., Jonson, C. L., Cullen, F. T., & Burton, V. S.Jr. (2021). Public Support for Policies to Reduce School Shootings: A Moral-Altruistic Model. *Journal of Research in Crime and Delinquency, 58*(3), 269–305. DOI: 10.1177/0022427820953202

Bushman, B. J. (2018). Narcissism, fame seeking, and mass shootings. *The American Behavioral Scientist, 62*(2), 229–241. DOI: 10.1177/0002764217739660

Christodoulou, E. (2024). The development of UNESCO's programmes for preventing violent extremism: Educational norms, institutional politics and declining legitimacy. *Compare: A Journal of Comparative Education, 54*(5), 750–769. DOI: 10.1080/03057925.2023.2252319

Cook, A. (2006). *Individual vs. systemic justice: using trust and moral outrage to predict reactions to vigilante murder* (Doctoral dissertation, University of Missouri-Columbia).

Coombs, W. T. (2017). *Revising situational crisis communication theory: The influences of social media on crisis communication theory and practice, Social media and crisis communication.* Routledge. DOI: 10.4324/9781315749068-3

Coombs, W. T., Holladay, S. J., & White, R. (2020). Corporate crises: Sticky crises and corporations. In *Advancing crisis communication effectiveness* (pp. 35–51). Routledge. DOI: 10.4324/9780429330650-5

Coombs, W. T., & Tachkova, E. R. (2023a). Integrating Moral Outrage in Situational Crisis Communication Theory: A Triadic Appraisal Model for Crises. *Management Communication Quarterly*, *37*(4), 798–820. DOI: 10.1177/08933189221151177

Coombs, W. T., & Tachkova, E. R. (2023b). How Emotions Can Enhance Crisis Communication: Theorizing Around Moral Outrage. *Journal of Public Relations Research*, *36*(1), 1–17.

Dahmen, N. S., Abdenour, J., McIntyre, K., & Noga-Styron, K. E. (2018). Covering mass shootings: Journalists' perceptions of coverage and factors influencing attitudes. *Journalism Practice*, *12*(4), 456–476. DOI: 10.1080/17512786.2017.1326832

Darley, J. M., & Pittman, T. S. (2003). The psychology of compensatory and retributive justice. *Personality and Social Psychology Review*, *7*(4), 324–336. DOI: 10.1207/S15327957PSPR0704_05 PMID: 14650389

Dumančić, M., & Krolo, K. (2017). Dehexing postwar West Balkan masculinities: The case of Bosnia, Croatia, and Serbia, 1998 to 2015. *Men and Masculinities*, *20*(2), 154–180. DOI: 10.1177/1097184X15625314

Durkheim, E. (1997). *The Division of Labor in Society*. The Free Press. (Original work published 1893)

Eagar, C. (2023). Bullet points: In Serbia, the gun is a way of life. *Spectator (London, England)*, *352*(10160), 24–25.

Engelbrecht, C. (2023), *A Look at Gun Violence in a Nation Awash With Weapons, The New York Times*, A8-L.

Ford, B. Q., & Feinberg, M. (2020). Coping with Politics: The Benefits and Costs of Emotion Regulation. *Current Opinion in Behavioral Sciences*, *34*, 123–128. DOI: 10.1016/j.cobeha.2020.02.014

Fox, J. A., Sanders, N. E., Fridel, E. E., Duwe, G., & Rocque, M. (2021). The contagion of mass shootings: The interdependence of large-scale massacres and mass media coverage. *Statistics and Public Policy (Philadelphia, Pa.)*, *8*(1), 53–66. DOI: 10.1080/2330443X.2021.1932645

Garcia-Bernardo, J., Qi, H., Shultz, J. M., Cohen, A. M., Johnson, N. F., & Dodds, P. S. (2018). *Social Media Appears to Affect the Timing, Location, and Severity of School Shootings, USA Today*, 10 (1), p.100.

Haidt, J. (2012). *The Righteous Mind: Why Good People are Divided by Politics and Religion*. Pantheon.

Helfgott, J. B. (2015). Criminal behavior and the copycat effect: Literature review and theoretical framework for empirical investigation. *Aggression and Violent Behavior*, *22*(1), 46–64. DOI: 10.1016/j.avb.2015.02.002

Henderson, R. K., & Schnall, S. (2021). Social threat indirectly increases moral condemnation via thwarting fundamental social needs. *Scientific Reports*, *11*(1), 21709. DOI: 10.1038/s41598-021-00752-2 PMID: 34741054

Kahneman, D. (2011). *Thinking fast and slow*. Farrar, Straus and Giroux.

Kambam, P. R., Pozios, V. K., Bond, K. L., & Ostermeyer, B. K. (2020). The influence of media related to mass shootings. *Psychiatric Annals*, *50*(9), 393–398. DOI: 10.3928/00485713-20200811-01

Kellner, D. (2012). School shootings, crises of masculinities, and media spectacle: Some critical perspectives. In *School Shootings: Mediatized Violence in a Global Age* (pp. 299–330). Emerald Group Publishing Limited. DOI: 10.1108/S2050-2060(2012)0000007018

Langman, P. (2015). *School shooters: Understanding high school, college, and adult perpetrators*. Rowman and Littlefield. DOI: 10.5771/9781442233577

Langman, P. (2017). *Role Models, Contagions, and Copycats: An Exploration of the Influence of Prior Killers on Subsequent Attack*. Available at: https://schoolshooters .info/role-models- *contagious-and-copycats-exploration-influence-prior-killers-subsequent-attacks* (accessed 12 May 2024)

Lankford, A., & Madfis, E. (2017). Don't name them, don't show them, but report everything else: A pragmatic proposal for denying mass killers the attention they seek and deterring future offenders. *The American Behavioral Scientist*, *62*(2), 260–279. DOI: 10.1177/0002764217730854

Lee, H., Pickett, J. T., Burton, A. L., Cullen, F. T., Jonson, C. L., & Burton, V. S.Jr. (2022). Attributions as anchors: How the public explains school shootings and why it matters. *Justice Quarterly*, *39*(3), 497–524. DOI: 10.1080/07418825.2020.1769710

Ljepava, N., & Selakovic, M. (2019). Trends and challenges of political polling and research in the digital age. In Bulatovic, Lj. (Ed.), *Digitalni prostori–izazovi i očekivanja/Digital spaces–challenges and expectations* (pp. 117–140).

Lodewijkx, H. F., Kersten, G. L., & Van Zomeren, M. (2008). Dual pathways to engage in 'silent marches' against violence: Moral outrage, moral cleansing and modes of identification. *Journal of Community & Applied Social Psychology*, *18*(3), 153–167. DOI: 10.1002/casp.916

Lubsanov, A. E., & Yantranova, V. B. (2022). The School Shooting Phenomenon, in Chepak. A. (Ed.): *All-Russian Contest for the Student's Scientific Project in a Foreign Language* (pp.72-75).

Meindl, J. N., & Ivy, J. W. (2017). Mass Shootings: The Role of the Media in Promoting Generalized Imitation. *American Journal of Public Health*, *107*(3), 368–370. DOI: 10.2105/AJPH.2016.303611 PMID: 28103074

Mihailov, E., Voinea, C., & Vică, C. (2023). Is Online Moral Outrage Outrageous? Rethinking the Indignation Machine. *Science and Engineering Ethics*, *29*(2), 12. DOI: 10.1007/s11948-023-00435-3 PMID: 37000374

Miller, D. T., Effron, D. A., & Zak, S. V. (2011). From moral outrage to social protest: The role of psychological standing, in Bobocel, R.; Kay, A.C.; Zanna, M.P., and Olson, J.M. (Eds): *The Psychology of Justice and Legitimacy: The Ontario Symposium* Volume 11 *(Ontario Symposia on Personality and Social Psychology Series) 1st Edition* (pp.103-124). Psychology Press.

Muschert, G. W. (2002). *Media and massacre: The social construction of the Columbine story*. University of Colorado at Boulder.

O'Mara, E. M., Jackson, L. E., Batson, C. D., & Gaertner, L. (2011). Will moral outrage stand up? Distinguishing among emotional reactions to a moral violation. *European Journal of Social Psychology*, *41*(2), 173–179. DOI: 10.1002/ejsp.754

O'Toole, M. E., Beresin, P. E., Berman, A., Gorelick, S. M., Helfgott, J. B., & Tobin, C. (2014). Celebrities Through Violence: The Copycat Effect and the Influence of Violence in Social Media on Mass Killers. *Violence and Gender*, *1*(3), 107–116. DOI: 10.1089/vio.2014.1512

Pantelic, Z. (2022). *Youth Emigration and Political Distrust in Serbia*. Westminster Foundation for Democracy.

Pejuskovic, B., & Lecic-Tosevski, D. (2023). School Shooting in Serbia–A Huge Collective Trauma and Challenge for Public Mental Health. *World Social Psychiatry, 5*(2), 179–182. DOI: 10.4103/wsp.wsp_46_23

Pescara, K. L., and Raleigh, M. J. (2017). The Contagion Effect as it Relates to Public Mass Shootings and Suicides, *Journal of Campus Behavioral Intervention (J-BIT)*, 5, pp.35-45.

Puryear, C., Vandello, J. A., & Gray, K. (2024). Moral panics on social media are fueled by signals of virality. *Journal of Personality and Social Psychology, 127*(1), 84–103. DOI: 10.1037/pspa0000379 PMID: 38421751

Reeping, P. M., Jacoby, S., Rajan, S., & Branas, C. C. (2020). Rapid response to mass shootings: A review and recommendations. *Criminology & Public Policy, 19*(1), 295–315. DOI: 10.1111/1745-9133.12479

Roseman, I. J. (1984). Cognitive determinants of emotions: A structural theory. *Review of Personality and Social Psychology, 5*, 11–36.

Rozin, P., Lowery, L., Imada, S., & Haidt, J. (1999). The CAD triad hypothesis: A mapping between three moral emotions (contempt, anger, and disgust) and three moral codes (community, autonomy, and divinity). *Journal of Personality and Social Psychology, 76*(4), 574–586. DOI: 10.1037/0022-3514.76.4.574 PMID: 10234846

Selaković, M., & Ljepava, N. (2023). Online-Ignited Crises and Post-Crisis Image Restoration: Example of Flight 3411 Incident. *International Journal of Management Trends: Key Concepts and Research, 2*(1), 15–40. DOI: 10.58898/ijmt.v2i1.15-40

Silva, J. R., & Capellan, J. A. (2019). The media's coverage of mass public shootings in America: Fifty years of newsworthiness. *International Journal of Comparative and Applied Criminal Justice, 43*(1), 77–97. DOI: 10.1080/01924036.2018.1437458

Silva, J. R., & Greene-Colozzi, E. A. (2019). Fame-seeking mass shooters in America: Severity, characteristics, and media coverage. *Aggression and Violent Behavior, 48*, 24–35. DOI: 10.1016/j.avb.2019.07.005

Silver, J. (2017). Moral Foundations, Intuitions of Justice, and the Intricacies of Punitive Sentiment. *Law & Society Review, 51*(2), 413–450. DOI: 10.1111/lasr.12264

Silver, J., Horgan, J., & Gill, P. (2019). Shared struggles? Cumulative strain theory and public mass murderers from 1990 to 2014. *Homicide Studies, 23*(1), 64–84. DOI: 10.1177/1088767918802881

Skitka, L. J., Bauman, C. W., & Sargis, E. G. (2005). Moral conviction: Another contributor to attitude strength or something more? *Journal of Personality and Social Psychology*, *88*(6), 895–917. DOI: 10.1037/0022-3514.88.6.895 PMID: 15982112

Smokvina, T. K., & Kovacic, M. P. (2021). Media Ethics and Professionalism in the Digital Communication Environment: Responses of Media Self-regulation and Regulatory Bodies. *JAVNOST-THE PUBLIC*, *28*, S43–S61.

Stanko, B. (2000). Murder and Moral Outrage: Understanding violence. *CJM*, *42*(1), 4–5.

Strahler, J., & Ziegert, T. (2015). Psychobiological stress response to a simulated school shooting in police officers. *Psychoneuroendocrinology*, *51*(1), 80–91. DOI: 10.1016/j.psyneuen.2014.09.016 PMID: 25290348

Telegraf. (2023). NASLOVNE STRANE DANAŠNJIH NOVINA: Prelistavanje štampe za petak, 5. maj 2023. Godine. Telegraf.rs. Retrieved from https://www.telegraf.rs/vesti/srbija/3673286-naslovne-strane-sutrasnjih-novina

Van Zomeren, M. (2006). Social-psychological paths to protest: An integrative perspective. The Netherlands: Doctoral dissertation, University of Amsterdam.

Van Zomeren, M., & Lodewijkx, H. M. (2005). Motivated responses to 'senseless' violence: Explaining emotional and behavioural responses through person and position identification. *European Journal of Social Psychology*, *35*(6), 755–766. DOI: 10.1002/ejsp.274

Vulić, T., & Pavlović, D. (2015). Domestic violence in the Serbian daily press: from facts to sensationalism. *Facta Universitatis, Series: Philosophy, Sociology, Psychology and History*, 111-122.

Yelderman, L. A., Joseph, J. J., West, M. P., & Butler, E. (2019). Mass shootings in the United States: Understanding the importance of mental health and firearm considerations. *Psychology, Public Policy, and Law*, *25*(3), 212–223. DOI: 10.1037/law0000200

Yen, V. Y., & Salmon, C. T. (2017). Further explication of mega-crisis concept and feasible responses, in *SHS Web of Conferences* (Vol. 33, p. 00034). EDP Sciences. DOI: 10.1051/shsconf/20173300034

Chapter 12

Zygmunt Bauman's Approach to the Loss of Meaning of Morality in Postmodern Wars

Ulhak Cimen
Independent Researcher, Turkey

Ömer Alanka
https://orcid.org/0009-0007-6570-2504
Independent Researcher, Turkey

ABSTRACT

Zygmunt Bauman, one of the most important sociologists of the 20th and 21st centuries, emphasizes that in the new post-modern period, which he calls the liquid modern period, situations such as genocide, war, and massacre continue without any change and that this situation is in line with the basic principles of modernism. The thinker claims that the concept of dehumanization is currently valid and that the weak are destroyed as a result of people disconnected from morality using technology incredibly well. Postmodern massacres and wars are based on the absolute rightness of the strong with rational justifications. This study, which conducts a literature review and descriptive analysis, tries to understand postmodern wars from the perspective of Zygmunt Bauman.

DOI: 10.4018/979-8-3373-7589-2.ch012

INTRODUCTION

It is thought that inhumane situations such as genocide and massacres experienced in modern times are incompatible with the basic principles of modern civilization. Still, these situations are extremely consistent with the principles of modernism Because modernism, in its obsession with establishing order, managed to legitimize everything within the framework of its logic. In the postmodern process experienced after modernism, the same inhumanities continue both literally and figuratively. Even if there is a claim that these have been diluted and mitigated to some extent, the events experienced (Palestinian Genocide, Bosnian Massacre, Iraq War, Afghanistan Invasion, Uyghur Turk Genocide etc.) show that modern logic has not changed at all. The most morally horrifying aspect of genocides is not the possibility of being put behind barbed wire or being driven into gas chambers, but the glorification of the logical aspects of the reasons that you can carry out the same cruelty (Bauman, 2020, p. 83). It is more acceptable to be a victim of genocide than to be its perpetrator and implementer (Bauman & Donskis, 2020, p. 49). It is often asserted that inhumane situations such as genocide and massacres experienced in modern times are fundamentally incompatible with the basic principles of modern civilization. Yet, paradoxically, these very situations prove to be remarkably consistent with the underlying tenets of modernism itself. This is because modernism, in its relentless obsession with establishing order and achieving a perfectly rationalized society, inherently possessed the capacity to legitimize virtually anything within the strict framework of its own logic. In the postmodern process that followed modernism, these same inhumane acts tragically persist, both literally and figuratively. While there might be a prevalent claim that such brutalities have been diluted or mitigated to some extent in this new era, the devastating events we continue to witness – from the Palestinian genocide and the Bosnian massacre to the Iraq War, the Afghanistan invasion, and the ongoing Uyghur Turk genocide – starkly demonstrate that the underlying modern logic has remained disturbingly unchanged. The most profoundly morally horrifying aspect of genocides is not merely the chilling prospect of being herded behind barbed wire or driven into gas chambers. Rather, it is the deeply disturbing glorification of the very logical aspects and rationales that enable the perpetration of such extreme cruelty. In this warped ethical landscape, it becomes more acceptable to be a victim of genocide than to be its perpetrator and implementer.

Dehumanization of enemies is not a situation specific to the modern era. Dehumanization was experienced frequently, especially in wars before the modern era. Soldiers had to kill each other to prevent themselves from dying. There was a definite symmetry in the struggles that took place on the battlefield. Everyone would fight against each other in the same time, space and space. Human values would be suspended for a short time and the indefinability and worthlessness of the other would be declared on the battlefield. In order to show moral respect for one's own identity, one should not die, and therefore, one should exclude the other from the moral sphere on the battlefield. The dehumanization of the other during the fight with the enemy has been revolutionized by the modern system. Because the technology and rational organization under the control of the modern system have invented mass killings from a distance instead of symmetrical mutual struggle. Modern technologies of mass slaughter are rational in carrying out mass killings. Modern weapons completely render the moral identities of the other insignificant before destroying them. The fact that the Japanese were seen by the Americans as little yellow mice instead of humans or that other societies generally turned them into animals, can be given as examples of this situation (Bauman, 2014, p. 71). While the dehumanization of enemies is indeed not an invention of the modern era, having been a pervasive feature of warfare across civilizations as a means to suspend moral inhibitions during conflict, the modern system profoundly revolutionized its nature and scale. Pre-modern battles, characterized by direct, symmetrical engagement in shared time and space, required individuals to actively declare the "indefinability and worthlessness" of the opponent to justify the act of killing for self-preservation and the protection of their own moral identity. In this context, dehumanization was often a visceral, immediate, and personally enacted psychological mechanism, a temporary moral suspension for a specific encounter. However, modernity's pursuit of rational organization and technological advancement transformed this primal act into a systematic, bureaucratized process. Zygmunt Bauman argues that modern atrocities, unlike their pre-modern counterparts, are not primarily driven by intense hatred or unbridled passion, but by cold, calculated efficiency and moral disengagement. The invention of "mass killings from a distance" fundamentally alters the moral landscape of warfare. Technologies like long-range artillery, aerial bombing, and ultimately, nuclear weapons, create a vast physical and psychological distance between the perpetrator and the victim.

This technologically mediated moral distance is crucial to understanding Bauman's perspective. When the act of killing becomes remote and depersonalized, the need for direct, emotional dehumanization diminishes. Instead, the "moral identities" of the 'other' are rendered insignificant before destruction, not through a passionate declaration of worthlessness on the battlefield, but through bureaucratic classification, statistical aggregation, and the very design of the weaponry itself. The targets become abstract units, data points, or, as your examples powerfully illustrate, "little yellow mice" or "animals" – not because of a soldier's immediate hatred, but because of a systemic process that strips them of their individuality and humanity a priori.

Bauman emphasizes that this routinization and rationalization of violence make it easier for individuals to participate in atrocities without feeling directly responsible for the moral implications. The division of labor, the hierarchical command structure, and the technical nature of modern warfare allow moral actors to externalize their ethical duties onto the system. The act of pulling a trigger from afar, or pressing a button that unleashes devastation, lacks the direct, tangible feedback that would trigger moral revulsion in face-to-face combat. Thus, the meaning of morality is not so much lost, but diluted, diffused, and rendered irrelevant by the very mechanisms designed to make killing more efficient and less emotionally taxing for the perpetrators. This systemic dehumanization, therefore, becomes a cornerstone of modern warfare's capacity for unprecedented scale and cruelty, even in the absence of traditional, face-to-face moral confrontation.

The openness that America has offered to these people in the Syrian war, where a quarter of a million people have died (according to 2016 data), is not only a political show but also a dehumanization of people. America accepted only 2,500 Syrian refugees in the first 5 years of the war, which is only six out of ten thousand of the number of refugees at that time (Bauman, 2018, p. 83). While the dehumanization of enemies is indeed not an invention of the modern era, having been a pervasive feature of warfare across civilizations as a means to suspend moral inhibitions during conflict, the modern system profoundly revolutionized its nature and scale. Pre-modern battles, characterized by direct, symmetrical engagement in shared time and space, required individuals to actively declare the "indefinability and worthlessness" of the opponent to justify the act of killing for self-preservation and the protection of their own moral identity. In this context, dehumanization was often a visceral, immediate, and personally enacted psychological mechanism, a temporary moral suspension for a specific encounter. However, modernity's pursuit of rational organization and technological advancement transformed this primal act into a systematic, bureaucratized process. Zygmunt Bauman argues that modern atrocities, unlike their pre-modern counterparts, are not primarily driven by intense hatred or unbridled passion, but by cold, calculated efficiency and moral disengagement. The invention of "mass killings from a distance" fundamentally alters the moral landscape of warfare. Technologies like long-range artillery, aerial bombing, and ultimately, nuclear weapons, create a vast physical and psychological distance between the perpetrator and the victim.

This technologically mediated moral distance is crucial to understanding Bauman's perspective. When the act of killing becomes remote and depersonalized, the need for direct, emotional dehumanization diminishes. Instead, the "moral identities" of the 'other' are rendered insignificant before destruction, not through a passionate declaration of worthlessness on the battlefield, but through bureaucratic classification, statistical aggregation, and the very design of the weaponry itself. The targets become abstract units, data points, or, as your examples powerfully illustrate, "little yellow mice" or "animals" – not because of a soldier's immediate hatred, but because of a systemic process that strips them of their individuality and humanity a priori.

Bauman emphasizes that this routinization and rationalization of violence make it easier for individuals to participate in atrocities without feeling directly responsible for the moral implications. The division of labor, the hierarchical command structure, and the technical nature of modern warfare allow moral actors to externalize their ethical duties onto the system. The act of pulling a trigger from afar, or pressing a button that unleashes devastation, lacks the direct, tangible feedback that would trigger moral revulsion in face-to-face combat. Thus, the meaning of morality is not so much lost, but diluted, diffused, and rendered irrelevant by the very mechanisms designed to make killing more efficient and less emotionally taxing for the perpetrators. This systemic dehumanization, therefore, becomes a cornerstone of modern warfare's capacity for unprecedented scale and cruelty, even in the absence of traditional, face-to-face moral confrontation. Beyond the battlefield, the principles of dehumanization and moral distance manifest in the broader geopolitical sphere, particularly in how states and publics react to humanitarian crises. The perceived "openness" that a nation like America offered to Syrian refugees during a conflict that claimed over a quarter-million lives (according to 2016 data) serves as a stark illustration of this phenomenon. Far from genuine compassion, such gestures can often function as a mere political spectacle, designed to project an image of moral concern while simultaneously facilitating a deeper dehumanization of the very people they ostensibly aim to help.

Consider the chilling statistics: America accepted only 2,500 Syrian refugees in the first five years of the war, a meager six out of every ten thousand of the total refugee population at that time. This disproportionately low intake, especially against the backdrop of immense suffering, reveals a profound moral indifference masked by bureaucratic procedures and carefully managed public relations. Bauman would argue that in the liquid modern world, the suffering of distant others is often commodified and consumed as a fleeting image rather than a call to action. The vast numbers of displaced and dying become a mere statistical abstraction, reducing individual human tragedy to an impersonal data point.

This numerical devaluation contributes to what Bauman identifies as "adiaphorization": the process by which certain acts or individuals are pushed outside the realm of moral consideration, becoming ethically neutral or irrelevant. When only a tiny fraction of a massive suffering population is deemed worthy of tangible assistance, the vast majority are, by implication, deemed "adiaphoric" – not necessarily hated, but simply rendered morally invisible. Their plight becomes a matter of logistics, security concerns, or national interest, rather than a direct ethical imperative. This detachment allows powerful actors to maintain a veneer of humanitarianism while effectively insulating themselves from the moral demands that mass suffering would otherwise impose, thus contributing significantly to the loss of meaning of morality in contemporary conflicts.

The world that Bauman is trying to analyze is not the evil world where demons and monsters live. The world that the thinker evaluates is the universe of life where seemingly good and modern people; parents who would be very good role models in family life can actually turn the world into hell for the other (Bauman & Donskis, 2020, p. 16). Moreover, the most common form of evil is experienced by not seeing the suffering of the other, free from war and ideologies, by rejecting meaning, becoming insensitive, and moving away from moral responsibility. The world Zygmunt Bauman meticulously analyzes is not, as one might instinctively imagine, a simplistic realm inhabited solely by demonic figures or monstrous aberrations. Instead, the profound tragedy of the human condition, in Bauman's view, resides in the fact that the very universe of life he scrutinizes is populated by ostensibly "good" and "modern" individuals. These are often people who, in their private spheres, could be exemplary parents or upstanding citizens, yet possess the chilling capacity to transform the world into a living hell for the 'Other'. This paradox highlights Bauman's central argument: that the greatest atrocities are often committed not by inherently evil individuals, but by those operating within systems that enable moral disengagement.

Moreover, Bauman posits that the most insidious and pervasive form of evil in our contemporary world is manifested not through overt acts of aggression, but through a chilling indifference to the suffering of the 'Other'. This widespread moral insensitivity transcends the traditional boundaries of war and political ideologies. It emerges from a conscious rejection of meaning, a deliberate turning away from the moral implications of one's actions or inactions. When individuals become desensitized to the plight of others, and when moral responsibility is effectively shed or diffused across complex bureaucratic and technological structures, a vacuum is created where ethical considerations simply cease to hold sway. This state of moral detachment, characterized by the refusal to acknowledge the 'Other's' suffering, epitomizes the loss of meaning of morality in a postmodern context. It is a quiet, yet devastating, form of evil where empathy is sidelined, and the imperative to act morally is dissolved into a sea of apathy and instrumental rationality, allowing horrors to unfold silently and with disturbing regularity.

POSTMODERN MASSACRES

The most terrible, bloodiest, most brutal, most horrific massacres in history have been committed for the highest, noblest, most praiseworthy purposes (Bauman, 2014, p. 84). The discourse of creation is always placed in the background of the destruction carried out in massacres. The people who carry out the massacre are not monster-looking, psychopathic, murderous individuals. The people who commit the massacre are mostly friends, neighbors, ideal people who help the elderly cross the street, give their seats to the disabled and elderly on the bus, tolerate when someone passes in front of them in the checkout line at the supermarket, return the wallet they find to its owner, help the poor, etc. These people were helpful individuals who did not show any pathological symptoms before the war. If the monsters had only been committed by monsters, the world could have been made a safe place with the precautions that would be taken. In fact, the world is almost completely protected against the dangers that the creatures can create. Beyond this, there are scientists who can identify psychopaths and sociopaths and legal systems that can impose penal sanctions on them all over the world (Bauman, 2017, p. 185). Your assertion that the "most terrible, bloodiest, most brutal, most horrific massacres in history have been committed for the highest, noblest, most praiseworthy purposes" cuts to the very heart of Zygmunt Bauman's analysis of modern atrocities. It highlights a profound moral paradox: that the discourse of "creation," "progress," and "higher ideals" often serves as the chilling backdrop for widespread destruction. This isn't merely a historical curiosity; it's a fundamental characteristic of modern warfare and its accompanying dehumanization.

Bauman vehemently argues against the simplistic notion that perpetrators of massacres are inherently monstrous, psychopathic, or uniquely evil individuals. Indeed, as you astutely point out, they are often "friends, neighbors, ideal people" – individuals exhibiting all the hallmarks of social conformity and everyday morality: helping the elderly, showing tolerance, returning lost wallets, and aiding the poor. These are seemingly "helpful individuals who did not show any pathological symptoms before the war." This observation is crucial because it shatters the comfortable illusion that evil is easily identifiable and containable. If atrocity were solely the domain of "monsters," the world could, theoretically, be rendered safe through identification and isolation, as you suggest by referencing the capabilities of scientists to identify psychopaths and legal systems to penalize them.

However, Bauman's work forcefully reminds us that the true danger lies not in the easily detectable abnormality, but in the "normality" of the perpetrators. Modern systems, with their emphasis on rationality, efficiency, and bureaucratic procedure, possess an unparalleled capacity to "neutralize moral inhibitions" in ordinary people. The division of labor, the chain of command, and the technical depersonalization of victims allow individuals to compartmentalize their actions, separating their roles as dutiful functionaries from their personal moral compasses. The focus shifts from the moral implications of an act to its technical execution. In this framework, atrocities become "socially produced" rather than merely psychologically driven. It's not about individuals having a momentary lapse of judgment or succumbing to primitive urges; it's about a systemic environment that fosters moral blindness and the outsourcing of ethical responsibility. The "loss of meaning of morality" in postmodern wars, therefore, isn't about people becoming evil; it's about the erosion of the mechanisms that compel ordinary people to act morally, even when confronted with extreme circumstances.

The possibility that people we know as good can be monstrous is not understood before they reach certain positions. For example, if they had not been assigned to Abu Ghraib, we would always describe these people as good, compassionate, well-mannered and respectful. People who can always be described as having positive characters in their own social lives can engage in deviant behaviors such as torment, harassment and torture when assigned to a position. Therefore, it is necessary to re-evaluate the monster hypothesis in the liquid modern world. Because unfortunately, all the criteria and filters that evaluate the characteristics of normal people that distinguish them from monstrous people are flawed. It is also possible to see the more terrifying side of this situation in the Bosnian War. Moreover, the fact that Western civilization is blind and deaf to those who display monstrous behavior can also be evaluated as this civilization being complicit in monstrosity. Your insightful observation that "the possibility that people we know as good can be monstrous is not understood before they reach certain positions" strikes at the very core of Zygmunt Bauman's critique of moral responsibility in the liquid modern world. The unsettling truth, powerfully illustrated by examples like Abu Ghraib, is that individuals we readily describe as "good, compassionate, well-mannered, and respectful" in their private lives can, when placed within specific institutional frameworks and roles, engage in "deviant behaviors such as torment, harassment, and torture." This reveals that the capacity for cruelty is not solely the domain of a pre-existing, identifiable "monster" but is often contingent on context and systemic enablement.

This necessitates a profound re-evaluation of the traditional "monster hypothesis." Bauman argues that our conventional "criteria and filters that evaluate the characteristics of normal people that distinguish them from monstrous people are flawed" precisely because they fail to account for the disembedding of morality from personal character in modern, and particularly liquid modern, systems. In the liquid modern world, moral responsibility becomes increasingly elusive and fragmented. It's not just about inherent character; it's about the structures that dilute individual accountability, allowing 'normal' people to commit 'monstrous' acts without perceiving themselves as monsters. The horrifying realities of the Bosnian War, where neighbors turned against neighbors with unfathomable brutality, further illuminate this terrifying fluidity of moral identity.

Moreover, the "blindness and deafness" of Western civilization to such monstrous behavior is, as you correctly posit, deeply problematic. Bauman would contend that this complicity is not accidental but symptomatic of a deeper malaise within modern societies. In a world saturated with information yet paradoxically starved of genuine moral engagement, suffering can be "aestheticized" or simply ignored when it occurs at a distance or involves an "Other" deemed outside one's immediate moral sphere. This detachment is exacerbated by the very fluidity of liquid modernity, which tends to erode stable moral communities and universal ethical frameworks. When collective responsibility is dissolved, and individual moral anchors are loosened, societies risk becoming tacit accomplices in atrocity, by their inaction, their turning away, or their rationalization of unspeakable acts. The very systems designed to bring order ironically create conditions where moral meaning is lost, and the capacity for monstrosity becomes a disturbing, ever-present potential within the 'normal'. Therefore, it is also necessary to re-examine the portrayal of good people. In this sense, the monstrosity of good people can become the norm, not an exception.

"Most, if not all, of these people, who had never before been known to commit acts of violence, let alone murder, and who would not have been suspected of such a tendency, committed murder, shooting men and women, elderly and children, innocent people who were unarmed and not accused of any crime, without the slightest intention of harming themselves or their comrades" (Bauman, 2017, p. 189).

These people who are hunting foreigners are the same people who eat at the same table, watch a match at the same stadium, dance the halay at the same wedding and shed tears at the same funeral in normal life. In this sense, it is necessary to stay away from certain convictions in order not to become a victim of these people when the time comes. There is no difference between the injustice done to one person and the injustice done to all people. The five hundred thousand children who died after the United States' intervention in Iraq were considered only as by-products of the price that had to be paid (Bauman, 2012, p. 117). Your poignant observation that "these people who are hunting foreigners are the same people who eat at the same table, watch a match at the same stadium, dance the halay at the same wedding and shed tears at the same funeral in normal life" underscores a chilling aspect of Zygmunt Bauman's analysis: the proximity of the monstrous to the mundane. This everyday familiarity makes it profoundly difficult for us to recognize the potential for cruelty within ostensibly "normal" social structures and interactions. It reinforces the imperative to "stay away from certain convictions" that might blind us to this lurking capacity, urging vigilance against the insidious mechanisms that transform ordinary individuals into perpetrators.

The core of this moral erosion lies in the abstraction of suffering. As you powerfully state, "there is no difference between the injustice done to one person and the injustice done to all people." Yet, modern warfare, particularly in its post-modern manifestation, thrives on precisely this distinction. The tragic fate of "the five hundred thousand children who died after the United States' intervention in Iraq," dismissed as "only as by-products of the price that had to be paid," serves as a stark, horrifying example. This isn't merely a callous disregard for life; it's a symptom of a systemic process where human lives, especially those of the 'Other,' are quantified, categorized, and ultimately rendered invisible behind statistical data and strategic objectives.

Bauman would argue that this statisticalization of suffering creates an immense moral distance. When victims are reduced to numbers, they lose their individual stories, their humanity, and their capacity to evoke empathy. The ethical imperative to respond to their suffering is dissolved in a sea of impersonal data. The complex web of cause and effect in modern conflicts, often mediated by technology and distant decision-making, further obscures direct moral responsibility. Perpetrators, far removed from the immediate consequences of their actions, can justify immense casualties as "collateral damage" or unavoidable "by-products" of a larger, supposedly noble goal. This detachment is not a failure of individual morality in isolation, but a product of the rationalizing forces of modernity that persist in the postmodern landscape, transforming ethical dilemmas into mere logistical problems and effectively robbing morality of its meaning in the context of large-scale violence.The understanding that officially approves breaking an egg to make an omelet has ignored the fact that life can be lived without eating an omelet. What needs to be done when one wants to eat an omelet is the same as what needs to be done when one wants to fight. Just as there is no negative emotion when an egg is broken, there is no negative emotion when five hundred thousand children die. Western society classifies all other humanity as servants if they serve them. There is no harm in killing those who refuse service. However, the person who survives the death of other people has survived the death of his own humanity (Bauman, 2012, p. 118). Denying human dignity means killing the cause of humanity. A world that is compatible with human dignity seems impossible for now. People/societies that do not respect the dignity of others - who have essentially lost their own dignity - consider holding power as an absolute right. In the world of oppressors, things are not going well for the oppressed. There seems to be no clue for a situation that will change with the efforts of the oppressed. In the hunter society, it has become normal to reject responsibility for the other and to turn the other into material. This situation can be seen from international relations to family relations. The question of what will I gain from this job and what does it have to do with me are the most decisive questions of the postmodern era (Bauman & Obirek, 2018, p. 72). In what Bauman refers to as the "hunter society" – a metaphor capturing the ruthless, competitive, and constantly shifting nature of liquid modernity – it has indeed become disturbingly normal to reject responsibility for the 'Other' and to transform them into mere "material" or disposable resources. This cold, calculating detachment is not confined to grand geopolitical conflicts; it pervades every stratum of society, from the highest echelons of international relations down to the most intimate family dynamics. The pervasive questions of "what will I gain from this job?" and "what does it have to do with me?" are, as you rightly point out, the most decisive and defining inquiries of the postmodern era. These questions encapsulate a profound shift from a society of producers, bound by long-term commitments and collective responsibilities, to a society of consumers, driven by immediate gratification, fleeting connections, and self-serving utility.

This pervasive utilitarian calculus effectively hollows out the space for genuine moral concern. When every interaction, every relationship, is filtered through the lens of individual gain and personal relevance, the suffering of those outside one's immediate sphere of interest becomes easily dismissible. As Zygmunt Bauman meticulously documented, the logic of the consumer market has infiltrated not only economic transactions but also social and ethical relations. In this framework, people are often treated as commodities – to be acquired, used, and discarded when they no longer serve a purpose. This makes it tragically easy to overlook or rationalize immense injustices. The profound statement that "there is no difference between the injustice done to one person and the injustice done to all people" stands in stark contrast to the dominant postmodern calculus, where the mass deaths, such as the five hundred thousand children who perished after the United States' intervention in Iraq, are casually reclassified as mere "by-products of the price that had to be paid." This chilling transformation of human lives into abstract statistics or regrettable externalities is a quintessential manifestation of morality's diminished meaning, reduced to a mere footnote in the pursuit of strategic, self-serving objectives. The imperative to look away, to remain unburdened by the suffering of the 'Other', becomes a deeply ingrained social norm, allowing horrific realities to persist unaddressed.The moral responsibility that should be felt for the welfare and dignity of the other has evolved into a logic that is incompatible with morality, such as "what does it have to do with me?" In the struggle that an individual gives for his own interests, the pain and damage that befalls the other has mostly been evaluated as collateral damage. This situation is not limited to individual struggles. The costs of this situation can be heavier in international relations. For example, the deaths of a hundred thousand Iraqis, although they have nothing to do with the war, have been trivialized as collateral damage. However, collateral damage is actually known in advance as planned and foreseen damage, but it is not considered valuable enough to prevent the achievement of the goal. What the soldier does is to control events from a distance, that is, to fulfill the duty professionally. The appropriateness of performance is measured by excluding moral references. The only ethical principle that is valid in the fulfillment of professional performance is whether the hierarchical command line is strictly followed (Bauman, 2018, p. 148). In war-related evaluations, the success of the war is determined by whether military casualties are suffered along with compliance with the command hierarchy. The health of the soldiers in your ranks is much more important than the number of enemy deaths.

The deaths of half a million children in the Iraq war do not indicate a value because they are seen as secondary/collateral casualties (Bauman, 2018, pp. 150–151). Political spokespeople and public relations experts express such situations in military assessments as inevitable side effects of war targeting enemy forces. The West's high-tech professional armies do not need living people. The West, which mentally directs the world as it wishes, no longer needs territorial conquest and annexation. What is aimed at is short-term, highly effective globalizing coups that can provide remote control and domination. In fact, the wars of the modern era stem from the fact that local powers do not open their doors to the free market and do not offer their resources to the West for exploitation (Bauman, 2019, p. 137). The elimination of all situations that will prevent the freedom to make global profits is the most important desire of modern morality. The West does not need to kill many people in order to maintain its hegemony in the lands it has acquired. It does not matter whether the bombs to be dropped are residential areas or military bases. The death of civilians, whether intentionally or accidentally, is not a situation to be regretted by the West. The value of others does not correspond to anything in the context of humanity. The immorality of a civilization that claims to be a moral entity in this sense is an example of cognitive dissonance (Bauman, 2019, p. 183).

POSTMODERN MASSACRES AND GLOBALIZATION

It is necessary to see that there are state structures that support the economic and cultural area controlled by the minority class. Although it is claimed that global commercial powers do not have a nation, in reality these powers are in close contact with some states. These states will get their share by causing the poor to fight the poor and by mediating the rich to become richer when necessary (Bauman, 2016, p. 112). The seemingly abstract forces of global commercial powers are, as you rightly identify, far from detached entities without national allegiance. On the contrary, it's crucial to recognize the enduring presence of state structures that actively support the economic and cultural spheres controlled by a minority elite class. While it's often claimed that these global commercial giants transcend national borders, the reality is that they maintain intimate, symbiotic relationships with specific states. These states, rather than being neutral arbiters, become active participants in a global system designed to accumulate wealth for the few, often at the expense of the many.

This complex alliance allows these states to secure their "share" by orchestrating conflicts where "the poor fight the poor," effectively turning human suffering into a profitable enterprise. They also serve as crucial intermediaries, facilitating the flow of capital and resources that enable "the rich to become richer." In Zygmunt Bauman's "liquid modern" framework, this dynamic reflects the profound shift in sovereignty and power. While traditional states might appear to retain their territorial authority, their actual power is increasingly diffused and intertwined with the interests of global capital. The state, in this context, is not merely a passive bystander to economic globalization; it becomes a facilitator and enabler of its most morally questionable outcomes. It provides the necessary legal frameworks, military protection for strategic resources, and diplomatic leverage to ensure that the global commercial machine runs smoothly, often lubricating it with the blood of the marginalized. Thus, the "loss of meaning of morality" in postmodern wars is not just about individual desensitization or technological distance; it's also about how state apparatuses actively co-opt and legitimize these processes, transforming ethical considerations into mere calculations of profit and geopolitical advantage. The suffering of the "Other," particularly the economically disenfranchised, becomes a justifiable "cost of doing business" within this ruthless global matrix. Global economic powers will want it; intermediary states will implement the policy of divide, fragment and rule, and eventually the task of removing the debris will be left to the regional people. In Bauman's words, the global system will constantly need regional turmoil so that it does not fear anything. Global powers are the only power that determines the rules of the game in the world. When the rules of the game are not played in the determined way, these powers can apply various sanctions on states (Bauman, 2017, p. 267). These sanctions are mostly economic. States that implement their own policies outside the global are seen as criminals on the grounds that they bypass global financial markets and global free trade. These states are not given credit; their debts are not postponed, in other words, they are subject to economic sanctions in some way. Moreover, global powers can collapse the currency values of countries that they treat as lepers with various credit ratings they determine. Global powers/ multinationals that have an influence all over the world and somehow manage the world are trying to organize the ideal world in a way where states do not exist or cannot be powerful even if they do. Sometimes, economic sanctions alone are not enough. In addition to economic sanctions, military sanctions are also applied to teach a lesson to those who may show potential for resistance in the future. In other words, if the economic sanctions of global powers and the show of force they display in this way are not enough to make countries come to their senses, in other words, if the rules determined in the new world order are not followed, what needs to be done is to put military forces into action (Bauman, 2017, p. 268). Military forces arc coming into play in a different way than traditional war strategies. Previously, the face-to-face warfare has been replaced by an aircraft/jet show. In fact, there is no wrong in using various gases without being accountable to anyone in terms of fluid moral values. Or, more devilishly, it seems more logical to create conflict environments between various ethnic groups, structures and communities within countries and manage and monitor the events from a distance.

The inconsistent and directionless world has become a place where the strong can destroy the weak with an excuse. In this world, which we can describe as the third world, rich countries can declare war on poor countries for any reason to eliminate their concerns and cannot explain this with any rational reason. Bauman explains the new world disorder with the concept of secondary barbarism (Bauman, 2013, p. 38). The inconsistent and directionless world you describe has indeed devolved into a perilous arena where the strong can relentlessly destroy the weak under the flimsiest of pretexts. This chaotic landscape, often characterized as a "third world" (though more accurately reflecting a global condition of instability rather than a specific geographical region), allows affluent nations to declare war on poorer countries for any perceived slight or manufactured concern. Crucially, these acts of aggression are often undertaken without any coherent or rationally defensible explanation, highlighting a profound deficit in moral and logical accountability. This very condition of global instability and arbitrary power projection is precisely what Zygmunt Bauman meticulously unpacks through his concept of secondary barbarism. Unlike primary barbarism, which he might associate with a lack of civilization or primitive urges, secondary barbarism arises within or from the very heart of advanced, rationalized, and technologically sophisticated societies. It is a barbarism that does not reject reason but rather co-opts and distorts it, using its tools—efficiency, calculation, and detachment—to achieve morally reprehensible ends. In this context, the rich declare war on the poor not out of a raw, emotional hatred, but out of a cold, strategic calculation of self-interest, resource acquisition, or the elimination of perceived threats (which can be as abstract as "eliminating concerns"). The "inconsistent and directionless" nature of this world is, paradoxically, a product of a hyper-rationalized pursuit of individual and national gain, where universal moral principles are abandoned in favor of fluid, self-serving justifications. The weak become expendable "material," as previously discussed, and their destruction is not an regrettable accident but an integral, often unacknowledged, part of a global order where moral meaning has been comprehensively stripped away in the pursuit of power and profit.

Some states that cooperate with global trade conditions and create supply-demand imbalances in the world legitimize all kinds of ways to strengthen their economic power with their own justifications. These states conduct reconnaissance wars before occupying a place. In reconnaissance wars, states make preliminary preparations to determine the determination and durability of the enemy state; the resources and military power in the hands of the enemy (mostly with the help of border states) before seizing enemy lands (Bauman, 2018, p. 137). Reconnaissance units identify the strengths and weaknesses of the enemy and transmit all necessary information to the authorities before the main operation. Especially in asymmetric wars, the new form of military operations almost eliminates the possibility of confronting the enemy. In new tactics aimed at striking and killing from a distance, the characteristics of the target are given an inhumane quality and the military action is adiabatic, leaving the human aspect of the war out of consideration (Bauman, 2018, p. 148). In other words, the ethical evaluations and moral qualities of military operations are removed from war.

CONCLUSION

The education system of the rigid modern era viewed educators as the launchers of ballistic missiles and taught them how to keep the bullets stationary in the air. The factor that led to the extreme success of ballistic missiles in the metaphorical context of the education system in the early modern era was that the path to be followed and the goal to be reached were always rigid and fixed. The loss of traditional value of ancient educational wisdom, in other words, the transition from ballistic missiles to smart missiles, occurred in the liquid modern era (Bauman, 2020, p. 24). In this period, institutional loyalties, emotional ties with the workplace, and focusing on more than necessary areas of work, in other words, lifelong business relationships, have become illogical. Commercial situations, product designs, the understanding of competition, the functioning of capital and everything related to knowledge in general have changed and become short-lived, so to speak. Especially in postmodern wars, smart missiles that have replaced ballistic missiles have eliminated moral sensitivities and turned people into adiaphores. What ultimately happened is that moral sensitivities disappeared in wars and oppressions did not create any sensitivities. Modern civilization has been able to show humanity how much genocide it can do with its reasonable arrangements, forms, practices, solidarities and scientific nature. Modern civilization, which has taken technology and scientific reason to its own side, has been able to use the monopoly morality, which can be absolutely right, in accordance with its own purposes by creating other components that serve its purpose. Modern, technological, rational, bureaucratic forms of action and their legitimization have been able to reproduce generally accepted forms of thought.

"Western civilization articulated its struggle for hegemony in terms of the struggle of humanity against barbarity, of reason against ignorance, of objectivity against prejudice, of progress against corruption, of truth against superstition, of science against magic, of rationality against excitement. It described the history of its own rise as the gradual but steady replacement of the domination of man over nature by the domination of nature over man. It presented its own achievement as a definite advance in man's freedom of action, his creative potential, and his security. It defined freedom and security as its own mode of social order: modern society, the Western type, was defined as civilized society, and a civilized society was understood

as a state in which most of the natural ugliness and corruption had been largely eliminated or at least suppressed" (Bauman, 2016, p. 149).

According to this logic, the image of the civilized Western society was a delicate, gentlemanly, well-mannered and gentlemanly society where violence had disappeared. For Bauman, the discourse of non-violence in modern civilization is completely illusory; it is an indicator of self-deception and the tendency to become God. Civilization has not eliminated violence and the vile and immoral nature that feeds violence, on the contrary, it has made violence bureaucratic, scientific, rational and centralized. When the means of oppression became centralized, it became immensely powerful by getting rid of the structures that had previously controlled it. Ultimately, since bureaucratic violence, unlike individual violence, is abstract, the morality of the actions taken has been removed from the patterns of value judgment.

REFERENCES

Bauman, Z. (2012). Akışkan Aşk İnsan İlişkilerinin Kırılganlığına Dair, (Çev. Işık Ergüden). Versus Kitap, İstanbul.

Bauman, Z. (2013). Postmodernizm ve Hoşnutsuzlukları. (Çev. İsmail Türkmen). Ayrıntı Yayınları. İstanbul.

Bauman, Z. (2014). Bu Bir Günlük Değildir. (Çev: Didem Kizen). Jaguar Kitap, İstanbul.

Bauman, Z. (2014). Modernlik ve Müphemlik. (Çev: İsmail Türkmen). Ayrıntı Yayınları. İstanbul.

Bauman, Z. (2016). Cemaatler Güvenli Olmayan Bir Dünyada Güvenlik Arayışı. (Çev.: Nurdan Soysal). Say Yayınları, İstanbul.

Bauman, Z. (2016). Modernite ve Holokaust. (Çev: Süha Sertabipoğlu). Alfa Basım Yayım Dağıtım. İstanbul.

Bauman, Z. (2017). Akışkan Modernite. (Çev: Sinan Okan Çavuş). Can Sanat Yayınları, İstanbul.

Bauman, Z. (2017). *Akışkan Modern Dünyadan 44 Mektup* (Çev: Pelin Siral). Habitus Yayıncılık, İstanbul.

Bauman, Z. (2018). Kuşatılmış Toplum. (Çev.: Akın Emre Pilgir). Ayrıntı Yayınları, İstanbul.

Bauman, Z. (2018). *Retrotopya. (Çev.: Ali Karatay).* Sel Yayıncılık.

Bauman, Z. (2019). *Borçlu Zamanlarda Yaşamak Citlali Rovirosa-Madrazo ile Söyleşi.* (Çev: Akın Emre Pilgir). Ayrıntı Yayınları, İstanbul.

Bauman, Z. (2020). *Akışkan Korku* (Çev. Cumhur Atay). Ayrıntı Yayınları, İstanbul.

Bauman, Z. (2020). Eğitim Üzerine Riccardo Mazzeo ile Söyleşi. (Çev.: Akın Emre Pilgir). Ayrıntı Yayınları, İstanbul.

Bauman, Z., & Lyon, D. (2016). Akışkan Gözetim. (Çev: Elçin Yılmaz). Ayrıntı Yayınları. İstanbul.

Bauman, Z., & Stanislaw, O. (2018). Dünyaya ve Kendimize Dair. (Çev. Burcu Halaç). Ayrıntı Yayınları, İstanbul.

Bauman, Zygmunt ve Donskis, Leodinas. (2020). *Ahlaki Körlük Akışkan Modernlikte Duyarlılığın Yitimi.* (Çev. Akın Emre Pilgir), Ayrıntı Yayınları, İstanbul.

Compilation of References

Açıkalın, Ş. N., & Baykız, T. (2024). Değişim ve dönüşümün eşiğinde siber savaş. *Muhafazakâr Düşünce Dergisi, 20*(67), 200–220.

Agamben, G. (1998). *Homo Sacer: Sovereign Power and Bare Life*. Stanford University Press., https://search.alexanderstreet.com/view/work/bibliographic_entity|bibliographic_details|4714514

Agee, J. (2000). Agee on film: criticism and comment on the movies. In *Agee on Film: Criticism and Comments on the Movies*. Modern Library., https://openlibrary.org/books/OL45445M/Agee_on_film

Aghamohseni, K. (2024). Forbidden Music in Iran: The Tension between Communism and Imperialism. *Interdisciplinary Researches of Art, 2*(1), 81–102. DOI: 10.22124/ira.2024.28037.1024

Agnew, J., & Corbridge, S. (1995). *Mastering space: Hegemony, territory and international political economy*. Routledge.

Ahmad, M., Mekala, H., Lone, J., Robinson, K., & Shah, K. (2020). Addressing mass shootings in a new light. *Cureus, 12*(7). Advance online publication. DOI: 10.7759/cureus.9298 PMID: 32832295

Akargün, S. E. (1993). *Harp Tarihi Ders Kitabı-1*. Kara Harp Okulu Matbaası.

Akarsu, B. (1998). *Bedia, Wilhelm von Humboldt'da Dil-Kültür Bağlantısı*. İnkılâp Yayınları.

Akhavan, P., Ashraph, S., Barzani, B., & Matyas, D. (2020). What justice for the Yazidi Genocide?: Voices from below. *Human Rights Quarterly, 42*(1), 1–47. DOI: 10.1353/hrq.2020.0000

Akın, A. (2019). Stratejik Analiz. https://www.stratejikanaliz.com/hibrit-savas/

Al Jazeera. (2025). Have India and Pakistan started a drone war? https://www.aljazeera.com/news/2025/5/8/have-india-and-pakistan-started-a-drone-war

Alford, M. (2016). The Political Impact of the Department of Defense on Hollywood Cinema. *Quarterly Review of Film and Video, 33*(4), 1–16. DOI: 10.1080/10509208.2015.1086614

Allcott, H., & Gentzkow, M. (2017). Social Media and Fake News in the 2016 Election. *The Journal of Economic Perspectives, 31*(2), 211–236. DOI: 10.1257/jep.31.2.211

Allen, L. (2004). Music and Politics in Africa. *Social Dynamics, 30*(2), 1–19. DOI: 10.1080/02533950408628682

Allison, S. T., & Goethals, G. R. (2011). *Heroes: What they do and why we need them.* Oxford University Press.

Al-Mahadin, S. (2018). The politics of the untranslated. *Journal of African Cultural Studies, 30*(3), 307–322. DOI: 10.1080/13696815.2018.1427557

Althusser, L. (1971). Ideology and Ideological State Apparatuses (Notes towards an Investigation) (B. Brewster, Trans.). In *Lenin and Philosophy and Other Essays* (pp. 127–186). Monthly Review Press.

Althusser, L. (1971). Ideology and ideological state apparatuses (Notes towards an investigation). In Althusser, L. (Ed.), *Lenin and Philosophy and Other Essays* (pp. 86–127). Monthly Review Press.

Ambodo, T., Saputro, G. E., & Komariah, U. N. (2021). Multi-Agent Drone as Loitering Munition System to Support Military Operations Carrying Out Policy Mandate. *International Journal of Arts and Social Science, 4*(5), 2581–7922.

American Historical Association. (n.d.). The story of propaganda. https://www.historians.org/resource/the-story-of-propaganda/

Amin, A. (2025). Sino-Indian rivalry in Indo-Pacific and Bangladesh's geo-strategic hedging in response to its security threats. *Discover Global Society, 3*(1), 1–20. DOI: 10.1007/s44282-025-00144-1

Anand, C. (Director). (1964). *Haqeeqat* [Film].

Anand, C. (Director). (1973). *Hindustan ki Kasam* [Film].

Anderegg, M. A. (1991). *Inventing Vietnam: The war in film and television.* Temple University Press.

Anderson, Benedict. (2006). Imagined communities: reflections on the origin and spread of nationalism.

Anderson, B. R. O. (2006). *Imagined communities: Reflections on the origin and spread of nationalism* (Rev. ed). Verso.

Angelskår, S. (2005). *Policing Fantasy: problems of genre in fantasy literature* (Master's thesis). http://urn.nb.no/URN:NBN:no-12249

Anisin, A. (2023). Do mass shootings in central and Eastern Europe differ from US mass shootings? Insights from the MSCEE Data Set, *Journal of Mass Violence Research*, pp.1-13.

Ansen, D. (1992). Wars Domestic and Military: Three striking independent films—Irish, American and Australian—enliven the scene. *Newsweek, 119*(19), 77. https://www.proquest.com/magazines/wars-domestic-military/docview/1879164987/se-2?accountid=14511

Antebi, L., & Adar, O. (2024). *FPV Drones: From the Ukrainian Battlefield to the Middle East*. Institute for National Security Studies.

Aravind, S. G. (2023). The gender of war: National masculinities and Hindutva in Bollywood war cinema. In *Historicizing Myths in Contemporary India* (pp. 81–101). Routledge India., https://www.taylorfrancis.com/chapters/edit/10.4324/9781003363149-5/gender-war-aravind DOI: 10.4324/9781003363149-5

Arendt, H. (1963). *Eichmann in Jerusalem: A report on the banality of evil*. Viking Press.

Armenta, A. (2017). *Protect, serve, and deport: The rise of policing as immigration enforcement*. University of California Press. DOI: 10.1525/luminos.33

Aroesti, R. (2014, February 28). *Barkhad Abdi: Captain Phillips star's rise from limo driver to Oscar's contender*. The Guardian. https://www.theguardian.com/film/2014/feb/28/barkhad-abdi-captain-phillips-oscar-contender

Aruz, J., Graff, S. B., & Rakic, Y. (2014). *Assyria to Iberia at the dawn of the Classical Age*. Metropolitan Museum of Art.

Assmann, A. (2006). History, memory, and the genre of testimony. *Poetics Today, 27*(2), 261–273. DOI: 10.1215/03335372-2005-003

Atay, T. (2021). Din Hayattan Çıkar. Ankara: İletişim Yayınları. 6th publication.

Athique, A. (2012). Aggression and Transgression on the India-Pakistan Border. In *Filming the Line of Control* (pp. 21–39). Routledge India., https://api.taylorfrancis.com/content/chapters/edit/download?identifierName=doi&identifierValue=10.4324/9780203150351-3&type=chapterpdf

Audergon, A. (2004). Collective trauma: The nightmare of history. *Psychotherapy and Politics International*, *2*(1), 16–31. DOI: 10.1002/ppi.67

Augé, M. (1995). *Non-places: Introduction to an anthropology of supermodernity*. Verso.

Auslander, H. B. (1981). "If Ya Wanna End War and Stuff, You Gotta Sing Loud"—A Survey of Vietnam-Related Protest Music. *Journal of American Culture*, *4*(2), 108–113. DOI: 10.1111/j.1542-734X.1981.0402_108.x

Avieson, B., Giles, F., & Joseph, S. (Eds.). (2017). *Mediating memory: Tracing the limits of memoir*. Routledge., DOI: 10.4324/9781315107349

Awad, S. H., Wagoner, B., & Glaveanu, V. (2017). *The street art of resistance*. Springer Singapore., DOI: 10.1007/978-3-319-63330-5

Awasthy, V. (2025). Financial imperialism and British colonial power in India: An interwar history of economic strategies and fallouts. In Mandala, V. R. (Ed.), *Exploring power and authority in Indian history across the ages* (pp. 217–243). Springer., DOI: 10.1007/978-981-96-1679-4_12

Aydın, M. (2011). *Güncel kültürde temel kavramlar*. Açılımkitap.

Babovic´, M., Katarina G., & Vukovic, O.(2010). Mapiranje porodicˇnog nasilja prema zˇenama u Centralnoj Srbiji [Mapping of Domestic Violence against Women in Central Serbia]. Ministry of Labor and Social Policy: Project against Sexual and Gender Based Violence. Belgrade, Serbia: SeConS

Bailey-Fakhoury, C. (2017). Learning through collective testimony: African American motherwork, womanism, and praxis. *Journal of the Motherhood Initiative for Research and Community Involvement.*

Baker, K., & Raney, A. A. (2007). Equally super?: Gender-role stereotyping of superHeroes in children's animated programs. *Mass Communication & Society*, *10*(1), 25–41. DOI: 10.1080/15205430709337003

Balali-Mood, M., Moshiri, M., & Etemad, L. (2013). Medical aspects of bio-terrorism. *Toxicon*, *69*, 131–142. DOI: 10.1016/j.toxicon.2013.01.005 PMID: 23339855

Bamberg, M. (2008). Considering counter narratives. In M. Bamberg, M. Andrews & J. Benjamins (Eds.), *Considering counter-narratives: Narrating, resisting, making sense* (pp. 351-371). John Benjamins Publishing Company. https://doi.org/DOI: 10.1075/sin.4.43bam

Bandura. (1977). *Social Learning Theory,* New York: Englewood Cliffs.

Bandura, A. (1999). Moral disengagement in the perpetration of inhumanities. *Personality and Social Psychology Review, 3*(3), 193–209. DOI: 10.1207/s15327957pspr0303_3 PMID: 15661671

Banerjee, S. (2006). Armed masculinity, Hindu nationalism and female political... - Google Scholar. *International Feminist Journal of Politics, 8*(1), 62–83. DOI: 10.1080/14616740500415482

Banga, B., & Kapoor, V. (2020). Representation of Political Ideology through Bollywood Movies: A Critical Study of the Movie "Uri-The Surgical Strike.". *Journal of Xi'an University of Architecture & Technology, XII*(III), 2899–2905.

Barakat, S. (2007). Postwar reconstruction and the recovery of cultural heritage: critical lessons from the last fifteen years. *Cultural heritage in postwar recovery,* 2(1), 26-37.

Baron, J. (2006). The archive effect: Found footage and the audiovisual experience of history. *Film Quarterly, 60*(2), 48–57. DOI: 10.1525/fq.2006.60.2.48

Bar-Tal, D. (1990). Causes and consequences of delegitimization: Models of conflict and ethnocentrism. *The Journal of Social Issues, 46*(1), 65–81. DOI: 10.1111/j.1540-4560.1990.tb00272.x

Bartels, D. M. (2008). Principled moral sentiment and the flexibility of moral judgment and decision making. *Cognition, 108*(2), 381–417. DOI: 10.1016/j.cognition.2008.03.001

Bartky, S. L. (2015). *Femininity and domination: Studies in the phenomenology of oppression*. Routledge., DOI: 10.4324/9780203825259

Basinger, J. (2003). *The World War II Combat Film: Anatomy of a Genre*. Wesleyan University Press., https://www.weslpress.org/9780819566232/the-world-war-ii-combat-film/

Bates, M. J. (1998). *The Wars We Took to Vietnam: Cultural Conflict and Storytelling*. University of California Press., DOI: 10.1525/9780520917521

Batson, C. D., Kennedy, C. L., Nord, L. A., Stocks, E. L., Fleming, D. A., Marzette, C. M., Lishner, D. A., Hayes, R. E., Kolchinsky, L. M., & Zerger, T. (2007). Anger at unfairness: Is it moral outrage? *European Journal of Social Psychology, 37*(1), 1272–1285. DOI: 10.1002/ejsp.434

Baudrillard, J. (2006). The Gulf War Did Not Take Place. In *Hollywood and War, The Film Reader* (1st ed.). Routledge., DOI: 10.4324/9781003416661-30

Bauman, Z. (2023). Modernity and the Holocaust. In W. Longhofer & D. Winchester (Eds.), *Social Theory Re-Wired* (pp. 300-318). Routledge. https://doi.org/DOI: 10.4324/9781003320609-38

BBC News Turkish. (2022). Donald Trump'ın Twitter hesabı yeniden açıldı, https://www.bbc.com/turkce/articles/c0jg0vqk5p7o

BBC News. (2017). Putin's 'chef', the man behind the troll factory. https://www.bbc.com/news/world-europe-39419560

BBC News. (2025). India and Pakistan drone conflict: What we know so far. https://www.bbc.com/news/articles/cwy6w6507wqo

BBC. (2023). *Srbija i protesti: Desetine hiljada ljudi u šetnji i blokade mostova, opozicija najavila novi skup za 27. maj*. BBC. Retrieved from https://www.bbc.com/serbian/lat/srbija-65648580

Beard, M. (2013). Cyberwar and just war theory. Applied ethics, risk, justice and liberty, 1-12.

Benavides-Vanegas, F. S. (2005). From Santander to Camilo and Ché: Graffiti and Resistance in Contemporary Colombia. *Social Justice (San Francisco, Calif.), 32*(1 (99)), 53–61. https://www.jstor.org/stable/29768290

Benton, L. (2010). *A search for sovereignty: Law and geography in European empires, 1400–1900*. Cambridge University Press.

Benton, L. (2014). *Law and colonial cultures: Legal regimes in world history, 1400–1900*. Cambridge University Press.

Berman, L. E. A. (1978). Sibling Loss as an Organizer of Unconscious Guilt: A Case Study. *The Psychoanalytic Quarterly, 47*(4), 568–587. DOI: 10.1080/21674086.1978.11926858 PMID: 715121

Bernard, N. (2020) 'Corona' günlerinde her 10 kişiden altısı yalan haber paylaşıyor. https://www.diken.com.tr/podcast-corona-gunlerinde-her-10-kisiden-altisi-yalan-haber-paylasiyor/

Bernard, A. (2012). *Gods and mortals: Ancient myths and modern life*. Profile Books.

Bernard-Donals, M. (2001). Beyond the question of authenticity: Witness and testimony in the Fragments controversy. *PMLA, 116*(5), 1302–1315. DOI: 10.1632/pmla.2001.116.5.1302

Beta (2023). *Naslovne strane za četvrtak, 4. maj 2023. Godine*. Krstarica. Retrieved from https://www.krstarica.com/vesti/srbija/naslovne-strane-za-cetvrtak-4-maj-2023-godine/

Bevan, R. (2007). *The destruction of memory: Architecture at war*. Reaktion Books.

Beverley, J. (2004). *Testimonio: On the politics of truth*. University of Minnesota Press.

Bharucha, R. (1999). *In the name of the secular: Contemporary cultural activism in India*. Oxford University Press., https://cir.nii.ac.jp/crid/1130282272976718720

Bhutia, U. (2017). *The Role of Media in the Emergence of Hindu Nationalism in Indian Politics* [Sikkim University]. https://dspace.cus.ac.in/jspui/bitstream/1/5985/1/Ugen%20Bhutia-History-PhD.pdf

Bigelow, K. (Director). (2008). The hurt locker [Film]. Summit Entertainment.

Bigelow, K. (Director). (2012). Zero dark thirty [Film]. Columbia Pictures.

Bingöl, O. (2017). Hybrid war and its strategic implications to Turkey. *Gazi Akademik Bakış, 11*(21), 107–132. DOI: 10.19060/gav.379591

Blackmore, H. (2009). *Pirates and emperors, old and new: International terrorism in the real world*. University of California Press.

Blackstone, A. M. (2003). *Gender roles and society*. ABC-CLIO.

Blasco, R., Peris, J. F., & Rosell, J. (2010). Several different strategies for obtaining animal resources in the late Middle Pleistocene: The case of level XII at Bolomor Cave (Valencia, Spain). *Comptes Rendus. Palévol, 9*(4), 171–184. DOI: 10.1016/j.crpv.2010.05.004

Bloom, P. B. N., Kimhi, S., Fachter, S., Shamai, M., & Canetti, D. (2020). Coping with moral threat: Moral judgment amid war on terror. *The Journal of Conflict Resolution, 64*(2-3), 231–260. DOI: 10.1177/0022002719854209

Boggs, C., & Pollard, T. (2006). Hollywood and the Spectacle of Terrorism. *New Political Science, 28*(3), 335–351. DOI: 10.1080/07393140600856151

Bond, S. (2024, September 3). China is pushing divisive messages using fake U.S. voters, NPR. https://www.npr.org/2024/09/03/nx-s1-5096151/china-tiktok-x-fake-voters-influence-campaign

Borins, S. (2011). Online Adaydan Online Başkana. In Devran, Y. (Ed.), *Seçim Kampanyalarında Geleneksel Medya.* İnternet ve Sosyal Medyanın Kullanımı, Başlık Yayın Grubu.

Bose, S., & Ray Chaudhury, A. B. (2023). Crossing currents: India-China contention and the Teesta conundrum in Bangladesh. *Observer Research Foundation.* https://www.orfonline.org/expert-speak/crossing-currents-india-china-contention-and-the-teesta-conundrum-in-bangladesh

Botstein, L. (2022). Music in Time of War. *The Musical Quarterly, 105*(3-4), 235–244. DOI: 10.1093/musqtl/gdac011

Bourazi, A. H. N. (2024). *Trauma through the lens of attachment and resilience.*

Boyce, M. (2001). *Zoroastrians: Their Religious Beliefs and Practices.* Psychology Press.

Boyer, Pascal. (2001). Religion Explained: The Evolutionary Origins of Religious Thought.

Brabeck, K. (2003). IV. Testimonio: A strategy for collective resistance, cultural survival and building solidarity. *Feminism & Psychology, 13*(2), 252–258. DOI: 10.1177/0959353503013002009

Brancaleone, D. (2021). Framing the real. *AMPS Proceedings, 21*, 1–16.

Brennan, H. (2001). *The story of Irish dance.* Roberts Rinehart.

Breuilly, J. (1985). Reflections on Nationalism. *Philosophy of the Social Sciences, 15*(1), 65–75. DOI: 10.1177/004839318501500106

Brincat, S. (2023). Introduction: Che Guevara and world politics. *Globalizations, 20*(8), 1426–1446. DOI: 10.1080/14747731.2023.2274649

Brockington, J. L. (1998). *The Sanskrit Epics.* BRILL. DOI: 10.1163/9789004492677

Bronfen, E. (2012). *Specters of war : Hollywood's engagement with military conflict.* Rutgers University Press., https://www.jstor.org/stable/j.ctt5hj3rx

Brownmiller, S. (2005). *Against Our will: Men, women and rape.* Fawcett.

Buchanan, J., & Buchanan, E. (2013). *Impossible odds: The kidnapping of Jessica Buchanan and her dramatic rescue by SEAL Team Six.* Atria Books.

Buelens, G. (2006). Reciting Shells. Dada *and* Dada *in* & Dadaists *on* the First World War. *arcadia*, 41(2), 275-295. https://doi.org/DOI: 10.1515/ARCA.2006.019

Buffetrille, K. (2020). The Pilgrimage to Mount Kawakarpo: A Metaphor for Bardo?. , (67), 99-122.

Bulut, M., & Sezener, A. (2019). Dumuzi-Enkimdu'dan, Kabil-Habil'e Yerleşik Hayatın Serüveni. *Turkish Studies*, *14*(6), 3071–3081. DOI: 10.29228/Turkish-Studies.23495

Burch, P. W. (2022). Receptivity, Simultaneity: The Thin Red Line as Ecological Cinematic Poesis. *Film-Philosophy*, *26*(2), 242–266. DOI: 10.3366/film.2022.0197

Burden, M. C. (2006). *The blog of war: Front-line dispatches from soldiers in Iraq and Afghanistan*. Simon & Schuster.

Burlando, A., Cristea, A., & Lee, S. (2015). The trade consequences of maritime insecurity: Evidence from Somali piracy. *Review of International Economics*, *23*(3), 525–557. DOI: 10.1111/roie.12183

Burton, A. L., Pickett, J. T., Jonson, C. L., Cullen, F. T., & Burton, V. S.Jr. (2021). Public Support for Policies to Reduce School Shootings: A Moral-Altruistic Model. *Journal of Research in Crime and Delinquency*, *58*(3), 269–305. DOI: 10.1177/0022427820953202

Bushman, B. J. (2018). Narcissism, fame seeking, and mass shootings. *The American Behavioral Scientist*, *62*(2), 229–241. DOI: 10.1177/0002764217739660

Butler, J. (2004). *Precarious life: The powers of mourning and violence*. Verso.

Buzan, B., & Hansen, L. (2009). *The Evolution of International Security Studies*. Cambridge University Press. DOI: 10.1017/CBO9780511817762

Çakı, C. (2018). *Propaganda. M. Karaca ve C. Çakı içinde*. İletişim ve Propaganda, Eğitim Yayınevi.

Calcara, A., Gilli, A., Gilli, M., Marchetti, R., & Zaccagnini, I. (2022). Why drones have not revolutionized war: The enduring hider-finder competition in air warfare. *International Security*, *46*(4), 130–171. DOI: 10.1162/isec_a_00431

Caldwell, C. (1998, October). Spielberg at War. *Commentary (New York, N.Y.)*, *106*(4), 48–51. https://link-gale-com.libproxy.ucl.ac.uk/apps/doc/A21191193/AONE?u=ucl_ttda&sid=bookmark-AONE&xid=db2fdd6b

Çalışkan, A. (2023). Siber savaş: Bilgi krizi mi yoksa güvenliği mi? *SAVSAD Savunma ve Savaş Araştırmaları Dergisi*, *33*(1), 1–32. DOI: 10.54078/savsad.1188851

Callamari, P., & Reveron, D. (2003). Chine's Use of Perception Management. *International Journal of Intelligence and CounterIntelligence*, *16*(1), 1–15. DOI: 10.1080/713830380

Campbell, J. (1968). *The Hero with a thousand faces*. Princeton University Press.

Campbell, J. (2008). *The Hero with a Thousand Faces* (3rd ed.). New World Library., https://newworldlibrary.com/product/the-hero-with-a-thousand-faces

Carel, H., & Kidd, I. J. (2023). Institutional opacity, epistemic vulnerability, and institutional testimonial justice. In Altanian, M., & Baghramian, M. (Eds.), *Testimonial Injustice and Trust* (pp. 273–296). Routledge., DOI: 10.4324/9781003396789-20

Carey, J. (1989). A Cultural Approach to Communication. In *Communication as Culture: Essays on Media and Society* (pp. 13–36). Unwin Hyman.

Caruth, C. (1996). *Unclaimed experience: Trauma, narrative, and history*. Johns Hopkins University Press. DOI: 10.1353/book.20656

Caruth, C. (Ed.). (1995). *Trauma: Explorations in memory*. Johns Hopkins University Press. DOI: 10.56021/9781421413525

Castells, M. (2020). İnternet Galaksisi: İnternet, İş Dünyası ve Toplum Üzerine Düşünceler, (T. Asrak Hasdemir, E. Öztürk, H. Çakmak, A. Yıldız, S. K. Özdemir, B. T. Yıldırım, M. D. Aydın, N. Yıldız, A. Aydın, A. A. Yıldız, & A. E. Yıldız, Trans.). Phoenix Yayınevi.

Castrillo, V. U., Manco, A., Pascarella, D., & Gigante, G. (2022). A Review of Counter-UAS Technologies for Cooperative Defensive Teams of Drones. *Drones (Basel)*, *6*(3), 65. DOI: 10.3390/drones6030065

Cetorelli, V., Sasson, I., Shabila, N., & Burnham, G. (2017). Mortality and kidnapping estimates for the Yazidi population in the area of Mount Sinjar, Iraq, in August 2014: A retrospective household survey. *PLoS Medicine*, *14*(5), e1002297. DOI: 10.1371/journal.pmed.1002297

Chatterjee, P. (1989). Colonialism, nationalism, and colonialized women: The contest in India. *American Ethnologist*, *16*(4), 622–633. DOI: 10.1525/ae.1989.16.4.02a00020

Chatterjee, P. (1996). Whose Imagined Community? In Balakrishnan, G. (Ed.), *Mapping Nation* (pp. 214–225). Verso., https://journals.sagepub.com/doi/10.1177/030582989102000030601

Chatterjee, P. (2012). Nationalism Today. *Rethinking Marxism*, *24*(1), 9–25. DOI: 10.1080/08935696.2012.635023

Chaudhari, D. D., & Pawar, A. V. (2021). Propaganda analysis in social media: A bibliometric review. *Information Discovery and Delivery, 49*(1), 57–70. DOI: 10.1108/IDD-06-2020-0065

Chen, P., Mao, C., & Wu, G. (2016). *Shielding effectiveness assessment of enclosure by means of EMP excitation.* In *2016 Asia-Pacific International Symposium on Electromagnetic Compatibility (APEMC)* (Vol. 1, pp. 273–275). IEEE. DOI: 10.1109/APEMC.2016.7523032

Cheterian, V. (2021). ISIS genocide against the Yazidis and mass violence in the Middle East. *British Journal of Middle Eastern Studies, 48*(4), 629–641. DOI: 10.1080/13530194.2019.1683718

Chion, M. (1994). *Audio-vision: Sound on screen* (Gorbman, C., Trans.). Columbia University Press.

Chohan, U. W. (2024, July 15). Modern economic warfare and its implications. *SSRN.* https://ssrn.com/abstract=4894995

Choi, S., Kwon, O.-J., Oh, H., & Shin, D. (2020). Method for Effectiveness Assessment of Electronic Warfare Systems in Cyberspace. *Symmetry, 12*(12), 2107. DOI: 10.3390/sym12122107

Chomsky, N. (1957). *Syntactic structures.* Mouton. DOI: 10.1515/9783112316009

Chopra, A. (2024). Iran 'news' sites, hackers target Trump ahead of US election. AL-MONITOR. https://www.al-monitor.com/originals/2024/09/iran-news-sites-hackers-target-trump-ahead-us-election

Chopra, Y. (Director). (2004). *Veer Zara* [Film].

Chouliaraki, L. (2013). *The ironic spectator: Solidarity in the age of post-humanitarianism.* John Wiley & Sons.

Chouliaraki, L. (2016). Authoring the self: Media, voice and testimony in soldiers' memoirs. *Media, War & Conflict, 9*(1), 58–75. DOI: 10.1177/1750635216636509

Christie, N. (1986). The ideal victim. In Palgrave Macmillan UK eBooks (pp. 17–30). https://doi.org/DOI: 10.1007/978-1-349-08305-3_2

Christodoulou, E. (2024). The development of UNESCO's programmes for preventing violent extremism: Educational norms, institutional politics and declining legitimacy. *Compare: A Journal of Comparative Education, 54*(5), 750–769. DOI: 10.1080/03057925.2023.2252319

Çığ, M. İ. (1997). Kur'an İncil ve Tevrat'ın Sümer'deki Kökeni. İstanbul: Kaynak Yayınları. 9. Basım.

Cinema. (2013, May 9). *The Herald Scotland.* https://www.heraldscotland.com/life_style/arts_ents/13103812.film-consultant-real-life-experience-piracy-high-seas/

Clarke, J. (2025, May 24). What tariffs has Trump announced and why? *BBC News.* https://www.bbc.com/news/articles/cn93e12rypgo

Clarke, R. A., & Knake, R. K. (2010). Cyber war: The next threat to national security and what to do about it.

Clarke, R. A., & Knake, R. K. (2014). Cyber war. Old Saybrook: Tantor Media, Incorporated.

Clasen, M. (2017). *Why horror seduces.* Oxford University Press. DOI: 10.1093/oso/9780190666507.001.0001

Clausewitz, C. V. (2011). *Savaş Üzerine* (Koçak, S., Trans.). Doruk Yayıncılık.

Claverie, B., & Du Cluzel, F. (2022). Cognitive warfare: The advent of the concept of 'cognitics' in the field of warfare. In B. Claverie, B. Prébot, N. Buchler, & F. Du Cluzel (Eds.), *Cognitive warfare: The future of cognitive dominance* (pp. 1–7). NATO Collaboration Support Office. https://hal.science/hal-03635889

Clifford, R. (2023). Good grief: changing attitudes to childhood grief in children's literature.

Clifford, J., & Said, E. W. (1980). Orientalism. *History and Theory*, *19*(2), 204–223. DOI: 10.2307/2504800

Cohen, E. A. (1999). The Thin Red Line: Let it Slip Away. *SAIS Review (Paul H. Nitze School of Advanced International Studies)*, *19*(2), 237–241. DOI: 10.1353/sais.1999.0031

Cohen, J. (2001). Questions of credibility: Omissions, discrepancies and errors of recall in the testimony of asylum seekers. *International Journal of Refugee Law*, *13*(3), 293–303. DOI: 10.1093/ijrl/13.3.293

Cole, R. (1998). *The Encyclopedia of Propaganda.* M.E. Sharpe.

Connolly, S. (2018). *The spatial cinema* (Doctoral dissertation, University of Kent).

Cook, A. (2006). *Individual vs. systemic justice: using trust and moral outrage to predict reactions to vigilante murder* (Doctoral dissertation, University of Missouri-Columbia).

Coombs, W. T. (2017). *Revising situational crisis communication theory: The influences of social media on crisis communication theory and practice, Social media and crisis communication*. Routledge. DOI: 10.4324/9781315749068-3

Coombs, W. T., Holladay, S. J., & White, R. (2020). Corporate crises: Sticky crises and corporations. In *Advancing crisis communication effectiveness* (pp. 35–51). Routledge. DOI: 10.4324/9780429330650-5

Coombs, W. T., & Tachkova, E. R. (2023a). Integrating Moral Outrage in Situational Crisis Communication Theory: A Triadic Appraisal Model for Crises. *Management Communication Quarterly, 37*(4), 798–820. DOI: 10.1177/08933189221151177

Coombs, W. T., & Tachkova, E. R. (2023b). How Emotions Can Enhance Crisis Communication: Theorizing Around Moral Outrage. *Journal of Public Relations Research, 36*(1), 1–17.

Coxall, M. (2013). *Human Manipulation: A Handbook*. Cornelio Books.

Cozma, R. (2014). Were the Murrow Boys Warmongers? The relationship between sourcing, framing, and propaganda in war journalism. *Journalism Studies, 16*(3), 433–448. DOI: 10.1080/1461670X.2014.882098

Crabtree, M. L. (2007). Forbid Not the Little Ones: The Spirituality of Children in the Celtic Christian Tradition. *Christian Education Journal, 4*(2), 238–251. DOI: 10.1177/073989130700400204

Craze, J. (2016). Mercenarism and state formation: The rise and fall of a private military market in the Horn of Africa. *African Affairs, 115*(461), 645–667.

Cronin, M., & Adair, D. (2004). *The wearing of the green: a history of St Patrick's Day*. Routledge. DOI: 10.4324/9780203007143

Crowe, T. K., Sánchez, V., Howard, A., Western, B., & Barger, S. (2017). Veterans transitioning from isolation to integration: A look at veteran/service dog partnerships. *Disability and Rehabilitation, 40*(24), 2953–2961. DOI: 10.1080/09638288.2017.1363301 PMID: 28805082

Cubilié, A., & Good, C. (2003). Introduction: The future of testimony. *Discourse (Berkeley, Calif.), 25*(1), 4–18. DOI: 10.1353/dis.2004.0003

Cullinane, J. P. (2003). *An Coimisiún le Rincí Gaelacha (Irish Dancing Commission)*. Its Origins and Evolution. John Cullinane.

Dahmen, N. S., Abdenour, J., McIntyre, K., & Noga-Styron, K. E. (2018). Covering mass shootings: Journalists' perceptions of coverage and factors influencing attitudes. *Journalism Practice, 12*(4), 456–476. DOI: 10.1080/17512786.2017.1326832

Daiya, K. (2008). Re-Gendering the Nation: Masculinity, Romance, and Secular Citizenship. *Violent Belongings: Partition, Gender, and National Culture In.*

Dalley, S. (1991). Gilgamesh in the Arabian Nights. *Journal of the Royal Asiatic Society, 1*(1), 1–17. DOI: 10.1017/S1356186300000031

Damanhoury, K. E., & Saleh, F. (2024). Mediated Clash of Civilizations: Examining the Proximity-Visual Framing Nexus in Al Jazeera Arabic and Fox News' Coverage of the 2021 Gaza War. *Digital Journalism (Abingdon, England), 13*(1), 136–158. DOI: 10.1080/21670811.2024.2332614

Dane, B. O. (1994). Overview Of Cultural Attitudes Toward Death And Grief. AIDS and the New Orphans: Coping with Death, 13.

Darke, J. L. (1990). Sexual aggression: Achieving power through humiliation. In W.L. Marshall., D.R. Laws, & H. E. Barbaree (Eds.), *Handbook of sexual assault: Issues, theories, and treatment of the offender* (pp. 55-72). Springer US. https://doi.org/DOI: 10.1007/978-1-4899-0915-2_4

Darley, J. M., & Pittman, T. S. (2003). The psychology of compensatory and retributive justice. *Personality and Social Psychology Review, 7*(4), 324–336. DOI: 10.1207/S15327957PSPR0704_05 PMID: 14650389

Das, B., & Sarangi, I. (2017). Voiceless victims of war: An absurd truth. *Rupkatha Journal on Interdisciplinary Studies in Humanities, 9*(1). Advance online publication. DOI: 10.21659/rupkatha.v9n1.22

Dascal, M. (2008). Dichotomies and types of debate. In *Controversy and confrontation: Relating controversy analysis with argumentation theory* (pp. 27–49). John Benjamins Publishing Company. DOI: 10.1075/cvs.6.03das

Das, V. (2006). *Life and words: Violence and the descent into the ordinary.* University of California Press. DOI: 10.1525/9780520939530

Davidson, T., & Pierson, R. R. (2001). Voices from the margins: Subaltern women speak... and rewrite history. *Journal of Women's History, 13*(2), 169–179. DOI: 10.1353/jowh.2001.0045

De Beauvoir, S. (2010). *The Second Sex.* Knopf.

de Souza, P. (1999). *Piracy in the Graeco-Roman world.* Cambridge University Press.

Deaville, J. (2018). Vietnam: Music of Patriotism and Protest. In *Music and War in the United States* (211-227). Routledge.

Delaperrière, M. (2014). Testimony as a literary problem. *Teksty Drugie, 2*, 42–54.

Deleuze, G., & Guattari, F. (1987). *A Thousand Plateaus: Capitalism and Schizophrenia* (1st ed.). University of Minnesota Press., https://www.upress.umn.edu/9780816614028/a-thousand-plateaus/

Demuyakor, J. (2021). The propaganda model in the digital age: A review of literature on the effects of social media on news production. *Shanlax International Journal of Arts. Science and Humanities*, *8*(4), 232–239. DOI: 10.34293/sijash.v8i4.3598

Dennis, E. (2020). *How Hinduism Rooted Itself in Film*. https://scholarlycommons.obu.edu/scholars_day_conference/2020/christian_studies/1/

Deppe, C., & Schaal, G. S. (2024). Cognitive warfare: A conceptual analysis of the NATO ACT cognitive warfare exploratory concept. *Frontiers in Big Data*, *7*, 1452129. Advance online publication. DOI: 10.3389/fdata.2024.1452129 PMID: 39552633

Derman, G. S. (2021). Perceptıon Management in the Media. *International Journal of Social and Economic Sciences*, *11*(1), 64–78.

DeRosa, A. (2014). The end of futurity: Proleptic nostalgia and the war on terror. *Literature, Interpretation, Theory*, *25*(2), 88–107. DOI: 10.1080/10436928.2014.904709

Detue, F., & Lacoste, C. (2021). What testimony does to literature. *Synthesis: An Anglophone Journal of Comparative Literary Studies*, *13*(13), 22–36. DOI: 10.12681/syn.27559

DeTurk, S. (2015). The "Banksy effect" and street art in the Middle East. *Street Art & Urban Creativity,* 1(2), 22-30. https://doi.org/DOI: 10.25765/sauc.v1i2.25

Deutscher, G. (2013). *Dilin Aynasından Kelimeler Dünyamızı Nasıl Renklendirir, Çev: Cemal Yardımcı*. Metis Bilim.

Devakumar, Dr. T. (2025). Between the Headlines: A Comparative Study of Indian and International Television News Perspectives on the Pre-War India-Pakistan Tensions. *INTERNATIONAL JOURNAL OF SCIENTIFIC RESEARCH IN ENGINEERING AND MANAGEMENT, 09*(06), 1–9. DOI: 10.55041/IJSREM50221

Dhar, A. (Director). (2019). *Uri: The Surgical Strike* [Film].

Dibattista, M. (2000). Steven Spielberg, Robert Rodat and Frank Darabont (1998), Saving Private Ryan; Terence Malick (1998), The Thin Red Line. *Rethinking History*, *4*(2), 223–228. DOI: 10.1080/13642520050074885

Dick, B. F. (2022). The Star-Spangled Screen: The American World War II Film. In *The Star-Spangled Screen*. Lexington, Kentucky: The University Press of Kentucky. DOI: 10.2307/j.ctv2ks7284

Dictionary of Military and Associated Terms. (2020). Perception Management. https://www.dtic.mil/doctrine/jel/doddict/index.html

Dittmar, L., & Michaud, G. (1990). *From Hanoi to Hollywood: The Vietnam War in American film*. Rutgers University Press.

Dodge, Toby. (2005). Iraq's future: the aftermath of regime change.

Doğan, İ. (2014). Propaganda Aracı Olarak İnternet: Kayseri İli Merkez Seçmeni Üzerine Bir Alan Araştırması, [Unpublished graduate dissertation]. Selcuk University Social Sciences Institute, Konya.

Doğan, A., & Ertan, T. (2016). Siyasal İletişim Aracı Olarak ABD Başkanlık Seçimlerinde Twitter'ın Kullanımı: Hillary Clinton ve Donald Trump Örneği. *Atatürk İletişim Dergisi*, *11*, 139–159.

Doherty, T. (1998). Saving Private Ryan. *Cinéaste (New York, N.Y.)*, •••, 68–71. https://link.gale.com/apps/doc/A53709342/AONE?u=ucl_ttda&sid=bookmark-AONE&xid=0aab3062

Doja, A. (2019). Politics of mass rapes in ethnic conflict: A morphodynamics of raw madness and cooked evil. *Crime, Law, and Social Change*, *71*(5), 541–580. DOI: 10.1007/s10611-018-9800-0

Dostie-Goulet, E. (2009). Social Networks And The Development Of Political İnterest. *Journal of Youth Studies*, *12*(4), 405–421. DOI: 10.1080/13676260902866512

Dracopoulou, S., & Doxiadis, S. (1988). In Greece, lament for the dead, denial for the dying. *The Hastings Center Report*, *18*(4), 15. DOI: 10.2307/3563249 PMID: 3220738

Drumbl, M., & Mouthaan, S. (2021). 'A Hussy Who Rode on Horseback in Sexy Underwear in Front of the Prisoners': The Trials of Buchenwald's Ilse Koch. *International Criminal Law Review*, *21*(2), 280–312. DOI: 10.1163/15718123-bja10047

Dumančić, M., & Krolo, K. (2017). Dehexing postwar West Balkan masculinities: The case of Bosnia, Croatia, and Serbia, 1998 to 2015. *Men and Masculinities*, *20*(2), 154–180. DOI: 10.1177/1097184X15625314

Durfey, N., & Sajal, S. (2022). A comprehensive survey: Cybersecurity challenges and futures of autonomous drones. In *Proceedings of the 2022 Intermountain Engineering, Technology and Computing Conference* (pp. 1–6). IEEE. DOI: 10.1109/IETC54973.2022.9796881

Durkheim, E. (1997). *The Division of Labor in Society*. The Free Press. (Original work published 1893)

Dutro, E. (2008). 'That's why I was crying on this book': Trauma as testimony in responses to literature. *Changing English, 15*(4), 423–434. DOI: 10.1080/13586840802493076

Dutta, J. P. (Director). (1997). *Border* [Film].

Dwyer, O. J., & Alderman, D. H. (2008). Memorial landscapes: Analytic questions and metaphors. *GeoJournal, 73*(3), 165–178. DOI: 10.1007/s10708-008-9201-5

Dzhakupova, I. (2024). Greenberg, The Cold War, and French Existentialism: How Abstract Expressionism became Constructed as Inherently Masculine. *Groundings Undergraduate Journal, 15*, 175–191. DOI: 10.36399/GroundingsUG.15.139

Eagar, C. (2023). Bullet points: In Serbia, the gun is a way of life. *Spectator (London, England), 352*(10160), 24–25.

Eastwood, C. (Director). (2014). American sniper [Film]. Warner Bros. Pictures.

Easwaran, E. (1989). Classics of Indian spirituality.

Eberwein, R. (2010). *The Hollywood War Film*. The Hollywood War Film., DOI: 10.1002/9781444315103

Edy, J. (2006). *Troubled pasts: News and the collective memory of social unrest*. Temple University Press.

Ehrenhaus, P. (2001). Why We Fought: Holocaust Memory in Spielberg's Saving Private Ryan. *Critical Studies in Media Communication, 18*(3), 321–337. DOI: 10.1080/07393180128089

Elden, S. (2004). Understanding Henri Lefebvre. *Continuum*.

Elden, S. (2013). *The birth of territory*. University of Chicago Press. DOI: 10.7208/chicago/9780226041285.001.0001

Elden, S. (2021). Terrain, politics, history: Thinking the volume. *Political Geography, 87*, 102431. Advance online publication. DOI: 10.1177/2043820620951353#

Elektratig (2014). Did Cambyses Have Cats Nailed to His Soldiers' Shields? Elektratig. https://elektratig.blogspot.com/2014/01/did-cambyses-have-cats-nailed-to-his.html

Elish, M. C. (2017). Remote Split: A History of US Drone Operations and the Distributed Labor of War. *Science, Technology & Human Values, 42*(6), 1100–1131. https://www.jstor.org/stable/26580327. DOI: 10.1177/0162243917731523

Ellison, M. (1994). African-American Music and Muskets in Civil War New Orleans. *Louisiana History: The Journal of the Louisiana Historical Association*, 35(3), 285-319. https://www.jstor.org/stable/4233124

Ellis, S. (2015). Nelson Mandela, the South African Communist Party and the origins of Umkhonto we Sizwe. *Cold War History, 16*(1), 1–18. DOI: 10.1080/14682745.2015.1078315

El-Masri, S. (2018). Prosecuting ISIS for the sexual slavery of the Yazidi women and girls. *International Journal of Human Rights, 22*(8), 1047–1066. DOI: 10.1080/13642987.2018.1495195

Elshtain, J. B. (1998). Women and War: Ten Years On. [suspicious link removed]. *Review of International Studies, 24*(4), 447–460. DOI: 10.1017/S0260210598004471

Emarketer. (2024). 2024 Political Ad Spending Will Jump Nearly 30% vs. 2020. Emarketer. https://www.emarketer.com/press-releases/2024-political-ad-spending-will-jump-nearly-30-vs-2020/

Encyclopedia. (n.d.). Disinformation. https://encyclopedia.pub/entry/29164

Engelbrecht, C. (2023), *A Look at Gun Violence in a Nation Awash With Weapons, The New York Times*, A8-L.

Engle, K. (2005). Feminism and its (dis) contents: Criminalizing wartime rape in Bosnia and Herzegovina. *The American Journal of International Law, 99*(4), 778–816. DOI: 10.2307/3396669

Enloe, C. (2014). *Bananas, beaches and bases: Making feminist sense of international politics*. Univ of California Press. DOI: 10.1525/9780520957282

Ertem, Y. E. (2019). Sosyal Medyada Dezenformasyon. [Unpublished graduate dissertation]. Marmara University Social Sciences Institute, İstanbul.

Eslen, N. (2003). *Tarih Boyu Savaş ve Strateji*. Q Matris Yayınları.

Fahim, M., & Islam, M. N. (2024). Mapping the Air Time of Eastern & Western Media on Conflict and War: A Comparative Study of BBC, DW, TRT and Al Jazeera on the Coverage of Second Nagorno-Karabakh War & the Aftermath. *Communicatio Socialis, 37*(1), 79–98.

Fanon, F. (2000). The Wretched of the Earth. In Badmington, N. (Ed.), *Posthumanism* (pp. 23–26). Macmillan Education UK., DOI: 10.1007/978-1-137-05194-3_4

Farkas, J., & Schwartz, S. A. (2018). Please like, comment and share our campaign! How social media managers for Danish political parties perceive user-generated content. *Nordicom Review, 39*(2), 19–33. DOI: 10.2478/nor-2018-0008

Fassin, D. (2011). *Humanitarian reason: a moral history of the present.* University of California Press.

Faulkner, P. (2011). *Knowledge on trust.* Oxford University Press., DOI: 10.1093/acprof:oso/9780199589784.001.0001

Felman, S., & Laub, D. (1992). *Testimony: Crises of witnessing in literature, psychoanalysis, and history.* Taylor & Francis.

Ferguson, R. B. (2018). War is not part of human nature. *Scientific American, 319*(3).

Fernández-Morales, M. (2024). Survivor memory and rape memoir: Chanel Miller's Know my name. *Life Writing, 21*(2), 401–418. DOI: 10.1080/14484528.2023.2196367

Ferro, M. (1988). *Cinema and History.* Wayne State University Press., https://wsupress.wayne.edu/9780814319055/

Festinger, L. (1957). *A theory of cognitive dissonance.* Stanford University Press. DOI: 10.1515/9781503620766

Field, N. P., & Bonanno, G. A. (2001). The Role of Blame in Adaptation in the First 5 Years Following the Death of a Spouse. *American Behavioral Scientist,* 44(5), 764-781. https://doi.org/ (Original work published 2001)DOI: 10.1177/00027640121956485

Figueroa-Caballero, A. (2019). The criminalization of immigration in the United States: Examining the effects of news coverage linking undocumented immigrants with criminality. *Communication Monographs, 86*(1), 112–132. DOI: 10.1080/03637751.2018.1505049

Filkins, D. (2008). *The Forever War: NATIONAL BOOK CRITICS CIRCLE AWARD WINNER.* Vintage.

Fimi, D. (2017). *Celtic myth in contemporary children's fantasy: Idealization, identity, ideology.* Springer. DOI: 10.1057/978-1-137-55282-2

Fiske, A. P., & Rai, T. S. (2015). *Virtuous violence: Hurting and killing to create, sustain, end, and honour social relationships.* Cambridge University Press.

Flomenhaft, D. (2007). *The forgotten ones: The grief experience of adult siblings of World Trade Center victims* (Doctoral dissertation, New York University, School of Social Work).

Flynn, D., & O'Brien, E. (Eds.). (2018). *Representations of loss in Irish literature*. Springer. DOI: 10.1007/978-3-319-78550-9

Foertsch, S. (2022). Metamodernity, American transcendentalism and transhumanism in Japanese Anime.

Fomin, M. (2018). Hunting the deer in Celtic and Indo-European mythological contexts. Celtic myth in the, 21, 73-87.

Ford, B. Q., & Feinberg, M. (2020). Coping with Politics: The Benefits and Costs of Emotion Regulation. *Current Opinion in Behavioral Sciences*, *34*, 123–128. DOI: 10.1016/j.cobeha.2020.02.014

Forsythe, D. P. (2005). International humanitarianism in the contemporary world: Forms and issues. *Human Rights & Human Welfare*, *5*(1), 57. DOI: 10.56902/HRHW.2005.5.1.57

Foucault, M. (1986). Space, knowledge and power. In Rainbow, P. (Ed.), *The Foucault reader* (pp. 239–256). Pantheon Books.

Fox, J. A., Sanders, N. E., Fridel, E. E., Duwe, G., & Rocque, M. (2021). The contagion of mass shootings: The interdependence of large-scale massacres and mass media coverage. *Statistics and Public Policy (Philadelphia, Pa.)*, *8*(1), 53–66. DOI: 10.1080/2330443X.2021.1932645

Frank, A. (1995). The wounded storyteller: Body, illness. *Ethics*. Advance online publication. DOI: 10.7208/chicago/9780226260037.001.0001

Fredericks, S. C. (1978). Problems of fantasy. *Science Fiction Studies*, *5*(Part 1), 33–44. https://www.jstor.org/stable/4239155. DOI: 10.1525/sfs.5.1.0033

Freud, S. (1953). The uncanny (J. Strachey, Ed. & Trans.). In J. Strachey (Ed. & Trans.), The standard edition of the complete psychological works of Sigmund Freud (Vol. 17, pp. 217–252). London: The Hogarth Press. (Original work published 1919)

Freud, S. (1915). Repression. In *The standard edition of the complete psychological works of Sigmund Freud* (Vol. 14, pp. 141–158). Hogarth Press.

Freud, S. (1930). *Civilization and its discontents*. Hogarth.

Fricker, M. (2008). Forum on Miranda Fricker's "Epistemic injustice: Power and the ethics of knowing". *Theoria*, *23*(1), 69–71. DOI: 10.1387/theoria.7

Friedlander, S. (2014). *Nazi Germany and the Jews: The years of persecution: 1933-1939*. Hachette UK.

Fritzsche, P. (2001). The case of modern memory. *The Journal of Modern History, 73*(1), 87–117. DOI: 10.1086/319880

Furstenau, M., & MacAvoy, L. (2007). Terrence Malick's Heideggerian Cinema: War and the Question of Being in The Thin Red Line. In *The Cinema of Terrence Malick : Poetic Visions of America* (2nd ed., pp. 179–191). London : WallFlower Press. https://search-ebscohost-com.libproxy.ucl.ac.uk/login.aspx?direct=true&AuthType=ip,shib&db=nlebk&AN=781073&site=ehost-live&scope=site&custid=s8454451&ebv=EB&ppid=pp_179

Gabbert, J. (1986). Piracy in the early Hellenistic period: A career open to talents. *Greece & Rome*, 33(2), 156–163.

Gabriele, J. (2024). *Underwater communication technologies for submarines: Experimentation and innovations* [Honours thesis, University of New Haven]. University of New Haven Digital Commons. https://digitalcommons.newhaven.edu/honorstheses/80/

Gabriel, K. (1998). Manning the Border: Gender and War in 'Border'. *Economic and Political Weekly*, •••, 828–832.

Gallagher, M. (2020). Bibliotherapy as a Counselling Intervention for Bereavement, Loss and Grief. Irish Association for Counselling and Psychotherapy, 20(1).

Galland, D., & Gronning, M. (2019). The social production of spatial knowledge. *Environment & Planning A, 51*(2), 287–304.

Ganti, T. (2004). *Bollywood: A guidebook to popular Hindi cinema*. Routledge. https://www.taylorfrancis.com/books/mono/10.4324/9780203643945/bollywood-tejaswini-ganti

Garcia-Bernardo, J., Qi, H., Shultz, J. M., Cohen, A. M., Johnson, N. F., & Dodds, P. S. (2018). *Social Media Appears to Affect the Timing, Location, and Severity of School Shootings, USA Today*, 10 (1), p.100.

Garfield, A. (2002). The Offence of Strategic Influence: Making the Case for Perception Management Operations. *Journal of Information Warfare, 1*(3), 30–39.

Garland, C. (2018). *Understanding trauma: A psychoanalytical approach*. Routledge. DOI: 10.4324/9780429484575

Garofalo, R. (1982). Music and Musket: Bands and Bandsmen of the American Civil War: By Kenneth E. Olson. Westport, Connecticut: Greenwood Press, 1981. 299 pages. Illustrations, bibliography. Hard cover, $27.50. *Music Educators Journal, 68*(5), 23–25. DOI: 10.2307/3397736

Garwood, A. (1996). The Holocaust and the power of powerlessness: Survivor Guilt an Unhealed Wound. *British Journal of Psychotherapy, 13*(2), 243–258. DOI: 10.1111/j.1752-0118.1996.tb00880.x

Gates, P. (2005). "Fighting the Good Fight:" The Real and the Moral in the Contemporary Hollywood Combat Film. *Quarterly Review of Film and Video, 22*(4), 297–310. DOI: 10.1080/10509200590475788

Gaufman, E. (2016). Security threats and public perception. In New security challenges. https://doi.org/DOI: 10.1007/978-3-319-43201-4

Geist, V. (1998). Deer of the world: their evolution, behaviour, and ecology (Vol. 272). Mechanicsburg: Stackpole books.

Gencer, B. (July 26, 2023) Elon Musk ile Grimes oğlu X Æ A-12'nin adını değiştirdi, NTV Haber, https://www.ntv.com.tr/galeri/n-life/magazin/elon-musk-ile-grimes-oglu-x-a-12nin-adini-degistirdi,8zWSDE6xW0iR5wsw_IIrhw/4wVqA5KjvE2jcQEJ2V1NVg

Genel, M. G. (2012). Siyasal İletişim Kampanyalarında Sosyal Medyanın Kullanımı (12 Haziran 2011 Seçimleri "Twitter" Örneği, The Turkish Online Journal of Design. *Art and Communication, TOJDAC, 2*(4), 23–31. DOI: 10.7456/10204100/005

Geppert, A. C., Brandau, D., & Siebeneichner, T. (2018). Militarizing Outer Space Astroculture, Dystopia and the Cold War. *Spaceflight.*

Gerasimov, V. (2016). The value of science is in the foresight: New challenges demand rethinking the forms and methods of carrying out combat operations. *Military Review, 96*(1), 23–29. https://www.armyupress.army.mil/portals/7/military-review/archives/english/militaryreview_20160228_art008.pdf

Gerber, T. P., & Zavisca, J. (2016). Does Russian Propaganda Work? *The Washington Quarterly, 39*(2), 79–98. DOI: 10.1080/0163660X.2016.1204398

Gerhart, J. I., Hall, B. J., Russ, E. U., Canetti, D., & Hobfoll, S. E. (2014). Sleep disturbances predict later trauma-related distress: Cross-panel investigation amidst violent turmoil. *Health Psychology, 33*(4), 365–372. DOI: 10.1037/a0032572

Gerstein, J., & Woodruff Swan, B. (2024. Three Iranians charged in hacking of Trump presidential campaign. Politico. https://www.politico.com/news/2024/09/27/iranians-charged-trump-campaign-hack-00181454

Gibbons, M. B. (1992). A child dies, a child survives: The impact of sibling loss. *Journal of Pediatric Health Care, 6*(2), 65–72. DOI: 10.1016/0891-5245(92)90123-L PMID: 1573530

Gilbert, G. (2004). Robert Motherwell's World War Two collages: Signifying war as topical spectacle in abstract expressionist art. *Oxford Art Journal, 27*(3), 311–337. DOI: 10.1093/oaj/27.3.311

Gilbert, S. (2007). Singing Against Apartheid: ANC Cultural Groups and the International Anti-Apartheid Struggle*. *Journal of Southern African Studies, 33*(2), 421–441. DOI: 10.1080/03057070701292848

Glenn, R. W. (2009). Thoughts on "Hybrid" Conflict. Small Wars Journal, https://smallwarsjournal.com/wp-content/uploads/2022/02/188-glenn.pdf

Glover, H. (1984). Survival guilt and the Vietnam veteran. *The Journal of Nervous and Mental Disease, 172*(7), 393–397. DOI: 10.1097/00005053-198407000-00003 PMID: 6726209

Goodman, A., Bergbower, H., Perrotte, V., & Chaudhary, A. (2020). Survival after sexual violence and genocide: Trauma and healing for Yazidi women in Northern Iraq. *Health (Irvine, Calif.), 12*(06), 612–628. DOI: 10.4236/health.2020.126046

Gopinath, S., & Deshmukh, R. (2023). *Historicizing Myths in Contemporary India: Cinematic Representations and Nationalist Agendas in Hindi Cinema.* Taylor & Francis. DOI: 10.4324/9781003363149

Gounaris, B. C. (2017). The privatization of war: Privateers in the Mediterranean and beyond. *Mediterranean Historical Review, 32*(2), 135–152.

Gouni, O., & Verny, T. (2023). The human toll of collective trauma: The ravages of war and persecution. *International Journal of Prenatal & Life Sciences*, 1–40. https://doi.org/DOI: 10.24946/IJPLS/20231010

Graves, R. (2025). *Goodbye to All That.* Aegitas.

Greengrass, P. (Director). (2013). *Captain Phillips* [Film]. Columbia Pictures.

Greitens, S. C., Lee, M., & Yazici, E. (2020). Counterterrorism and preventive repression: China's changing strategy in Xinjiang. *International Security, 44*(3), 9–47. DOI: 10.1162/isec_a_00368

Grigore, L., & Cristescu, C. (2024). The Use of Drones in Tactical Military Operations in the Integrated and Cybernetic Battlefield. *Land Forces Academy Review, 29*(2), 269–273. DOI: 10.2478/raft-2024-0029

Güler, H. (2017). Rusya Araştırmaları Enstitüsü. http://www.rusen.org/hibrit-savas-nedir/

Gültekin, M. (2022). Hibrit Savaş, Sosyal Medya ve Propaganda. İslami Analiz. https://islamianaliz.com/makale/12496108/mucahit-gultekin/hibrit-savas-sosyal-medya-ve-propaganda

Gültekin, M. (2024). *Algı yönetimi ve Manipülasyon: Kanmanın ve Kandırmanın Psikolojisi*. Pınar Yayınları.

Gupta, A., & Dubey, M. A. (2024). *Reforms in Indian military since 2014* [Research report]. ResearchGate. https://www.researchgate.net/publication/380034381_Reforms_in_Indian_Military_Since_2014

Gürkan, H. (2015). *Karşı Sinema*. Es Yayınları.

Haidt, J. (2012). *The Righteous Mind: Why Good People are Divided by Politics and Religion*. Pantheon.

Hakimi, M., Reid, N. L., & Witten, S. (2015). An introduction: Adapting to a rapidly changing world. *American Society of International Law Proceedings, 109*, 1–7. DOI: 10.5305/procannmeetasil.109.2015.0001

Hall, S. (1993). Culture, community, nation. *Cultural Studies, 7*(3), 349–363. DOI: 10.1080/09502389300490251

Hallwood, P., & Miceli, T. J. (2015). *Maritime piracy and its control: An economic analysis*. Palgrave Macmillan. DOI: 10.1057/9781137461506

Hansen, S. J. (2009). *Piracy in the greater Gulf of Aden: Myths, misconceptions and remedies. Norwegian Institute for Urban and Regional Research*. NIBR. DOI: 10.7577/nibr/rapport/2009/29

Hanson, H., & O'Rawe, C. (2010). *The femme fatale: Images, histories, contexts*. Palgrave Macmillan. DOI: 10.1057/9780230282018

Harding, A. (2016). *The mayor of Mogadishu: A story of chaos and redemption in the ruins of Somalia*. St. Martin's Press.

Harris, P. L., & Koenig, M. A. (2006). Trust in testimony: How children learn about science and religion. *Child Development, 77*(3), 505–524. DOI: 10.1111/j.1467-8624.2006.00886.x

Harris, P. L., & Lane, J. D. (2014). Infants understand how testimony works. *Topoi, 33*(2), 443–458. DOI: 10.1007/s11245-013-9180-0

Hasian, M. (2001). Nostalgic Longings, Memories of the "Good War," and Cinematic representations in Saving Private Ryan. *Critical Studies in Media Communication, 18*(3), 338–358. DOI: 10.1080/07393180128083

Hazou, R. (2022). Arts and conflict: Banksy, art and Palestinian solidarity. In *Tu Rangaranga Rights, responsibilities and global citizenship in Aotearoa New Zealand* (pp. 223–236). Massey University Press.

Hedges, C. (2002). *War is a force that gives us meaning*. PublicAffairs.

Heebøll-Holm, T. (2013). *Ports, piracy, and maritime war: Piracy in the English Channel and the Atlantic, c.1280–c.1330*. Brill. DOI: 10.1163/9789004248168

Heidegger, M. (1978). *Being and time*. Wiley-Blackwell., https://www.wiley.com/en-cn/Being+and+Time-p-9780631197706

Heidel, A. (1949). *Gilgamesh Epic and Old Testament Parallels* (Vol. 136). University of Chicago Press.

Heiney, S. P., Darr-Hope, H., Meriwether, M. P., & Adams, S. A. (2017). Healing by Creating: Patient Evaluations of Art-Making Program. *The Journal of Creative Behavior*, *51*(1), 35–44. DOI: 10.1002/jocb.84

Helfgott, J. B. (2015). Criminal behavior and the copycat effect: Literature review and theoretical framework for empirical investigation. *Aggression and Violent Behavior*, *22*(1), 46–64. DOI: 10.1016/j.avb.2015.02.002

Hen, K., Tsui, H., & Lam, D. (Directors). (2021). The battle at Lake Changjin [Film]. Bona Film Group.

Henderson, R. K., & Schnall, S. (2021). Social threat indirectly increases moral condemnation via thwarting fundamental social needs. *Scientific Reports*, *11*(1), 21709. DOI: 10.1038/s41598-021-00752-2 PMID: 34741054

Herman, E. S., & Chomsky, N. (1988). *Manufacturing consent: The political economy of the mass media*. Pantheon Books.

Herman, E. S., & Chomsky, N. (1999). *Medya Halka Nasıl Evet Dedirtir*. Minerva Yayınları.

Herre, B., Guirao, L. R., & Roser, M. (2016). War and Peace. https://ourworldindata.org/war-and-peace

Hersh, S. M. (1970). *My Lai 4: A report on the massacre and its aftermath*. Random House.

Heuvelmans, B. (2003). *In the wake of sea serpents*. Kegan Paul International.

Hickman, M. J. (2015). *Diaspora space: celebrating St. Patrick's Day in New York and London. Consuming St Patrick's Day*. Cambridge Scholars Publishing., https://dictionary.cambridge.org/dictionary/english/warhttps://en.wikipedia.org/wiki/Riverdance:_The_Animated_Adventurehttps://kids-in-mind.com/r/riverdance-the-animated-adventure-parents-guide-movie-review-rating.htm#:~:texthttps://www.imdb.com/title/tt10841088/ https://www.maxim.org.nz/article/faq6-where-is-euthanasia-legal/ https://www.thenumbers.com/movies/franchise/Riverdance#tab=summary&franchise_movies_overview=od5https://wordvis.com/

High, A. C. & Solomon, D. H. (2011). Locating Computer-Mediated Social Support Within Online Communication Environments.

Hill, A., & White, A. (2011). Painting Peace? Murals and the Northern Ireland Peace Process. *Irish Political Studies*, *27*(1), 71–88. DOI: 10.1080/07907184.2012.636184

Hirschberger, G. (2018). Collective trauma and the social construction of meaning. *Frontiers in Psychology*, *9*, 1441. DOI: 10.3389/fpsyg.2018.01441

Hirsch, M. (2012). *The generation of postmemory: Writing and visual culture after the Holocaust*. Columbia University Press. DOI: 10.7312/hirs15652

Hirsch, M. (2017). Touching memory: Postmemory, art, and the Holocaust. *Esprit (Paris, France)*, (10), 42–61. DOI: 10.3917/espri.1710.0042

Hobbes, T. (1997). Leviathan. Touchstone.

Hobbes, T. (2018). LEVIATHAN (Complete Edition). e-artnow.

Hobbes, T. (1651). *Leviathan*. Penguin Books.

Hodgkins, J. (2002). In the Wake of Desert Storm: A Consideration of Modern World War II Films. *The Journal of Popular Film and Television*, *30*(2), 74–84. DOI: 10.1080/01956050209602842

Hoffman, F. G. (2007). Conflict in the 21 st Century: The Rise of Hybrid Wars. Potomac Institute for Policy Studies. https://www.potomacinstitute.org/images/stories/publications/potomac_hybridwar_0108.pdf

Holat, O. (2021). Yeni Medya ve Siber Savaş Kavramları Bağlamında Stuxnet Saldırısı Örneğinin İncelenmesi. *Abant Kültürel Araştırmalar Dergisi*, *6*(11), 105–121.

Holm, E., & de Bruijn, M. (2016). Pirates and the discursive construction of the Somali coast. In Biersteker, T. J. (Eds.), *International Political Sociology*. Routledge.

Homer, . (2007). *The Iliad*. University of Michigan Press.

Homer. (n.d.). *The Odyssey* (Book XII). (R. Fagles, Trans.). Penguin Classics.

Horowitz, M. C. (2018). Artificial Intelligence, International Competition, and the Balance of Power. *Texas National Security Review*, *1*(3), 36–57.

Hossain, M. A., Hasan, M. M., & Wasantha, N. (2025). The power of social media in driving political change: A case study on Bangladesh's regime change in 2024. *International Journal of the Sociology of Law*, *2*(1), 257–275. DOI: 10.62951/ijsl.v2i1.405

Howard, S. (Director). (2005). Jarhead [Film]. Universal Pictures.

Hudson, A. (2002). Advocacy by UK-based development NGOs. *Nonprofit and Voluntary Sector Quarterly*, *31*(3), 402–418. DOI: 10.1177/0899764002313005

Hughes, C. W. (2003). North Korea: Coming in from the cold? *Journal of East Asian Studies (Seoul).*

Human Rights Watch. (2017). *Flawed justice: Accountability for ISIS crimes in Iraq.* https://www.hrw.org/report/2017/12/05/flawed-justice/accountability-isis-crimes-iraq

Huo, J. (2024). Foreign İnfluence Efforts Reached A Fever Pitch During the 2024 Election. NPR. https://www.npr.org/2024/11/09/nx-s1-5181965/2024-election-foreign-influence-russia-china-iran

Hutson, S. P., Hall, J. M., & Pack, F. L. (2015). Survivor guilt: Analyzing the concept and its contexts. *Advances in Nursing Science*, *38*(1), 20–33. DOI: 10.1097/ANS.0000000000000058 PMID: 25635503

Ibrahim, A. (2024). Employment of FPV Drones in Russia-Ukraine War: Lessons and Future Outlook. *Modern Diplomacy.* https://moderndiplomacy.eu/2024/12/14/employment-of-fpv-drones-in-russia-ukraine-war-lessons-and-future-outlook/

Ibrahim, F., Rhode, S., & Daseking, M. (2023). Cognitive & psychological warfare. *The Defence Horizon Journal.* https://tdhj.org/blog/post/cognitive-psychological-warfare/

Ibrahim, H., Ertl, V., Catani, C., Ismail, A. A., & Neuner, F. (2018). Trauma and perceived social rejection among Yazidi women and girls who survived enslavement and genocide. *BMC Medicine*, *16*(1), 1–11. DOI: 10.1186/s12916-018-1140-5

İldem, T. (2021). Dezenformasyona Karşı Toplumsal Dayanıklılık. Ekonomi ve Dış Politika Araştırmalar Merkezi. https://edam.org.tr/wpcontent/uploads/2021/03/210305-Dezenformasyon-Ildem.pdf

Imperial War Museums. (1915). Women of Britain say 'Go!' Retrieved February 1, 2023, from https://www.iwm.org.uk/collections/item/object/205020521

International Criminal Court. (2015). *Statement of the prosecutor of the International Criminal Court, Fatou Bensouda, on the alleged crimes committed by ISIS.* https://www.icc-cpi.int/news/statement-prosecutor-international-criminal-court-fatou-bensouda-alleged-crimes-committed-isis

Irish Childhood Bereavement Network. (2017). *Standards for Supporting Bereaved Children & Young People – A Framework for Development.* ICBN.

Isaak, J., & Hanna, M. J. (2018). User Data Privacy: Facebook, Cambridge Analytica, and Privacy Protection. *Computer, 51*(8), 56–59. DOI: 10.1109/MC.2018.3191268

Ivanova, N., & Ryzhchenko, O. (2019). Meta-Genre Of Fantasy In The Context Of Modern Science. *Research Trends In Modern Linguistics And Literature, 2*, 23–37. DOI: 10.29038/2617-6696-2019.2.23.37

Iyer, J. B., & Das, S. (2025). Leveraging history to invoke nationalism: From the annals of history to social engineering of present and future in Hindi cinema. *National Identities, 27*(1–2), 155–174. DOI: 10.1080/14608944.2024.2365914

Jachec, N. (1991). "The Space between Art and Political Action": Abstract Expressionism and Ethical Choice in Postwar America, 1945-1950. *Oxford Art Journal, 14*(2), 18–29. https://www.jstor.org/stable/1360522. DOI: 10.1093/oxartj/14.2.18

Jackson, R. (2005). *Writing the War on Terrorism: Language*. Politics and Counter-terrorism.

Jacob, J. F. R. (1997). *Surrender at Dacca: Birth of a Nation*. Manohar Publishers.

Jaffal, Z. (2020). Rape as genocide crime in International Criminal Law-The case of Yazidi women in Iraq. *International Journal of Criminal Justice Sciences, 15*(2), 230–247.

Jara, J. (1984). *An unfinished song: the life of Victor Jara*. Ticknor and Fields.

Jardim, T. (2023). *Ilse Koch on Trial: Making the "Bitch of Buchenwald"*. Harvard University Press., DOI: 10.2307/jj.362392

Johnson, E. (2024). How Far Does the Russian Military Presence Stretch? *Florida Political Chronicle, 31*(2).

Johnson, J. C. (2014). *Spaces of detention: Spatiality and the organization of immigration detention*. Routledge.

Johnson, J. C. (2020). *Seafarers, statecraft, and the rise of maritime frontiers in the ancient Mediterranean*. Oxford University Press.

Jomini, A. H. (2013). *Savaş Sanatı* (Büyükonat, A. T., Trans.). Doruk Yayımcılık.

Jonason, P. K., Webster, G. D., Schmitt, D. P., Li, N. P., & Crysel, L. (2012). The Antihero in Popular Culture: Life History Theory and the Dark Triad Personality Traits. *Review of General Psychology, 16*(2), 192–199. DOI: 10.1037/a0027914

Jones, N. (2015). The production of cinematic space. Retrieved from https://www.academia.edu/11910555, Accessed date: 21.01.2025

Jones, A. (2002). Equivocal Masculinity: New York Dada in the Context of World War I. *Art History, 25*(2), 162–205. DOI: 10.1111/1467-8365.00312

Jones, G. (2008). *Killing Monsters: Our Children's Need For Fantasy, Heroism, and Make-Believe Violence*. Basic Books.

Jones, S. (2019). Testimony through culture: Towards a theoretical framework. *Rethinking History, 23*(3), 257–278. DOI: 10.1080/13642529.2019.1620909

Jonsson, O. (2020). Putting Modern War in Perspective, in Oscar Jonsson (Ed.), Modern Warfare: New Technologies and Enduring Concepts. Stockholm Free World Forum. https://www.academia.edu/44001543/Modern_Warfare_New_Technologies_and_Enduring_Concepts

Jordan, D., Kiras, J. D., Lonsdale, D. J., Speller, I., Tuck, C., & Walton, C. D. (2008). *Understanding Modern Warfare*. Cambridge University Press.

Josan, A., & Voicu, C. (2015). Hybrid Wars in the Age of Asymmetric Conflicts. *Review of the Air Force Academy, 1*(28), 49–52.

Joshi, M. (2022). *China's military-civil fusion strategy: The US response and implications for India*. Observer Research Foundation. https://www.orfonline.org/research/china-s-military-civil-fusion-strategy-the-us-response-and-implications-for-india

Jowet, G. S., & O'donnel, V. (1999). *Propaganda and Persuasion*. Sage Publications.

JUNG, C. G. (1959). Collected Works of C.G. Jung, Volume 9 (Part 2): Aion: Researches into the Phenomenology of the Self (G. ADLER & R. F. C. HULL, Eds.). Princeton University Press. [suspicious link removed]

JUNG, C. G. (1969). Collected Works of C.G. Jung, Volume 9 (Part 1): Archetypes and the Collective Unconscious (G. ADLER & R. F. C. HULL, Eds.). Princeton University Press. [suspicious link removed]

Kahneman, D. (2011). *Thinking fast and slow*. Farrar, Straus and Giroux.

Kaldor, M. (2013). In defence of New Wars. *Stability, 2*(1), 1–16. DOI: 10.5334/sta.at

Kambam, P. R., Pozios, V. K., Bond, K. L., & Ostermeyer, B. K. (2020). The influence of media related to mass shootings. *Psychiatric Annals, 50*(9), 393–398. DOI: 10.3928/00485713-20200811-01

Kant, I. (1795). *Perpetual Peace: A Philosophical Sketch.* Essays on Kant's Cosmopolitan Ideal.

Kaplan, E. A. (2005). *Trauma culture: The politics of terror and loss in media and literature.* Rutgers University Press.

Kara, M. (2013) Siber Saldırılar, Siber Savaşlar ve Etkileri, [Unpublished graduate dissertation]. İstanbul Bilgi University.

Karagöz, K. (2018). Post-Truth Çağında Yayıncılığın Geleceği. *Trt-Akademi, 3*(6), 678–709.

Karakaya, S. (2006). Türk Medyasında Haberin Manipülasyonu ve Dezenformasyon. In Türkoğlu, N. (Ed.), *Medya Okuryazarlığı* (pp. 176–178). Marmara Üniversitesi İletişim Fakültesi.

Kathju, J. (2025). Indian media fuels panic with disinformation. *Al Jazeera Media Institute.* https://institute.aljazeera.net/en/ajr/article/indian-media-fuels-panic-disinformation

Katoch, P. C. (2023). Brain warfare. *SP's Naval Forces.* https://www.spsnavalforces.com/experts-speak/?id=606&h=Brain-Warfare

Kaura, V. (2020). India's Pakistan policy: From 2016 'surgical strike' to 2019 Balakot 'airstrike'. *The Round Table, 109*(3), 317–327. DOI: 10.1080/00358533.2020.1760499

Kearney, R. (1988). *Transitions: Narratives in modern Irish culture.* Manchester University Press.

Keeling, M. E., Ozuna, S. M., Kintzle, S., & Castro, C. A. (2018). Veterans' Civilian Employment Experiences: Lessons Learnt From Focus Groups. *Journal of Career Development,* 46(6), 692-705. https://doi.org/ (Original work published 2019)DOI: 10.1177/0894845318776785

Kellner, D. (2009). The Persian Gulf TV War Revisited. https://pages.gseis.ucla.edu/faculty/kellner/essays/gulftvwarrevisited.pdf

Kellner, D. (2012). School shootings, crises of masculinities, and media spectacle: Some critical perspectives. In *School Shootings: Mediatized Violence in a Global Age* (pp. 299–330). Emerald Group Publishing Limited. DOI: 10.1108/S2050-2060(2012)0000007018

Kelly, L. (2013). *Surviving sexual violence.* John Wiley & Sons.

Kelly, R. J. (2007). *The foraging spectrum: Diversity in hunter-gatherer lifeways.* ISD. DOI: 10.2307/j.ctv2sx9gc9

Kelly, R. L. (2013). *The lifeways of hunter-gatherers: the foraging spectrum.* Cambridge University Press. DOI: 10.1017/CBO9781139176132

Kelman, H. C. (1973). Violence without moral restraint: Reflections on the dehumanization of victims and victimizers. *The Journal of Social Issues, 29*(4), 25–61. DOI: 10.1111/j.1540-4560.1973.tb00102.x

Kerner, A. (2011). *Film and the Holocaust: New perspectives on dramas, documentaries, and experimental films.* A&C Black. DOI: 10.5040/9781628928716

Keskin, F. (1998). *Uluslararası Hukukta Kuvvet Kullanma: Savaş, Karışma ve Birleşmiş Milletler.* Mülkiyeliler Birliği Vakfı Yayınları.

Kidd, I. J., Medina, J., & Pohlhaus, G.Jr., (Eds.). (2017). *The routledge handbook of epistemic injustice* (1st ed.). Routledge., DOI: 10.4324/9781315212043

Kilcullen, D. (2019). The evolution of unconventional warfare. *Scandinavian Journal of Military Studies, 2*(1), 61–71. DOI: 10.31374/sjms.35

Kim, Y. (2003). Kidnapping Politics in East Asia. *Journal of East Asian Studies (Seoul).*

Kingston, R. (2014). *The pirate's nest: The rise and fall of Blackbeard's pirate haven.* University of North Carolina Press.

Kirkegaard, A. (2017). Silencing artists: Reflections on music censorship in the case of Mahsa Vahdat. *Researching music censorship*, 28-51.

Kizilhan, J. I. (2025). Mental health and collective trauma among Yazidi Genocide Survivors. *Journal of Mental Health Disorders, 5*(1), 47–49. DOI: 10.33696/mentalhealth.5.036

Kizilhan, J. I., Berger, T., Sennhauser, L., & Wenzel, T. (2023). The psychological impact of genocide on the Yazidis. *Frontiers in Psychology, 14*, 1074283. DOI: 10.3389/fpsyg.2023.1074283

Kleis Nielsen, R., & Graves, L. (2020). News You Don't Believe: Audience Perspectives on Fake News. *Journalism Studies, 21*(6), 826–841. DOI: 10.1080/1461670X.2020.1745667

Knowles, C. (2023). Articulating understanding: a phenomenological approach to testimony on gendered violence. In M, Altanian & M. Baghramian (Eds.), *Testimonial Injustice and Trust* (pp. 173-197). Routledge. https://doi.org/DOI: 10.4324/9781003396789-14

Kolhatkar, A. J. (2024). Escalation and air power. *Journal of Defence Studies, 18*(2), 101–120. https://www.idsa.in/wp-content/uploads/2024/08/18-2_AJ-Kolhatkar-07 .pdf

Korpe, M., Reitov, O., & Cloonan, M. (2006). Music censorship from Plato to the present. *Music and manipulation: On the social uses and social control of music*, 1, 239-263.

Kovacs, M. G. (1989). *The epic of Gilgamesh.* Stanford University Press.

Kubany, E. S., & Manke, F. P. (1995). Cognitive therapy for trauma-related guilt: Conceptual bases and treatment outlines. *Cognitive and Behavioral Practice, 2*(1), 27–61. DOI: 10.1016/S1077-7229(05)80004-5

Kübler-Ross, E., & Kessler, D. (2014). *On grief and grieving: Finding the meaning of grief through the five stages of loss.* Simon and Schuster.

Kuhn, U., & Williams, H. (2025). Governing the impact of emerging technologies: Actors, technologies, and regulation. *The Journal of Strategic Studies, 48*(3), 559–570. DOI: 10.1080/01402390.2025.2472213

Kulasekaran, V. (Director). (2021). *Shershaah* [Film].

Kurubaş, E. (2017). Dünya Politikasının Dönüşümü ve Uluslararası İlişkiler Çağı. In Kardaş, Ş., & Balcı, A. (Eds.), *Uluslararası İlişkilere Giriş* (7th ed.). Küre Yayınları.

Kwint, M., Breward, C., & Aynsley, J. (1999). *Material memories: Design and evocation.* Berg.

Lacan, J. (1977). *Écrits: A selection.* W. W. Norton & Company.

Lacan, J. (1998). *The four fundamental concepts of psychoanalysis* (Vol. 11). W. W. Norton & Company.

LaCapra, D. (2001). *Writing history, writing trauma.* Johns Hopkins University Press.

Lakho, M. B., Siddiqui, H. A., Bakhtawar, S., & Soomro, M. A. (2025). Media Narratives and Public Perception during Cross-Border Tensions: A Comparative Study of Pakistan and India. *Review Journal of Social Psychology & Social Works, 3*(2), 748–765. DOI: 10.71145/rjsp.v3i2.234

Landsberg, A. (2004). *Prosthetic memory: The transformation of American remembrance in the age of mass culture*. Columbia University Press.

Landsman, S., & Bredin, J. (1987). The triumph of justice. *Michigan Law Review, 85*(5/6), 1095. DOI: 10.2307/1289035

Langman, P. (2017). *Role Models, Contagions, and Copycats: An Exploration of the Influence of Prior Killers on Subsequent Attack.* Available at: https://schoolshooters .info/role-models- *contagious-and-copycats-exploration-influence-prior-killers-subsequent-attacks* (accessed 12 May 2024)

Langman, P. (2015). *School shooters: Understanding high school, college, and adult perpetrators*. Rowman and Littlefield. DOI: 10.5771/9781442233577

Lankford, A., & Madfis, E. (2017). Don't name them, don't show them, but report everything else: A pragmatic proposal for denying mass killers the attention they seek and deterring future offenders. *The American Behavioral Scientist, 62*(2), 260–279. DOI: 10.1177/0002764217730854

Lanoszka, A. (2016). Russian Hybrid Warfare and Extended Deterrence in Eastern Europe. *International Affairs, 92*(1), 175–195. DOI: 10.1111/1468-2346.12509

Larson, E. M. (1717). *Wallenberg's account of the Kraken.* [Primary Source].

Latif Ega, Abdi . "The Great Somali Apologia." *Africasacountry.com*, 26 Mar. 2019, africasacountry.com/2019/03/the-great-somali-apologia. Accessed 11 July 2025.

Laub, D. (1991). Truth and testimony the process and the struggle. *The American Imago, 48*(1), 75–91.

Le Bon, G. (2014). *Kitleler Psikolojisi* (Can, H., Trans.). Tutku Yayınevi.

le Roux-Kemp, A. (2014). Struggle Music: South African Politics in Song. *Law and Humanities, 8*(2), 247–268. DOI: 10.5235/17521483.8.2.247

Lee, R. A. (2009). Protest Music as Alternative Media During the Vietnam War Era. *War and the media: Essays on news reporting, propaganda and popular culture,* 24-40.

Lee, S. (2025). *Cinematic Landscapes and Geopolitical Imaginaries*. Numberanalytics.com. https://www.numberanalytics.com/blog/cinematic-landscapes-geopolitical -imaginaries

Lee, H., Pickett, J. T., Burton, A. L., Cullen, F. T., Jonson, C. L., & Burton, V. S.Jr. (2022). Attributions as anchors: How the public explains school shootings and why it matters. *Justice Quarterly, 39*(3), 497–524. DOI: 10.1080/07418825.2020.1769710

Lefebvre, H. (1974). La production de l'espace. *Anthropos: Studie z Oboru Anthropologie, Paleoethnologie, Paleontologie a Kvartérní Geologie.*

Lefebvre, H. (1991). *The production of space* (Nicholson-Smith, D., Trans.). Blackwell. (Original work published 1974)

Lerner, M. J. (1980). The Belief in a Just World. In *The Belief in a Just World. Perspectives in Social Psychology*. Springer., DOI: 10.1007/978-1-4899-0448-5_2

Levine, M. G. (2006). *The belated witness: Literature, testimony, and the question of Holocaust survival.* Stanford University Press., DOI: 10.1515/9781503616820

Lewis, D., & Madon, S. (2004). Information systems and nongovernmental development organizations: Advocacy, organizational learning, and accountability. *The Information Society, 20*(2), 117–126. DOI: 10.1080/01972240490423049

Leys, R. (2007). *From guilt to shame: Auschwitz and after* (Vol. 8). Princeton University Press.

Lind, W. S. (2004). Understanding Forth Generation War. U.S.Army Combined Arms Center. https://www.hsdl.org/?view&did=482203

Lindsey, L. L. (2020). *Gender: Sociological perspectives.* Routledge., DOI: 10.4324/9781315102023

Lind, W. S., Nightengale, K., Schmitt, J. F., Sutton, J. W., & Wilson, G. I. (1989). The Changing Face of War: Into the Fourth Generation. *The Marine Corps Gazette, 73*(10), 22–26.

Lind, W. S., Nightengale, K., Schmitt, J. F., Sutton, J. W., & Wilson, G. I. (2001). The Changing Face Of War: Into The Fourth Generation. *The Marine Corps Gazette, 85*(11), 65–68.

Linebarger, P. M. A. (1954). *Psychological warfare* (2nd ed.). Duell, Sloan and Pearce.

Lippmann, W. (1922). *Public opinion.* Harcourt, Brace.

Lister, A. M., & Stuart, A. J. (2019). The extinction of the giant deer Megaloceros giganteus (Blumenbach): New radiocarbon evidence. *Quaternary International, 500*, 185–203. DOI: 10.1016/j.quaint.2019.03.025

Litz, B. T., Stein, N., Delaney, E., Lebowitz, L., Nash, W. P., Silva, C., & Maguen, S. (2009). Moral injury and moral repair in war veterans: A preliminary model and intervention strategy. *Clinical Psychology Review, 29*(8), 695–706. DOI: 10.1016/j.cpr.2009.07.003 PMID: 19683376

Livingstone, S., & Das, R. (2013). *Interpretation/Reception* (pp. 9780199756841–0134) [Dataset]. DOI: 10.1093/obo/9780199756841-0134

Livingstone, S. (2019). Reception Studies. In *The Blackwell Encyclopedia of Sociology* (pp. 1–3). John Wiley & Sons, Ltd., DOI: 10.1002/9781405165518.wbeosr031.pub3

Ljepava, N., & Selakovic, M. (2019). Trends and challenges of political polling and research in the digital age. In Bulatovic, Lj. (Ed.), *Digitalni prostori–izazovi i očekivanja/Digital spaces–challenges and expectations* (pp. 117–140).

Lodewijkx, H. F., Kersten, G. L., & Van Zomeren, M. (2008). Dual pathways to engage in 'silent marches' against violence: Moral outrage, moral cleansing and modes of identification. *Journal of Community & Applied Social Psychology, 18*(3), 153–167. DOI: 10.1002/casp.916

Lowenthal, D. (1989). Nostalgia Tells It Like It Wasn't. In Chase, M., & Shaw, C. (Eds.), *The Imagined Past: History and Nostalgia* (1st ed., pp. 18–32). Manchester University Press., https://www.amazon.co.uk/Imagined-Past-History-Nostalgia/dp/0719028752

Lubsanov, A. E., & Yantranova, V. B. (2022). The School Shooting Phenomenon, in Chepak. A. (Ed.): *All-Russian Contest for the Student's Scientific Project in a Foreign Language* (pp.72-75).

Luhmann, N. (1990). The Cognitive Programs of Constructivism and a Reality That Remains Unknown. In Krohn, W., Küppers, G., & Nowotny, H. (Eds.), *Selforganization: Portrait of Scientific Revolution* (pp. 64–85). Springer., https://doi.org/https://doi.org/10.1007/978-94-017-2975-8_5 DOI: 10.1007/978-94-017-2975-8_5

Mabima, J. (2018). *Social networking sites as a tool of psychological operations: A case study. SSRN.* https://doi.org/DOI: 10.2139/ssrn.3261039

Mac Ginty, R. (2021). Temporality and Contextualisation in Peace and Conflict Studies: The Forgotten Value Of War Memoirs And Personal Diaries. *Cooperation and Conflict, 57*(2), 191–209. DOI: 10.1177/00108367211027605

Machado, L. (2016). *A palimpsest-image: Place, space and film geographies* (Doctoral dissertation, University of Westminster).

MacKenzie, J. M. (1984). *Propaganda and empire: The manipulation of British public opinion, 1880-1960* (Rev. ed.). Manchester University Press.

MacKinnon, C. A. (2017). Rape, genocide, and women's human rights. In Lattimer, M. (Ed.), *Genocide and Human Rights* (pp. 133–144). Routledge., DOI: 10.4324/9781351157568-4

Madlingozi, T. (2010). On transitional justice entrepreneurs and the production of victims. *Journal of Human Rights Practice*, *2*(2), 208–228. DOI: 10.1093/jhuman/huq005

Mandela, N. (2008). *Long walk to freedom: The Autobiography of Nelson Mandela*. Little, Brown.

Maniratnam, G. R. (Director). (1992). *Roja* [Film].

Manovich, L, (2001). The Language of New Media. Massachusetts Institute of Technology.

Mansoor, P. R. (2012). Introduction: Hybrid Warfare in History. in Peter R. Mansoor ve Williamson Murray (Ed.), Hybrid Warfare: Fighting Complex Opponents from the Ancient World to the Present. Cambridge University Press.

Marche, G. (2012). Expressivism and resistance: Graffiti as an infrapolitical form of protest against the war on terror. *Revue Française d'Études Américaines*, *131*(1), 78–96. DOI: 10.3917/rfea.131.0078

Marizka, A. (2024). *Retracing The Past Through Historical Fantasy Novel: The Poppy War* (Doctoral dissertation, UNIMED).

Marković, S., & Bulut, T. (2017). Attractiveness of the female body: Preference for the average or the supernormal? *Psihologija (Beograd)*, *50*(3), 403–426. DOI: 10.2298/PSI1703403M

Markus, H. (2021). Maritime masculinity and the myth of *Captain Phillips*. *Journal of Cultural Studies*, *15*(1), 78–93.

Marusic, B. (2015). *Evidence and agency: Norms of belief for promising and resolving*. OUP Oxford., DOI: 10.1093/acprof:oso/9780198714040.001.0001

Masero, A. (2010). The Changes in Irish Dance Since Riverdance.

Maslin, J. (1998). Saving Private Ryan. *The New York Times*, E1–E28. https://archive.nytimes.com/www.nytimes.com/library/film/072498ryan-film-review.html

Massey, D. (2013). *Space, place and gender*. University of Minnesota Press.

McAdams, D. P., & McLean, K. C. (2013). Narrative Identity. Current Directions in Psychological Science, 22(3), 233-238. https://doi.org/ (Original work published 2013)DOI: 10.1177/0963721413475622

McAuliffe, C., & Iveson, K. (2011). Art and crime (and other things besides…): Conceptualising graffiti in the city. *Geography Compass*, *5*(3), 128–143. DOI: 10.1111/j.1749-8198.2011.00414.x

McCombs, M. (2002, June). The agenda-setting role of the mass media in the shaping of public opinion. In *Mass Media Economics 2002 Conference, London School of Economics*: https://sticerd.lse.ac.uk/dps/extra/McCombs.pdf

McCombs, M., & Valenzuela, S. (2020). *Setting the agenda: Mass media and public opinion*. John Wiley & Sons.

McElroy, P. C. (2011). On redemption, the value of putting bad things to good use. Stanford Digital Repository. https://purl.stanford.edu/gq597rc8146

McIvor, C. (2012). Essences of social change: City Fusion, interculturalism and the Dublin St. Patrick's Day festival in post-Celtic Tiger Ireland. *Public*, *23*(45), 180–191. DOI: 10.1386/public.23.45.180_1

McJunkin, B. A. (2023). Rape as indignity. *SSRN*, *109*(385). Advance online publication. DOI: 10.2139/ssrn.4347325

McMahon, L. (2019). *Animal Worlds: Film, Philosophy and Time* (1st ed.). Edinburgh University Press., DOI: 10.1515/9781474446402

McNeal, J. (2023). *Ambiguous Morality in the Human Condition as Reflected by Modern Villains from Popular Culture* (Doctoral dissertation). https://hdl.handle.net/2104/12591

McWhirter, C. (2012). Christian, *Battle Hymns: The Power and Popularity of Music in the Civil War* (Chapel Hill, NC, 2012; online edn. *North Carolina Scholarship Online*, *24*. Advance online publication. Retrieved May 15, 2025, from. DOI: 10.5149/9780807882627_mcwhirter

Mearsheimer, J. J. (2001). *The tragedy of great power politics*. W. W. Norton & Company.

Medhurst, M. J., Ivie, R. L., Wander, P., & Scott, R. L. (1997). *Cold War Rhetoric: Strategy, Metaphor, and Ideology*. Michigan State University Press. [suspicious link removed]

Meindl, J. N., & Ivy, J. W. (2017). Mass Shootings: The Role of the Media in Promoting Generalized Imitation. *American Journal of Public Health*, *107*(3), 368–370. DOI: 10.2105/AJPH.2016.303611 PMID: 28103074

Metz, C. (1991). *Film language: A Semiotics of the Cinema* (Taylor, M., Trans.). Univ. of Chicago Press.

Migeon, M. (2024). Peace and Conflict in Public Space: Gendered Murals Shaping Belfast. *Geopolitics*, *29*(5), 1604–1634. DOI: 10.1080/14650045.2024.2321164

Mihailov, E., Voinea, C., & Vică, C. (2023). Is Online Moral Outrage Outrageous? Rethinking the Indignation Machine. *Science and Engineering Ethics*, *29*(2), 12. DOI: 10.1007/s11948-023-00435-3 PMID: 37000374

Miladi, N. (2018). Urban Graffiti, Political Activism and Resistance. In *The Routledge Companion to Media and Activism* (pp. 241–249). Routledge. DOI: 10.4324/9781315475059-26

Miller, D. T., Effron, D. A., & Zak, S. V. (2011). From moral outrage to social protest: The role of psychological standing, in Bobocel, R.; Kay, A.C.; Zanna, M.P., and Olson, J.M. (Eds): *The Psychology of Justice and Legitimacy: The Ontario Symposium* Volume 11 *(Ontario Symposia on Personality and Social Psychology Series) 1st Edition* (pp.103-124). Psychology Press.

Miller, M. (2024). Influence Campaigns From Iran, China, Russia Ramping Up Ahead of Elections, Microsoft Finds. Politico. https://www.politico.com/news/2024/10/23/influence-iran-china-russia-elections-microsoft-00185075

Miller, M. K., Clark, J. D., & Jehle, A. (2015). *Cognitive Dissonance Theory (Festinger).* The Blackwell Encyclopedia of Sociology., DOI: 10.1002/9781405165518.wbeosc058.pub2

Miller, M. K., Rauch, J. A., & Kaplan, T. (2016) Gender Differences in Movie SuperHeroes' Roles, Appearances, and Violence. Ada: A Journal of Gender, New Media, and Technology, No. 10. DOI: 10.7264/N3HX19ZK

Mills, R. (2014). Letters of marque and reprisal. *Journal of Maritime Law and Commerce*, *45*(2), 201–215.

Ministry of External Affairs. (2025). *Statement by Foreign Secretary: OPERATION SINDOOR* [Press release]. Government of India. https://www.mea.gov.in/Speeches-Statements.htm?dtl/39473/Statement+by+Foreign+Secretary+OPERATION+SINDOOR

Misek, R. (2008). Exploding Binaries: Point-of-View and Combat in The Thin Red Line. *Quarterly Review of Film and Video*, *21*(1), 116–123. DOI: 10.1080/10509200601074710

Mishra, S., & Desai, H. (2025). The Media Framing of India's Nuclear Weapons Discourse. *International Journal of Nuclear Security*, *9*(1). Advance online publication. DOI: 10.7290/ijns09153822

Misra, A. (2011). The Missionary Position: Christianity and Politics of Religious Conversion in India. *Nationalism & Ethnic Politics*, *17*(4), 361–381. DOI: 10.1080/13537113.2011.622635

Mohammed, H. A. (2022). The last genocide against the Yazidi People. *Review of Middle East Studies*, *56*(1), 108–122. DOI: 10.1017/rms.2023.1

Monaco, J. (2009). *How to read a film: Movies, media, and beyond*. Oxford University Press.

Monteleone, E. (Director). (2002). El Alamein: The line of fire [Film]. Fiorella Film, RAI Cinema.

Moody, C. (2021). Iranian music censorship & international human rights law. *Brooklyn Journal of International Law*, *47*(1), 232–259.

Mora, R. A. (2014). Counter-narrative. *Qualitative Inquiry*, *8*(1), 23–44.

Morris, R. C. (Ed.). (2010). *Can the subaltern speak?: Reflections on the history of an idea*. Columbia University Press.

Mowery, N. (2021). *Even Objects Revolt Against Regimentation: Art and Censorship in Post-WWI Germany and the Rise of Dadaism. A Film and Thesis* (Doctoral dissertation, University of Oregon).

Mulgan, R. (2000). 'Accountability': An ever-expanding concept? *Public Administration*, *78*(3), 555–573. DOI: 10.1111/1467-9299.00218

Mulvey, L. (1989). Visual pleasure and narrative cinema. In *Visual and other pleasures* (pp. 14–26). Springer. DOI: 10.1007/978-1-349-19798-9_3

Muoghalu, C. O. (2012). Rape and women's sexual health in Nigeria: The stark realities of being female in a patriarchal world. *American Anthropologist*, *19*(1-2), 33–41.

Murad, N. (2018). *The last girl: my story of captivity, and my fight against the Islamic State*. Crown.

Murphy, M. (2011). *Small boats, weak states, dirty money: Piracy and maritime terrorism in the modern world.* Hurst.

Murray, H., Pethania, Y., & Medin, E. (2021). Survivor guilt: A cognitive approach. *Cognitive Behaviour Therapist*, *14*, e28. DOI: 10.1017/S1754470X21000246 PMID: 34557258

Murtaza, G., & Saud, A. (2025). India's Fifth-Generation Warfare against Pakistan and the Role of Indian Media in Shaping Public Perceptions of the China-Pakistan Economic Corridor (CPEC). *Journal of Social Sciences Review*, *5*(1), 249–256. DOI: 10.62843/jssr.v5i1.487

Musarat, Y. (2024). (De-)Legitimizing War: A Linguistic Analysis of Indian and Pakistani Civil and Military Conflict Discourses. *Critical Military Studies*, *10*(1), 61–82. DOI: 10.1080/23337486.2023.2188997

Muschert, G. W. (2002). *Media and massacre: The social construction of the Columbine story*. University of Colorado at Boulder.

Musolff, A. (2021). *National Conceptualisations of the Body Politic: Cultural Experience and Political Imagination*. Springer., DOI: 10.1007/978-981-15-8740-5

Muszyński-Sulima, W. (2023). Cold War in space: Reconnaissance satellites and US-Soviet security competition. *European Journal of American Studies*, *18*(2). Advance online publication. DOI: 10.4000/ejas.20427

Mutua, M. (2002). *Human rights: A political and cultural critique*. University of Pennsylvania Press., DOI: 10.9783/9780812204155

Myllylä, M. (2020). The Good, The Bad, and The Ugly Graffiti. Emotions In Technology Design: From Experience To Ethics, 87-104. https://doi.org/DOI: 10.1007/978-3-030-53483-7_6

Nabavi, T. (2012). *Razieh & Bijandi, Mohammad*. Bandura's Social Learning Theory & Social Cognitive Learning Theory.

Nag, H. (2023). *Drone Warfare Capacity Building in Securing India's National Interests*. Vivekananda International Foundation.

Nagle, J. (2005). "Everybody is Irish on St. Paddy's": Ambivalence and Alterity at London's St. Patrick's Day 2002. *Identities (Yverdon)*, *12*(4), 563–583. DOI: 10.1080/10702890500332733

Nagle, W. J. (2023). *Morality and modern warfare*. Highlyy Publishing LLP.

Nagy, M. (1948). The child's theories concerning death. *The Pedagogical Seminary and Journal of Genetic Psychology*, *73*(1), 3–27. DOI: 10.1080/08856559.1948.10533458 PMID: 18893204

Nasriddinov, D. (2023). Fantasy Genre and Its Scientific Interpretation in Theoretical Views. https://doi.org/DOI: 10.47689/2181-3701-vol1-iss2-pp56-62

Nasui, R. C. (2023). An Upgrade to the Gepard System. Possible Usage of the CA-94. *AFASTUD*, *2023*, 28.

NDTV. (2025). India-Pakistan tension live update: Operation Sindoor, Pahalgam terror attack, Jammu and Kashmir, Rajasthan, blackout in Punjab. https://www.ndtv.com/india-news/india-pakistan-tension-live-update-operation-sindoor-pahalgam-terror-attack-jammu-and-kashmir-rajasthan-blackout-in-punjab-8366542

Nelson, K. C. (2022). Malefaction and atrocities against women and girl child. *The Indian Journal of Law and Legal Research*, *4*(1).

Neureiter, M. (2016). Sources of media bias in coverage of the Israeli–Palestinian conflict: The 2010 Gaza flotilla raid in German, British, and US newspapers. *Israel Affairs*, *23*(1), 66–86. DOI: 10.1080/13537121.2016.1244381

New York Times. (2025). India and Pakistan Talked Big, but Satellite Imagery Shows Limited Damage. https://www.nytimes.com/interactive/2025/05/14/world/asia/india-pakistan-attack-damage-satellite-images.html

Nguyen, P. D. (2023). *The role of small and medium UAVs in recent conflicts: Lessons and forecasting trends in the development of UAV forces in the military* [Master's dissertation]. Canadian Forces College.

Nihlani, G. (Director). (1982). *Vijetha* [Film].

Nilmaola, S. (2017). Deconstructing cinematic piracy: Representing Somali pirates in *Captain Phillips. Critical Studies in Media Communication*, *34*(4), 347–362.

Nordmarker, A. (2016). Graffiti–For Joy and Confirmation. Motivational Aspects, Triggering and Inhibiting Factors, and Emotional Satisfactions In Graffiti: The Creative-Interactive Dimension Of Vandalism.

Norris, F. H., Stevens, S. P., Pfefferbaum, B., Wyche, K. F., & Pfefferbaum, R. L. (2008). Community resilience as a metaphor, theory, set of capacities, and strategy for disaster readiness. *American Journal of Community Psychology*, *41*(1-2), 127–150. DOI: 10.1007/s10464-007-9156-6

North Atlantic Treaty Organization. (2020). Game of drones: How new technologies affect deterrence, defence, and security. *NATO Review*. https://www.nato.int/docu/review/articles/2020/05/05/game-of-drones-how-new-technologies-affect-deterrence-defence-and-security/index.html

Ó Gráda, C. (2004). Ireland's Great Famine: an overview.

O'Brien, C., Gallagher, S., & Malafouris, L. (2025). Handmade Therapy: The Hedonic Impacts of Engaging in Pottery Making. *Topoi*, *44*(4), 1–14. DOI: 10.1007/s11245-025-10181-9

O'Brien, T. (1990). *The Things They Carried*. Houghton Mifflin.

O'Connor, S. (2019). *How Chinese companies facilitate technology transfer from the United States*. U.S.-China Economic and Security Review Commission. https://www.uscc.gov/sites/default/files/Research/How%20Chinese%20Companies%20Facilitate%20Tech%20Transfer%20from%20the%20US.pdf(WITA)

O'Mara, E. M., Jackson, L. E., Batson, C. D., & Gaertner, L. (2011). Will moral outrage stand up? Distinguishing among emotional reactions to a moral violation. *European Journal of Social Psychology, 41*(2), 173–179. DOI: 10.1002/ejsp.754

O'Toole, M. E., Beresin, P. E., Berman, A., Gorelick, S. M., Helfgott, J. B., & Tobin, C. (2014). Celebrities Through Violence: The Copycat Effect and the Influence of Violence in Social Media on Mass Killers. *Violence and Gender, 1*(3), 107–116. DOI: 10.1089/vio.2014.1512

Ohri, R., & Sharma, R. (2014, September 18). Maulvi in Army allegedly censured for using slogan that sends 'message of extremism.' *The Economic Times*. https://economictimes.indiatimes.com/news/politics-and-nation/maulvi-in-army-allegedly-censured-for-using-slogan-that-sends-message-of-extremism/articleshow/42747816.cms

Okulate, G. T., & Jones, O. B. E. (2006). Post-Traumatic Stress Disorder, Survivor Guilt and Substance Use–A Study Of Hospitalised Nigerian Army Veterans. *South African Medical Journal, 96*(2), 144–146. PMID: 16532084

Oleinik, A. (2023). War Propaganda Effectiveness: A Comparative Content-Analysis Of Media Coverage of the Two First Months of Russia's Invasion of Ukraine. *Atlantic Journal of Communication, 32*(4), 526–544. DOI: 10.1080/15456870.2023.2187801

Oliver, K. (2001). *Witnessing: beyond recognition*. University of Minnesota Press.

Olsen, J. A. (2010). *A History of Air Warfare*. University of Nebraska Press. https://muse.jhu.edu/book/28122

Outlook. (2025). Fact-check: Viral claim of Indian strike on Karachi port uses 2020 Gaza images. *Outlook India*. https://www.outlookindia.com/national/fact-check-viral-claim-of-indian-strike-on-karachi-port-uses-2020-gaza-images

Owen, A. S. (2002). Memory, war and American identity: *Saving Private Ryan* as cinematic jeremiad. *Critical Studies in Media Communication, 19*(3), 249–282. DOI: 10.1080/07393180216565

Öymen, O. (2014). *Bir Propaganda Silahı Olarak Basın*. Remzi Yayınevi.

Özdemir, İ. (2015). Propaganda Aracı Olarak Sosyal Medya. T.C Republic of Türkiye Ministry of Interior. https://www.icisleri.gov.tr/arem/propaganda-araci-olarak-sosyal-medya

Özdemiroğlu, P. (2013). Sahte Tweet'ler Bizi Yıldıramaz, Milliyet Gazetesi, https://www.milliyet.com.tr/ekonomi/sahte-tweet-ler-bizi-yildiramaz-1698125

Pagels, E. (1995). *The Origin of Satan: How Christians Demonized Jews, Pagans, and Heretics*. Vintage Books.

Paget, D., & Lipkin, S. N. (2009). 'Movie-of-the-Week' docudrama, 'historical-event' television, and the Steven Spielberg series *Band of Brothers. New Review of Film and Television Studies*, *7*(1), 93–107. DOI: 10.1080/17400300802603047

Pandit, S., & Chattopadhyay, S. (2017). Coverage of the surgical strike on television news in India: Nationalism, journalistic discourse and India–Pakistan conflict. *Journalism Studies*, *12*(2), 162–176.

Panofsky, E. (2012). Style and Medium in the Motion Pictures. In Cardullo, B. (Ed.), *Stage and Screen: Adaptation Theory from 1916 to 2000* (pp. 39–54). Continuum., DOI: 10.5040/9781628928983.ch-004

Pantelic, Z. (2022). *Youth Emigration and Political Distrust in Serbia*. Westminster Foundation for Democracy.

Parashar, S. (2018). Discursive (in)securities and postcolonial anxiety: Enabling excessive militarism in India. *Security Dialogue*, *49*(1–2), 123–135. DOI: 10.1177/0967010617746527

Paris, M. (2000). *From the Wright brothers to Top Gun: Aviation, nationalism, and popular cinema*. Manchester University Press.

Parker, G. (1988). *The Military Revolution: Military Innovation and the Rise of the West, 1500–1800*. Cambridge University Press.

Parrott, J. (2021). *Swimming through the Bardo: Death as a Means to Self-Discovery*. Pacifica Graduate Institute.

Parziale, A. E. (2013). *Representations of trauma in contemporary American literature and film: Moving from erasure to creative transformation*.

Patterson, O. (2018). *Slavery and social death: A comparative study, with a new preface*. Harvard University Press.

Peace, P. (1917). *A Philosophical Essay, translated with Introduction and Notes by M. Campbell Smith, with a Preface by L. Latta*. George Allen and Unwin.

Pearl, D., Intriligator, J., & Liu, W. (2025). Seamless integration: The evolution, design and future impact of wearable technology (Version 1) [Preprint]. *arXiv*. https://arxiv.org/abs/2502.05797v1

Pecht, E., & Tishler, A. (2015). The value of military intelligence. *Defence and Peace Economics*, *26*(2), 179–211. DOI: 10.1080/10242694.2014.886435

Pejuskovic, B., & Lecic-Tosevski, D. (2023). School Shooting in Serbia–A Huge Collective Trauma and Challenge for Public Mental Health. *World Social Psychiatry*, *5*(2), 179–182. DOI: 10.4103/wsp.wsp_46_23

Pembecioğlu, N. (2019). *Duyarsızlaşma, Kültürlerarası Anlayış ve Ölüm Üstüne: Bir Jindabyne Çözümlemesi. Sinema Kültür İletişim Okumaları*. Parşömen.

Pembecioğlu, N. (2019). *Policy Makers*. Educators & Parents Up For Media Literacy At The Crossroads Of Rights, Values And Identity, CiCea Conference.

Pereira, V., & Kunkolienkar, G. R. (2013). EMP (Electro-Magnetic Pulse) weapon technology along with EMP shielding & detection methodology. In *2013 Fourth International Conference on Computing, Communications and Networking Technologies (ICCCNT)* (pp. 1-5). IEEE. DOI: 10.1109/ICCCNT.2013.6726651

Peri, D. (2025). Pakistan used 400 drones, possibly from Turkiye: India. *The Hindu*. https://www.thehindu.com/news/national/pakistan-launched-300-400-turkish-drones-across-western-border-pak-ad-radar-hit-in-indian-retaliation/article69558073.ece

Periasamy, R. (Director). (2024). *Amaran* [Film].

Pescara, K. L., and Raleigh, M. J. (2017). The Contagion Effect as it Relates to Public Mass Shootings and Suicides, *Journal of Campus Behavioral Intervention (J-BIT)*, 5, pp.35-45.

Petzold, D. (1986). Fantasy Fiction and Related Genres. *Modern Fiction Studies*, *32*(1), 11–20. DOI: 10.1353/mfs.0.1205

Pevere, G. (2013, August 16). *A Hijacking: Hostage thriller turns focus to the cost of saving lives*. The Globe and Mail. https://www.theglobeandmail.com/arts/film/film-reviews/hostage-thriller-turns-focus-to-the-cost-of-saving-lives/article13802725/

Pew Research Center. (2024). How Americans Navigate Politics on TikTok, X, Facebook and Instagram. https://www.pewresearch.org/internet/2024/06/12/how-americans-navigate-politics-on-tiktok-x-facebook-and-instagram/

Pick, A. (2011). *Creaturely poetics : animality and vulnerability in literature and film*. Columbia University Press., https://cup.columbia.edu/book/creaturely-poetics/9780231147866/

Pieris, R. (1951). Speech and Society: A sociological approach to language. *American Sociological Review, 16*(4), 499–505. DOI: 10.2307/2088282

Popkin, J. D. (2003). Holocaust memories, historians' memoirs: First-person narrative and the memory of the Holocaust. *History & Memory, 15*(1), 49–84. DOI: 10.2979/his.2003.15.1.49

Potot-Warren, J. (2020). Identifying Genocide: The Yazidi massacre in the context of the convention on the prevention and punishment of genocide 1948. *The Student Journal of Professional Practice and Academic Research, 2*(1), 30–49. DOI: 10.19164/sjppar.v2i1.922

Potzsch, H. (2015). The emergence of iBorder: Bordering bodies, networks, and machines. *Environment and Planning. D, Society & Space, 33*(1), 101–118. DOI: 10.1068/d14050p

Prateek, G., Knopf, R., & Peterson, B. C. (2017). 779 Newspaper Articles Show How the Ruling and Opposition Parties Used the Narrative of the "Common Man" during Demonetisation. *Economic and Political Weekly, 52*(36). https://www.epw.in/engage/article/analysis-779-newspaper-articles-shows-how-ruling-and-opposition-parties-used

Proulx, J., Aubut, J., McKibben, A., & Côté, M. (1994). Penile responses of rapists and nonrapists to rape stimuli involving physical violence or humiliation. *Archives of Sexual Behavior, 23*, 295–310. DOI: 10.1007/BF01541565

Puryear, C., Vandello, J. A., & Gray, K. (2024). Moral panics on social media are fueled by signals of virality. *Journal of Personality and Social Psychology, 127*(1), 84–103. DOI: 10.1037/pspa0000379 PMID: 38421751

Puvimanasinghe, T. S., & Price, I. R. (2016). Healing through giving testimony: An empirical study with Sri Lankan torture survivors. *Transcultural Psychiatry, 53*(5), 531–550. DOI: 10.1177/1363461516651361

Raghavan, S. (Director). (2012). *Agent Vinod* [Film].

Raghavapudi, H. (Director). (2022). *Sita Ramam* [Film].

Raichle, M. E. (2015). The brain's default mode network. *Annual Review of Neuroscience, 38*(1), 433–447. DOI: 10.1146/annurev-neuro-071013-014030 PMID: 25938726

Raj, S. J., & Suresh, A. K. (2022). Surgical Strikes on Screen: Narrations of Terrorism and Military Cross-Border Violence in Bollywood Cinema. In Choe, S. (Ed.), *The Palgrave Handbook of Violence in Film and Media* (pp. 169–186). Springer International Publishing., DOI: 10.1007/978-3-031-05390-0_9

Rashi, T., & Schleifer, R. (2023). The Ethics of Psychological Warfare – Lessons from Israel. *Democracy and Security*, *19*(2), 199–210. DOI: 10.1080/17419166.2023.2210472

Rastovac, H. (2009). Contending with Censorship: The Underground Music Scene in Urban Iran. *The McNair Scholars Journal*, *273*, 59–82.

Reeping, P. M., Jacoby, S., Rajan, S., & Branas, C. C. (2020). Rapid response to mass shootings: A review and recommendations. *Criminology & Public Policy*, *19*(1), 295–315. DOI: 10.1111/1745-9133.12479

Regalado, A. (2024). Brainwashing, mind control, and the CIA: The history of Operation Midnight Climax. *MIT Technology Review*. https://www.technologyreview.com/2024/04/12/1090726/brainwashing-mind-control-history-operation-midnight-climax/

Reill, A. (2023, December 5). A Simple Way to Make Better Decisions. *Harvard Business Review*.

Reinke, J. A. (2001). What Are My Middle School Students Reading? *Michigan Reading Journal*, *33*(4), 6.

Renz, B. (2016). Russia and "Hybrid Warfare". *Contemporary Politics*, *22*(3), 283–300. DOI: 10.1080/13569775.2016.1201316

Reuters. (2024). Philadelphia Prosecutor Sues Elon Musk Group to Stop $1 Million Lottery for Voters. https://www.reuters.com/legal/philadelphia-da-sues-elon-musk-pac-stop-1-million-lottery-voters-cnbc-reports-2024-10-28/

Revkin, M. R., & Wood, E. J. (2021). The Islamic State's pattern of sexual violence: Ideology and institutions, policies and practices. *Journal of Global Security Studies*, *6*(2), ogaa038. Advance online publication. DOI: 10.1093/jogss/ogaa038

Richey, L. A., & Ponte, S. (2011). *Brand aid: Shopping well to save the world.* U of Minnesota Press.

Robinson, E. J., Einav, S., & Fox, A. (2013). Reading to learn: Prereaders' and early readers' trust in text as a source of knowledge. *Developmental Psychology*, *49*(3), 505–513. DOI: 10.1037/a0029494

Robinson, L. (2011). Soldiers' stories of the Falklands war: Recomposing trauma in memoir. *Contemporary British History*, *25*(4), 569–589. DOI: 10.1080/13619462.2011.623866

Roseman, I. J. (1984). Cognitive determinants of emotions: A structural theory. *Review of Personality and Social Psychology*, *5*, 11–36.

Ross, R. (2023a). *Divine Rivals*. Macmillan Publishers.

Ross, R. (2023b). *Ruthless Vows*. Macmillan Publishers.

Rothberg, M. (2014). Multidirectional memory in migratory settings: The case of post-Holocaust Germany. In C. De Cesari & A. Rigney, *Transnational Memory: Circulation, Articulation, Scales*, 123-145. https://doi.org/DOI: 10.1515/9783110359107.123

Rothkopf, J. (1999, April 11). Shellshock. *In These Times*, 28–29.

Roy-Chaudhury, R. (2025). India–Pakistan drone and missile conflict: Differing and disputed narratives. *International Institute for Strategic Studies*. https://www.iiss.org/online-analysis/online-analysis/2025/05/indiapakistan-drone-and-missile-conflict-differing-and-disputed-narratives/

Roy, K. (2022). *A Global History of Warfare and Technology: From Slings to Robots*. Springer Nature. DOI: 10.1007/978-981-19-3478-0

Rozin, P., Lowery, L., Imada, S., & Haidt, J. (1999). The CAD triad hypothesis: A mapping between three moral emotions (contempt, anger, and disgust) and three moral codes (community, autonomy, and divinity). *Journal of Personality and Social Psychology*, *76*(4), 574–586. DOI: 10.1037/0022-3514.76.4.574 PMID: 10234846

Rubarth, S. (2015). Masculinity and the sea in Greek myth. *Classical World*, *108*(3), 323–343.

Rubio-Marín, R. (Ed.). (2009). *The gender of reparations: unsettling sexual hierarchies while redressing human rights violations*. Cambridge University Press., DOI: 10.1017/CBO9780511596711

Ruhaak, R. (2019). Towards an Alternative Black Death Narrative for Ireland: Ecological and Socio-Economic Divides on the Medieval European Frontier. *Journal of the North Atlantic*, *2019*(39), 1–16. DOI: 10.3721/037.006.3901

Sadowski, D., & Becker, J. (2010). Beyond the "Hybrid" Threat: Asserting the Essential Unity of Warfare. *Journal Article*, *7*(1), 13.

Sagar, R. (Director). (1972). *Lalkar* [Film].

Said, E. W. (1978). *Orientalism*. Pantheon Books.

Saigal, R. (Director). (1950). *Samadhi* [Film].

Samuels, M. (2017). Erwin Rommel and German Military Doctrine, 1912–1940. War in History, 24(3), 308-335. https://doi.org/ (Original work published 2017) DOI: 10.1177/0968344515623982

Samuels, A. (1985). *Jung and the post-Jungians* (1st ed.). Routledge., DOI: 10.4324/9780203359297

San José, P. C. (2020). The rapes committed against the Yazidi women: a genocide?: A Study of the crime of rape as a form of genocide in International Criminal Law. *Comillas Journal of International Relations*, (18), 50–71. DOI: 10.14422/cir.i18.y2020.003

Sandars, N. K. (1960). The epic of Gilgamesh. http://www.aina.org/books/eog/eog.pdf

Sander, O. (2017). *Siyasi Tarih: İlk Çağlardan 1918'e* (33rd ed.). İmge Kitapevi.

Sandford, S. (2024). "On their own terms": Agency, advocacy and representation in refugee Webcomics. *Journal of Australian Studies*, *48*(2), 194–208. DOI: 10.1080/14443058.2024.2343827

Sandhya Devi, N. K. (2025). Unweaving the accustomed silence: A study of comfort women using How We Disappeared by Jing Jing Lee (2020). *Journal of Gender-Based Violence*, •••, 1–17. DOI: 10.1332/23986808Y2025D000000080

Sarsılmaz, N., & Yiğitler, K. (2012). How World War II Shaped the Art and Literature? *Anadolu Bil Meslek Yüksekokulu Dergisi*, (25), 50–58.

Sassoon, S. (1930). *Memoirs of an Infantry Officer*. Faber & Faber.

Saud, A., & Kazim, N. (2022). Disinformation and Propaganda Tactics: Impacts of Indian Information Warfare on Pakistan. *Journal of Indian Studies*, *8*(02), 335–354.

Scarpino, R., & Trainer, J. (2024). Sanctions by the numbers: 2023 year in review. *Center for a New American Security*. https://www.cnas.org/publications/reports/sanctions-by-the-numbers-2023-year-in-review

Scherr, R. (2015). Framing human rights: Comics form and the politics of recognition in Joe Sacco's Footnotes in Gaza. *Textual Practice*, *29*(1), 111–131. DOI: 10.1080/0950236X.2014.952771

Schoen, F., & Lamb, C. J. (2012). *Deception, Disinformation, and Strategic Communications: How One Interagency Group Made a Major Difference*. National Defense University Press.

Schönfeld, C. (2022). Memories of World War II in German Film after 1945. In Bragança, M., & Tame, P. (Eds.), *German Film after 1945". The Long Aftermath: Cultural Legacies of Europe at War, 1936-2016* (pp. 200–218). Berghahn Books., DOI: 10.1515/9781782381549-015

Schroefl, J., & Kaufman, S. J. (2014). Hybrid Actors, Tactical Variety: Rethinking Asymmetric and Hybrid War. *Studies in Conflict and Terrorism*, *37*(10), 862–880. DOI: 10.1080/1057610X.2014.941435

Selaković, M., & Ljepava, N. (2023). Online-Ignited Crises and Post-Crisis Image Restoration: Example of Flight 3411 Incident. *International Journal of Management Trends: Key Concepts and Research*, *2*(1), 15–40. DOI: 10.58898/ijmt.v2i1.15-40

Sellars, K. (2010). Imperfect justice at Nuremberg and Tokyo. *European Journal of International Law*, *21*(4), 1085–1102. DOI: 10.1093/ejil/chq070

Seo, H. (2014). Visual Propaganda in the Age of Social Media: An Empirical Analysis of Twitter İmages During the 2012 Israeli–Hamas Conflict. *Visual Communication Quarterly*, *21*(3), 50–161. DOI: 10.1080/15551393.2014.955501

Seth, A., De, S., Arya, A., Wilkinson, S., Singh, S., & Pal, J. (2022, April 8). Closed Ranks: The Discursive Value of Military Support for Indian Politicians on Social Media. *Proceedings of the 2022 International Conference on Information and Communication Technologies and Development*. Information and Communication Technologies and Development, Seattle. DOI: 10.1145/3572334.3572395

Shafto, P., Eaves, B., Navarro, D. J., & Perfors, A. (2012). Epistemic trust: Modeling children's reasoning about others' knowledge and intent. *Developmental Science*, *15*(3), 436–447. DOI: 10.1111/j.1467-7687.2012.01135.x

Shaheen, J. G. (2001). *Reel bad Arabs: How Hollywood vilifies a people*. Interlink Publishing Group.

Shakarian, P., Shakarian, J., & Ruef, A. (2013). *Introduction to Cyber-Warfare: A Multidisciplinary Approach*. Morgan Kaufmann Publishers – Elsevier.

Sharma, S., & Pal, P. (2023). The rise of spy universe in contemporary Hindi cinema: Masculinity and nation-building in War (2019). *Intersections: Gender and Sexuality in Asia and the Pacific*, *50*. https://www.academia.edu/download/113394978/The_Rise_of_Spy_Universe.pdf

Sharma, H., & Pegu, U. K. (2023). Is the liberal media dead?: A critical analysis of Modi's Government and the performance of Indian Media. *India Review*, *22*(3), 219–248. DOI: 10.1080/14736489.2023.2201538

Shelef, N. G., & Koo, A. Z.-X. (2022). Homelands and nationalism. *Nationalities Papers*, *50*(3), 417–429. DOI: 10.1017/nps.2021.38

Sherfudeen, C., Mahmood, K., & Hossain, M. J. (2024). Cultural Nationalism or National Culture as Portrayed in Indian Popular Culture: The Case of Indian Visual Media. *International Peer Reviewed Journal: AWEJ for Translation & Literary Studies*, 115–130.

Sherlock, J. I. (1999). Globalisation, western culture and Riverdance. In *Thinking identities: Ethnicity, racism and culture* (pp. 205–218). Palgrave Macmillan UK. DOI: 10.1057/9780230375963_10

Sherry, F. (1986). *Raiders and rebels: A history of the golden age of piracy*. Harper & Row.

Shirm, G. (2023). Testimony, reconstructed: the secondary witness and testimonial possession in Bart van Es' The cut out girl: A story of war and family, lost and found. *Textual Practice*, *37*(8), 1281–1301. DOI: 10.1080/0950236X.2022.2111702

Shnabel, N., & Nadler, A. (2008). A needs-based model of reconciliation: Satisfying the differential emotional needs of victim and perpetrator as a key to promoting reconciliation. *Journal of Personality and Social Psychology*, *94*(1), 116–132. DOI: 10.1037/0022-3514.94.1.116

Shoesmith, B. (1988). Swadeshi cinema: Cinema, politics and culture: The writings of D.G. Phalke. *Continuum (Perth)*, *2*(1), 44–50. Advance online publication. DOI: 10.1080/10304318809359350

Siegel, P. C. (2005). Perception Management: IO's Stepchild? *Low Intensity Conflict & Law Enforcement.*, *13*(2), 117–134. DOI: 10.1080/09662840500347314

Silva, J. R., & Capellan, J. A. (2019). The media's coverage of mass public shootings in America: Fifty years of newsworthiness. *International Journal of Comparative and Applied Criminal Justice*, *43*(1), 77–97. DOI: 10.1080/01924036.2018.1437458

Silva, J. R., & Greene-Colozzi, E. A. (2019). Fame-seeking mass shooters in America: Severity, characteristics, and media coverage. *Aggression and Violent Behavior*, *48*, 24–35. DOI: 10.1016/j.avb.2019.07.005

Silver, J. (2017). Moral Foundations, Intuitions of Justice, and the Intricacies of Punitive Sentiment. *Law & Society Review*, *51*(2), 413–450. DOI: 10.1111/lasr.12264

Silver, J., Horgan, J., & Gill, P. (2019). Shared struggles? Cumulative strain theory and public mass murderers from 1990 to 2014. *Homicide Studies*, *23*(1), 64–84. DOI: 10.1177/1088767918802881

Simmons, B. K. (1992). The effect of censorship on attitudes toward popular music. *Popular Music and Society*, *16*(4), 61–67. DOI: 10.1080/03007769208591497

Simonds, J. (2025). *Maritime risk, insurance, and seafarer safety: Navigating piracy in the Indian Ocean*. Routledge. DOI: 10.4324/9781032664736

Simon, T. W. (1996). Defining genocide. *Wisconsin International Law Journal*, *15*, 243.

Singer, M. (2004). Shame, Guilt, Self-Hatred and Remorse in the Psychotherapy of Vietnam Combat Veterans Who Committed Atrocities. *American Journal of Psychotherapy*, *58*(4), 377–385. DOI: 10.1176/appi.psychotherapy.2004.58.4.377 PMID: 15807083

Sinnerbrink, R. (2006). A Heideggerian Cinema? On Terrence Malick's The Thin Red Line. *Film-Philosophy*, *10*(3), 26–37. DOI: 10.3366/film.2006.0027

Skitka, L. J., Bauman, C. W., & Sargis, E. G. (2005). Moral conviction: Another contributor to attitude strength or something more? *Journal of Personality and Social Psychology*, *88*(6), 895–917. DOI: 10.1037/0022-3514.88.6.895 PMID: 15982112

Smith, I. S. (2014). *In lonely places: Film noir beyond the city*. McFarland.

Smokvina, T. K., & Kovacic, M. P. (2021). Media Ethics and Professionalism in the Digital Communication Environment: Responses of Media Self-regulation and Regulatory Bodies. *JAVNOST-THE PUBLIC*, *28*, S43–S61.

Sobel, D. M., & Kushnir, T. (2013). Knowledge matters: How children evaluate the reliability of testimony as a process of rational inference. *Psychological Review*, *120*(4), 779–797. DOI: 10.1037/a0034191

Socolow, M. J. (2007). "News is a Weapon": Domestic Radio Propaganda and Broadcast Journalism in America, 1939–1944. *American Journalism*, *24*(3), 109–131. DOI: 10.1080/08821127.2007.10678081

Soja, E. (1989). *Postmodern geographies*. Verso.

Somalinet. (2013, October 12). "Captain Phillips! - SomaliNet Forums." somalinet .com/forums/viewtopic.php?t=336546#gsc.tab=0.

Somasundaram, D. (2007). Collective trauma in northern Sri Lanka: A qualitative psychosocial-ecological study. *International Journal of Mental Health Systems*, *1*(1), 1–27. DOI: 10.1186/1752-4458-1-5

Somasundaram, D. (2014). Addressing collective trauma: Conceptualisations and interventions. *Intervention (Amstelveen, Netherlands)*, *12*(4), 43–60. DOI: 10.1097/WTF.0000000000000068

Sönmez, G. (2022). Rusya'nın 2014 Ukrayna Krizi'ndeki Hibrit Savaş Stratejisi. *Gazi Akademik Bakış*, *15*(30), 489–511. DOI: 10.19060/gav.1136936

Sorlin, P. (1994). *The film in history: Restaging the past*. Blackwell.

Sosa, E. (1994). Testimony and coherence. In Matilal, B. K., & Chakrabarti, A. (Eds.), *Knowing from words: Western and Indian philosophical analysis of understanding and testimony* (pp. 59–67). Springer Netherlands., DOI: 10.1007/978-94-017-2018-2_5

Sotirovic, M., & McLeod, J. M. (2004). Knowledge as Understanding: The Information Processing Approach to Political Learning. In Kaid, L. L. (Ed.), *Handbook of Political Communication Research* (pp. 357–394). Lawrence Erlbaum Associates.

Sperber, D., Clément, F., Heintz, C., Mascaro, O., Mercier, H., Origgi, G., & Wilson, D. (2010). Epistemic vigilance. *Mind & Language*, *25*(4), 359–393. DOI: 10.1111/j.1468-0017.2010.01394.x

Spiegal, S. (1991). *The pirate myth: Genealogies of an imperial figure*. University of California Press.

Spiegelman, A. (1986). *Maus: A survivor's tale*. Pantheon.

Spielberg, S. (Director). (1998). Saving private ryan [Film]. DreamWorks Pictures; Paramount Pictures.Stone, O. (Director). (1986). Platoon [Film]. Hemdale Film Corporation.

Spivak, G. C. (2023). Can the subaltern speak? In Cain, P. H., & Harrison, M. (Eds.), *Imperialism* (pp. 171–219). Routledge.

Staiger, J. (2005). *Media Reception Studies*. New York University Press., DOI: 10.18574/nyu/9780814786741.001.0001

Stanger, M. (2018, Mayıs 2). 6 People Who Were Literally Erased From History. Business Insider. http://uk.businessinsider.com/people-who-were-erased-from-history-2013-12/#nikolai-yezhov-joseph-stalins-head-of-secret-police-1

Stanko, B. (2000). Murder and Moral Outrage: Understanding violence. *CJM*, *42*(1), 4–5.

Starks, L. S. (2002). "Remember me": Psychoanalysis, cinema, and the crisis of modernity. *Shakespeare Quarterly*, *53*(2), 181–200. DOI: 10.1353/shq.2002.0025

Stefanovic, D., & Loizides, N. (2011). The way home: Peaceful return of victims of ethnic cleansing. *Human Rights Quarterly, 33*(2), 408–430. DOI: 10.1353/hrq.2011.0029

Steinberg, P. (2001). *The social construction of the ocean.* Cambridge University Press.

Steinberg, P., & Peters, K. (2015). Wet ontologies, fluid spaces: Giving depth to volume through oceanic thinking. *Environment and Planning. D, Society & Space, 33*(2), 247–264. DOI: 10.1068/d14148p

Steitkasimova, E. (2019). Dangerous femininity: Sirens and the patriarchal Athenian society. *Classical Mythology Review, 12*(3), 44–57.

Sternbach, H. (1991). The serotonin syndrome. *The American Journal of Psychiatry, 148*(6), 705–713. DOI: 10.1176/ajp.148.6.705

Strahler, J., & Ziegert, T. (2015). Psychobiological stress response to a simulated school shooting in police officers. *Psychoneuroendocrinology, 51*(1), 80–91. DOI: 10.1016/j.psyneuen.2014.09.016 PMID: 25290348

Strauss, S., Gourevitch, P., & Forges, A. D. (2000). Genocide in Rwanda. *African Studies Review, 43*(2), 126. DOI: 10.2307/524988

Strejilevich, N. (2006). Testimony: Beyond the language of truth. *Human Rights Quarterly, 28*(3), 701–713. DOI: 10.1353/hrq.2006.0038

Stylianou, P., & Zembylas, M. (2018). Dealing with the concepts of "grief" and "grieving" in the classroom: Children's perceptions, emotions, and behavior. *Omega, 77*(3), 240–266. DOI: 10.1177/0030222815626717 PMID: 29940830

Suk, J. (2009). *At home in the law: How the domestic violence revolution is transforming privacy.* Yale University Press.

Swinfen, A. (2019). *In defence of fantasy: A study of the genre in English and American literature since 1945.* Routledge., DOI: 10.4324/9780429321221

Sykes, C. J. (1999). *The end of privacy: The attack on personal rights at home, at work, on-line, and in court.* St. Martin's Press.

Sy, T., & Smith, J. (2013). The exceptional nation: Framing of the U.S. in *Captain Phillips. Journal of American Popular Culture, 12*(1), 58–72.

Szakacsi, I. (2016). Certain Aspects of Military Applications of Robotics in the Future Missions. *Defence Review, 1.*

Taddeo, M. (2012). Information Warfare: A Philosophical Perspective. *Philosophy & Technology*, *25*(1), 105–120. DOI: 10.1007/s13347-011-0040-9

Tajfel, H., & Turner, J. C. (2004). The Social Identity Theory of Intergroup Behavior. In Jost, J. T., & Sidanius, J. (Eds.), *Political psychology: Key readings* (pp. 276–293). Psychology Press., DOI: 10.4324/9780203505984-16

Tawil-Souri, H. (2014). Cinema as the Space to Transgress Palestine's Territorial Trap. *Middle East Journal of Culture and Communication*, *7*(2), 169–189. DOI: 10.1163/18739865-00702005

Taylor, P. M. (1998). *War and the Media: Propaganda and Persuasion in the Gulf War* (Revised). Manchester: Manchester University Press. https://manchesteruniversitypress.co.uk/9780719055508/

Taylor, D. (2018). Humiliation as a harm of sexual violence: Feminist versus neoliberal perspectives. *Hypatia*, *33*(3), 434–450. DOI: 10.1111/hypa.12427

Telegraf. (2023). NASLOVNE STRANE DANAŠNJIH NOVINA: Prelistavanje štampe za petak, 5. maj 2023. Godine. Telegraf.rs. Retrieved from https://www.telegraf.rs/vesti/srbija/3673286-naslovne-strane-sutrasnjih-novina

Teune, S. (2005, September). Art and the re-invention of political protest. In Paper presented at the *3rd ECPR Conference,* 8, 10.

Thankachan, K., Thomas, P. E., & Menon, V. A. (2025). Conflict Caught in Propaganda Nationalism: An Analysis of the India-Pakistan Newspaper Coverage of Mumbai, Uri, and Pulwama Attacks. In *Impacts of Leakage, Whistleblowing, and the Rise of Propaganda* (pp. 169-184). IGI Global Scientific Publishing.

The Hindu. (2025). Pahalgam terror attack: Operation Sindoor launch live updates. https://www.thehindu.com/news/national/pahalgam-terror-attack-operation-sindoor-launch-live-updates-may-7-2025/article69543511.ece

The SAIS Review. (2024). Social Media, Disinformation, and AI: Transforming the Landscape of the 2024 U.S. Presidential Political Campaigns. Johns Hopkins University. https://saisreview.sais.jhu.edu/social-media-disinformation-and-ai-transforming-the-landscape-of-the-2024-u-s-presidential-political-campaigns/

Thomas, T. H. (1930). Some War Memoirs. *The Journal of Modern History*, *2*(4), 629–640. DOI: 10.1086/235644

Thomson, J. E. (1994). *Mercenaries, pirates, and sovereigns: State-building and extraterritorial violence in early modern Europe*. Princeton University Press.

Ticktin, M. I. (2011). *Casualties of care: Immigration and the politics of humanitarianism in France*. Univ of California Press. DOI: 10.1525/9780520950535

Tigay, J. H. (2002). *The evolution of the Gilgamesh epic*. Bolchazy-Carducci Publishers.

Times of India. (2025). Seven dead, 38 injured as Pakistan Army targets civilian areas along LoC in J&K. https://timesofindia.indiatimes.com/india/seven-dead-38-injured-as-pakistan-army-targets-civilian-areas-along-loc-in-jk/articleshow/120952157.cms

TOI News Desk. (2025). Army foils Pakistan's attempts to send swarm drones across LoC. *The Times of India*. https://timesofindia.indiatimes.com/india/army-foils-pakistan-attempts-to-send-swarm-drones-across-loc/articleshow/121014854.cms

Toplu, M. (2018). Kamu Politikalarının Oluşumu ve Uygulanmasında Medya/Sosyal Medya'nın Rolü. *Journal of Communication Theory & Research.*, *46*, 346–376.

Traesar, L., & Saktiningrum, N. (2024). THE PORTRAYAL OF FEMALE VILLAINS: A REPRESENTATION STUDY ON CRUELLA. *Rubikon Journal of Transnational American Studies*, *11*(2), 327. DOI: 10.22146/rubikon.v11i2.94900

Traverso, A., & Broderick, M. (2010). Interrogating trauma: Towards a critical trauma studies. *Continuum (Perth)*, *24*(1), 3–15. DOI: 10.1080/10304310903461270

Traw, R. (1993). *Nothing in the Middle: What Middle Schoolers Are Reading.*

Traynor, M. (2021). Beyond the limits of imagination: Abductive inferences from imagined phenomena. *Synthese*, *199*(5), 14293–14315. DOI: 10.1007/s11229-021-03421-z

Tripathi, S. (2024). A force for good. *Index on Censorship*, *53*(4), 55–57. DOI: 10.1177/03064220241306623

Tufte, T. (2017). *Communication and social change: A citizen perspective*. John Wiley & Sons.

Türkay, Ş. (2013). Siber Savaş Hukuku ve Uygulanma Sorunsalı. *Journal of Istanbul University Law Faculty*, *71*(1), 1177–1227.

Twitter Inc. (2021). Permanent suspension of @realDonaldTrump. Twitter Blog. https://blog.twitter.com/en_us/topics/company/2020/suspension

Ulman, R. B., & Brothers, D. (2013). *The shattered self: A psychoanalytic study of trauma*. Routledge. DOI: 10.4324/9780203766408

Uluk, M. (2018). *Hakikat Sonrası Çağda Yeni Medya ve Yalan Haber*. Dorlion.

Unerman, J., & O'Dwyer, B. (2006). Theorising accountability for NGO advocacy. *Accounting, Auditing & Accountability Journal, 19*(3), 349–376. DOI: 10.1108/09513570610670334

University of Oregon School of Journalism and Communication. (2024). Media Influence on Politics: 7 Election Trends in 2024. https://journalism.uoregon.edu/news/media-influence-on-politics

Upadhyay, S. P., & Robinson, R. (2012). Revisiting communalism and fundamentalism in India. *Economic and Political Weekly*, •••, 35–57.

US. Marine Corps. (1997). Warfighting. https://www.marines.mil/Portals/1/Publications/MCDP%201%20Warfighting.pdf

Utma, S. (2022). *Medyada Gerçeklik Algısının Dönüşümü: Medya ve Dezenformasyon*. İKSAD Publishing House.

Vallar, C. (2009). *Pirates and plunder: A history*. Maritime Archives Press.

Van Creveld, M. (1991). *The Transformation of War*. The Free Press.

Van der Dennen, J. M. (1990). Origin and evolution of 'primitive'warfare. In *Sociobiology and Conflict: Evolutionary perspectives on competition, cooperation, violence and warfare* (pp. 149–188). Springer Netherlands. DOI: 10.1007/978-94-009-1830-6_9

Van der Kolk, B. (2014). The body keeps the score: Brain, mind, and body in the healing of trauma. *New York, 3*, 14-211.

Van Schaack, B. (2018). The Iraq investigative team and prospects for justice for the Yazidi genocide. *Journal of International Criminal Justice, 16*(1), 113–139. DOI: 10.1093/jicj/mqy002

van Schaik, M. (2023, July). Place-Identity and Post-war Rebuilding. In *World Congress of Architects* (95-105). Design for Resilient Communities. UIA 2023. Sustainable Development Goals Series. Cham: Springer International Publishing. https://doi.org/DOI: 10.1007/978-3-031-36640-6_7

Van Zomeren, M. (2006). Social-psychological paths to protest: An integrative perspective. The Netherlands: Doctoral dissertation, University of Amsterdam.

Van Zomeren, M., & Lodewijkx, H. M. (2005). Motivated responses to 'senseless' violence: Explaining emotional and behavioural responses through person and position identification. *European Journal of Social Psychology, 35*(6), 755–766. DOI: 10.1002/ejsp.274

Varlık, A. B. (2013). Savaşı Tanımlamak: Terminolojik Bir Yaklaşım. *Avrasya Terim Dergisi, 1*(2), 114–129.

Vélez-Serna, M., & Stauff, M. (2023, June 7). Ports: On the material and symbolic mediation of global capitalism - NECSUS. NECSUS. https://necsus-ejms.org/ports-on-the-material-and-symbolic-mediation-of-global-capitalism/

Verma, V. (2022). Non-Contact Warfare: A Strategy for Future. *CLAWS Journal, 15*(2), 57–70. https://core.ac.uk/download/pdf/555351808.pdf

Verma, R., & Roy, A. G. (2022). Ideological positioning in the representation of borders: An analysis of recent Hindi films. *India Review, 21*(3), 307–328. DOI: 10.1080/14736489.2022.2086405

Verpoest, L., Engelen, L., Heynickx, R., Schmidt, J., Uyttenhove, P., & Verstraete, P. (Eds.). (2020). *Revival after the Great War: Rebuild, remember, repair, reform.* Leuven University Press. DOI: 10.2307/j.ctv1b0fvbf

Vice, S. (2008). Holocaust poetry and testimony. *Critical Survey, 20*(2), 7–17. DOI: 10.3167/cs.2008.200202

Vidler, A. (1992). *The architectural uncanny: Essays in the modern unhomely.* MIT Press.

Virilio, P. (1989). *War and cinema: The logistics of perception.* Verso.

Virilio, P. (1989). *War and Cinema: The Logistics of Perception.* Verso., https://www.versobooks.com/en-gb/products/1264-war-and-cinema?_pos=1&_sid=ab635b484&_ss=r

Viswanath, G. (2014). *The" nation" in War: A Study of Military Literature and Hindi War Cinema.* Cambridge Scholars Publishing.

Vonnegut, K. (1969). *Slaughterhouse-Five.* Delacorte Press.

Vuillemin, J. (as cited in Grozholz, P.). (2007). History enters into the considerations of theoretical philosophy through mathematics. *Philosophy of Mathematics Review*, 28(3), 228–232.

Vulić, T., & Pavlović, D. (2015). Domestic violence in the Serbian daily press: from facts to sensationalism. *Facta Universitatis, Series: Philosophy, Sociology, Psychology and History*, 111-122.

Walker, J. (2005). *Trauma cinema: Documenting incest and the Holocaust.* University of California Press. DOI: 10.1525/9780520937932

Walsh, J. "Does Feel-Good "Captain Phillips" Film Unfairly Depict Somali Plight?" *MinnPost*, 16 Oct. 2013, www.minnpost.com/politics-policy/2013/10/does-feel-good-captain-phillips-film-unfairly-depict-somali-plight. Accessed 11 July 2025.

Walsh, J. P. (2020). Social Media and Moral Panics: Assessing The Effects Of Technological Change On Societal Reaction. *International Journal of Cultural Studies, 23*(6), 840–859. Advance online publication. DOI: 10.1177/1367877920912257 PMID: 40479194

Waltz, K. N. (2001). *Man, the State and War: A Theoretical Analysis* (3rd ed.). Columbia University Press.

Wardle, C. & Derakhshan, H. (2017). Information Disorder: Toward an Interdisciplinary Framework for Researchand Policymaking, Council of Europe Report DGI 09.

Warren, G. M. (2017). The human colonisation of Ireland in Northwest European context. Advances in Irish Quaternary Studies, 293-316.

Weine, S. (2006). *Testimony after catastrophe: Narrating the traumas of political violence*. Northwestern University Press.

Werner, G. G., Riemann, D., & Ehring, T. (2021). Fear of sleep and trauma-induced insomnia: A review and conceptual model. *Sleep Medicine Reviews, 55*, 101383. DOI: 10.1016/j.smrv.2020.101383

Wessells, M., & Strang, A. (2006). Religion as resource and risk. *A world turned upside down: Social ecological approaches to children in war zones*, In N. Boothby, A. Strang & M. Wessells (Eds.), *A World Turned Upside Down- Social Eclological Approaches to Children in War Zones* (199-222). Bloom-fied, CT: Kumarian Press. https://doi.org/DOI: 10.1515/9781565493179-012

WEST. R. (1984). On War (M. HOWARD & P. PARET, Eds.). Princeton University Press. https://doi.org/DOI: 10.2307/j.ctt7svzz

Wilkins, K. G. (2020). Advocating with accountability for social justice. In Tacchi, J., & Tufte, T. (Eds.), *Communicating for Change: Concepts to Think With* (pp. 53–62)., DOI: 10.1007/978-3-030-42513-5_5

Williams, A. L. (2002). *Film and nationalism*. Rutgers University Press.

Williams, T. (1988). Diagnosis and treatment of survivor guilt: The bad penny syndrome. In *Human adaptation to extreme stress: From the Holocaust to Vietnam* (pp. 319–336). Springer US., DOI: 10.1007/978-1-4899-0786-8_14

Wilson, G., Hill, M., & Kiernan, M. D. (2018). Loneliness and social isolation of military veterans: Systematic narrative review. *Occupational Medicine, 68*(9), 600–609. DOI: 10.1093/occmed/kqy160 PMID: 30551163

Wilson, K. (1990). Black Bands and Black Culture: A Study of Black Military Bands in the Union Army During the Civil War. *Australasian Journal of American Studies, 9*(1), 31–37. https://www.jstor.org/stable/41054165

Winter, J. (2018). War Memoirs, Witnessing and Silence. *War Stories: The War Memoir in History and Literature*, Berghahn Books, 27-47.

Woods, R. (2021). Testimony and its mediations in life writing. *Life Writing, 19*(3), 441–454. DOI: 10.1080/14484528.2021.1873718

Woolley, S., & Howard, P. (2019). Computational Propaganda Worldwide. in Samuel Woolley ve Philip Howard (Ed.), Computational Propaganda Political Parties, Politicians and Political Manipulation on Social Media (p. 3-20). Oxford University Press.

Wright, E. (2021). *PTSD and narration: Traumatic stories' impact on realism in film and literature* (Doctoral dissertation, University of Maryland, Baltimore County).

Wright, Q. (1942). *A study of war* (Vols. 1–2). University of Chicago Press.

Yalçınkaya, H. (2008). *Savaş Uluslararası İlişkilerde Güç Kullanımı*. İmge Kitabevi.

Yelderman, L. A., Joseph, J. J., West, M. P., & Butler, E. (2019). Mass shootings in the United States: Understanding the importance of mental health and firearm considerations. *Psychology, Public Policy, and Law, 25*(3), 212–223. DOI: 10.1037/law0000200

Yen, V. Y., & Salmon, C. T. (2017). Further explication of mega-crisis concept and feasible responses, in *SHS Web of Conferences* (Vol. 33, p. 00034). EDP Sciences. DOI: 10.1051/shsconf/20173300034

Yenal, S., & Akdemir, N. (2020). Uluslararası İlişkilerde Yeni Bir Kuvvet Çarpanı: Siber Savaşlar Üzerine Bir Vaka Analizi. ÇAKÜ. *Journal of Institute of Social Sciences, 11*(1), 414–450.

Yılmaz, M. E. (2010). *Savaş ve Uluslararası Sistem*. Nobel Yayın Dağıtım.

Young, A. (2012). Criminal Images: The Affective Judgment of Graffiti and Street Art. *Crime, Media, Culture, 8*(3), 297–314. DOI: 10.1177/1741659012443232

Young, A., & Omosun, F. (2024). How Does It Look from Where You Are? A Visual Media Framing Analysis of the 2022 War in Ukraine. *Digital Journalism (Abingdon, England), 13*(1), 18–36. DOI: 10.1080/21670811.2023.2296560

Zinn, H. (1998, October 10). Private Ryan Saves War. *The Progressive*, *62*, 38. https://link-gale-com.libproxy.ucl.ac.uk/apps/doc/A21200696/AONE?u=ucl_ttda&sid=bookmark-AONE&xid=15cedc36

Zukin, S. (1995). *The cultures of cities*. Blackwell.

About the Contributors

***

Ömer Alanka was born in Erzurum in 1981. He graduated from Ankara University Faculty of Communication in 2005. He completed his Master's degree in the Department of Journalism at Atatürk University Institute of Social Sciences in 2008. In 2009, he was appointed as a Research Assistant at Ege University Faculty of Communication. In 2012, he received his Ph.D. from Ege University Institute of Social Sciences, Department of Journalism, with his thesis titled "The Ontology of Image and Conceptual Photography." In 2012, he continued to work at Atatürk University's Faculty of Communication as a Research Assistant. In 2014, he was appointed as an Assistant Professor at Atatürk University's Faculty of Communication. In 2018, he was appointed as an Associate Professor in the field of Journalism and Media Studies. In 2022, he joined the Faculty of Communication at Hatay Mustafa Kemal University as an Associate Professor. In 2023, he was appointed as a Professor at the Faculty of Communication at Hatay Mustafa Kemal University. Since 2023, he has been serving as the Acting Dean of the Faculty of Communication at Hatay Mustafa Kemal University. In addition to his academic duties, Alanka has also held administrative positions such as Head of the Department of Information Technology, Head of the Department of Journalism, Vice Dean, and Dean.

Aida Alinia studied dramatic literature at Tehran Art University and is now pursuing her master's degree at Istinye University.

Sevda Bozkurt She was born in 1992 in Balıkesir. She graduated from Atatürk University, Faculty of Communication, Department of Radio, Television and Cinema in 2014. During this period, she worked in various advertising agencies and media organizations. In 2014, she began her master's studies in the Department of Radio, Television and Cinema at the Institute of Social Sciences, Atatürk University, and in 2018 she completed her master's thesis entitled "Internet Usage Habits of Individuals

Living in Kepsut District of Balıkesir." In the same year, she commenced her doctoral studies in the Department of Basic Communication Sciences at Atatürk University. In 2020, she began her academic career as a lecturer at the Gendarmerie and Coast Guard Academy in Ankara. In 2024, she completed her doctoral dissertation entitled "Social Media Literacy in the Context of Instagram Cookie Applications: The Case of Ankara." Sevda Bozkurt continues her research on local media, new media, culture, social media, and media and social media literacy.

Ülhak Çimen was born in Erzurum, Turkey. He completed his primary, secondary, and high school education in Erzurum. Having graduated from Atatürk University, Faculty of Communication, Department of Radio, Television and Cinema, Çimen is still an Associate Professor at the same institution.

Nebahat Akgün Çomak has been conducting research and teaching at the Galatasaray University, Türkiye. She has published a number of research articles, books and participated in many projects with both national and international recognition. Her research interests include, language, communication, mythology, narratives and representation studies.

Debadyuti Dasgupta is an independent researcher with a keen interest in military history. His passion spans military technology, hardware, and, most notably, the 1971 Bangladesh-Pakistan war, a subject he has explored in depth through his four published books. He has extended his passion for the 1971 war to the online space, where he has created various social media pages under the banner "1971 War Book Series." These platforms document the history of the 1971 war, shedding light on key incidents, personal accounts, and events from both the Eastern and Western fronts. In addition to writing, he is an avid weapon enthusiast and is currently engaged in constructing electric vehicles. His academic contributions include two books on engineering research projects and six research papers in the field of mechanical engineering. Mr. Dasgupta has also collaborated with the Engineering Corps where he contributed to the creation of four military history murals at the East Gate of Fort William, India.

Tannistha Dasgupta has a PhD in English Language Education (ELE) from The English and Foreign Languages University (EFLU), Hyderabad, India. Currently, she is a Senior Assistant Professor at the VIT-AP University, Amaravati, Andhra Predesh, India. She has developed content as a subject expert of video programmes of UGC-MHRD sponsored (Government of India) MOOC Courses for SwayamPrabha channel. Her research interests include language assessment, educational technology and online tools, learner motivation, and task-based language teaching. She has received the E-Teacher Scholarship for the Online Professional English Network

Program, sponsored by the U.S. Department of State and nominated by the U.S. Embassy, New Delhi. Recently, she received the English Language Centre Eastbourne Graham Smith Scholarship, awarded by IATEFL, UK.

Hasan Gürkan is an Associate Professor at Istinye University in Turkey and a research fellow at Western Caspian University in Azerbaijan. He was a Maria Zambrano International Fellow and a member of the ARPA research team at the University of Girona, Spain, from May 2022 to May 2024. Gürkan conducted postdoctoral research at the University of Vienna, supported by TÜBİTAK, from 2015 to 2016. He earned his Ph.D. in radio, television, and cinema from Istanbul University. He authored "Counter Cinema" and "Journalism Practices" and has published in journals like Feminist Media Studies and Studies in European Cinema. His research focuses on counter cinema, Hollywood cinema, alternative media, cinema and representation, transnational films, and gender studies.

Layal Hmaidi is a Psychology Lecturer at the American University in the Emirates. She holds a Master's degree in Psychology from the American University of Beirut. Her research examines the interdisciplinary juncture of Cognitive Psychology, Cross – Cultural Psychology, and Social Psychology. In addition to lecturing and conducting research, Layal Hmaidi serves as a Psychology examiner in the International Baccalaureate Organization.

Bidyut Bhusan Jena writes in English and Odia. Some of his poems and other pieces have appeared in various literary magazines, journals and newspapers like Muse India, Erothanatos, Rock Pebbles, The Hans India, The Eastern Times, The Sambad, the fall/winter 2021 issue of KNOT Magazine, Springfield, Colorado, Rock Pebbles' (one of the oldest literary magazines of India) anthology, India, The Antonym, a global online magazine dedicated to connecting Global Cultures through Words, Sound and Color, India and SMOKE Magazine (ISSUE 69), Liverpool, UK. His poetry collections – Pages (2019) and A Letterbox Across Time (2020) have been published by Writers Workshop, Kolkata, India and Hawakal Publishers, Kolkata, India respectively. Eight of his poems and a meditative prose piece have appeared in an anthology called Locktales (2021), India. Some of his poems and translations are in the 2023 anthology of Rock Pebbles and in the Penguin Random House India anthology on Odia Literature – The Big Book of Odia Literature respectively. His recent research publications are in The Journal of Poetry (Taylor & Francis Group), The Journal of Human Values (a Sage Journal), a Taylor & Francis anthology and Agathos A5 (a Web of Science Journal). Four of his research papers are under review in various international journals. He has received his MPhil and PhD from University of Hyderabad (UoH) and The English and Foreign Languages University (formerly CIEFL) respectively. Some of his areas of interest are Indic Studies, East-West Encounter, Comparative Philosophy, World Literatures, Comparative Literatures, Comparative Religions, Translation Studies (theory and practice) and Nature Studies. Currently, he teaches Indian Literatures, Epics, World Literatures, Philosophy, Creative Writing and an array of multidisciplinary courses at CHRIST Deemed to be University, Bangalore Central Campus, India.

Anushka Malik is a UGC-NET qualified doctoral scholar in the Department of English and Cultural Studies at CHRIST University (Central Campus), Bangalore. Her research focuses on representations of war and conflict in fantasy fiction, with a thematic emphasis on propaganda, history, and the psychological impact of war. She completed both her Bachelor's (2015–18) and Master's (2018–20) degrees in English at Panjab University, Chandigarh, and also holds a Bachelor's in Education (2020–22) from Guru Jambheshwar University, Hisar. She is fluent in English and Hindi and has earned the DELF A1 certification in French. When she's not immersed in academic work, she's usually lost in a fantasy novel, sipping her fifth cup of coffee for the day, or exploring new cafes and bakeries.

Sandhya Devi N.K. is working as an Assistant Professor at RV University, Bangalore. Her primary research areas are Queer Studies, Gender Studies, Cultural Studies, Partition Literature and World Literature. She has published several articles and chapters in the above-mentioned areas in IGI Global, Pertanika and several other reputed journals.

Harshini Nalam is a 3rd-year B.Sc. Psychology student at PES University with interests in research, clinical neuropsychology, personality psychology, and counselling. She has actively contributed to academic discourse by producing several conference proceedings and engaging with diverse scholarly platforms. Her areas of interest include brain–behavior relationships, therapeutic practices, and the exploration of Carl Jung's archetypes in understanding personality and the psyche. Passionate about research, publications, and conferences, Harshini aspires to pursue advanced studies and build a career that integrates scientific inquiry with clinical and counselling practice.

Renuprasad Patki teaches German courses at VIT-AP University. He is a researcher in the field of German studies, specifically contemporary German literature. His other research interests include foreign language teaching, GenAI in teaching-learning, and literary translations. He has been the recipient of several fellowships funded by DAAD, Germany including research at Georg August University, Göttingen, Germany. He recently published the Marathi translation of the German author Hermann Hesse's 'Wandering'.

Nilüfer Pembecioğlu is a professor at Istanbul University's Faculty of Communication, Radio, Television, and Cinema. She has published many articles on a national and international level about subjects like advertising, children, women, and citizenship, as well as language, linguistics, education, communication, journalism, and peace journalism. She has directed eight feature documentaries, thirty short films, and eight animations in addition to her academic research. Her areas of expertise included Cyberbullying, Social Discrimination and Exclusion, Digital Citizenship, Media Literacy, Systemic Family Therapy, Film Therapy, and Family Coaching. She primarily assists marginalized populations like women, refugees, gifted, deaf, and handicapped children.

Marko Selaković is manager, researcher and scholarly academic who is occupying various senior roles at SP Jain School of Global Management – Dubai campus (currently Deputy Director and Assistant Professor). He is a strategic management and communications professional with more than 20 years of top-level experience in Europe and the Gulf countries. He is specialized in strategic and international communications and development, stakeholder relations and crisis management. In addition to the academic positions, he is currently Chairman of Serbian Business Council in Dubai. Previously he was Head of Chamber of Commerce and Industry of Serbia Office to the UAE, Business Director of Expo 2020 Serbia, and Vice-president of the International Association of Business Communicators (IABC) Gulf Chapter. Selaković is author or co-author of numerous research papers related to the strategic, crisis, investor and internal communications.

Jessica Simonds is an Assistant Professor at the College of Security and Global Studies at the American University in the Emirates. Her research and teaching focus on maritime security, geopolitical strategy, and the intersections of risk, insurance, and global governance. She is the author of Maritime Risk, Insurance, and Seafarer Safety: Navigating Piracy in the Indian Ocean (Routledge), a critical study of how piracy is managed and represented in international systems. Dr. Simonds holds a PhD in International Relations and has extensive experience in curriculum development, particularly in the fields of terrorism studies, strategic thought, and security theory. Her work bridges academic research with real-world application, drawing on interdisciplinary methods to explore how narratives and policy interact in global security environments. She has presented her research at international conferences and actively supports emerging scholars through mentorship and thesis supervision.

Shruti Soudi holds a PhD in Positive Psychology from Karnatak University and has a background in counseling psychology. Her academic journey includes a Master's degree in Counseling Psychology and a Master of Philosophy in Health

Psychology from Karnatak University. Dr. Soudi has a diverse range of professional experiences. She has worked as a Mental Health Specialist at Inspiron Psychological Well-being Centre Pvt Ltd and as an Associate SME at Evelyn Learning Systems Pvt. Ltd. Additionally, she has been a Guest Associate Editor at Frontiers in Media Psychology since August 2023. Her research focuses on familial dynamics and parenting strategies, exploring their impact on children's emotional well-being, Bharata's Natya Shastra, Indian performing artists and she is passionate about the Indian Knowledge System and its contribution for mental health and wellbeing. She has also contributed to various studies on topics such as personality, coping with stress, spirituality, self-esteem, and forgiveness among musicians and students. Dr Soudi is currently an Associate Professor in the Department of Psychology in PES University.

Deepthi Krishna is an Assistant Professor at the School of Digital Media and Communication, Mahindra University, Hyderabad. Her research primarily focuses on film and media studies, with a specific emphasis on gender issues, including the representation of women and masculinity. Additionally, she explores the intersection of caste in media and the portrayal of spaces and cities, delving into how marginalisation is depicted and articulated in cinema and media narratives. Dr. Krishna has published numerous research articles and book chapters that contribute significantly to the discourse within media and film studies, highlighting critical socio-cultural themes.

Zitianxiao Xu is a graduate student in Film Studies at University College London. Her research interests encompass animation, cinematic materiality, and the representation of gender and identity in film, with a particular focus on queer and feminist screen studies.

Muhammad Yaqoub, Ph.D., is an assistant professor in the School of Literature, Arts and Media at Tongling University. His academic interests include film, drama, and television studies, film theory, glocalization, political science, and mass communication. Email: muhammadyaqoub@tlu.edu.cn ; muhammadyaqoub@ yahoo.com

Index

Symbols

A

B

C

D

E

F

G

H

I

J

M

N

O

P

R

S

T

V

W

Z

www.ingramcontent.com/pod-product-compliance
Lightning Source LLC
LaVergne TN
LVHW062325300926
849161LV00007B/1580